José A. C. Broekaert
Analytical Atomic
Spectrometry with Flames
and Plasmas

Analytical Atomic Spectrometry with Flames and Plasmas

José A. C. Broekaert

Analytical Atomic Spectrometry with Flames and Plasmas

Chemistry Library

Prof. Dr. José A. C. Broekaert
Universität Leipzig
Institut für Analytische Chemie
Linnéstraße 3
04103 Leipzig
Germany

Typesetting   Asco Typesetters, Hong Kong
Printing   betz-druck gmbH, D-64291
Darmstadt
Bookbinding   Wilhelm Osswald & Co., 67433
Neustadt

ISBN   3-527-30146-1

Library of Congress Card No.: applied for
A catalogue record for this book is available
from the British Library.
Die Deutsche Bibliothek – CIP
Cataloguing-in-Publication-Data
A catalogue record for this publication is
available from Die Deutsche Bibliothek

To my wife Paula and our daughters Ilse,
Sigrid and Carmen

Contents

Preface

Spectrochemical analysis is a powerful instrumental principle for the determination of the chemical elements and their species in a variety of sample types of different size, at widely different concentration levels and with very differing cost performance ratios and time consumption. In addition, not only monoelement but also multielement determinations are possible with widely differing precision and accuracy using the various different methods. The basic principles of spectrochemical analysis are related to the atomic and molecular structure and also to gas discharge physics as well as to instrumentation and measurement sciences. Therefore, research into spectrochemical analysis requires knowledge of the aforementioned disciplines to enable innovative developments of new methodologies to be achieved in terms of the improvement of power of detection, accuracy and cost performance ratios, these being the driving forces in analytical innovation. The development of analytical procedures also requires the analytical chemist to have a knowledge of the theory and the principles of the above mentioned disciplines. It is the aim of this monograph to bring together the theory and principles of todays spectrochemical methods that make use of flames and plasma sources. This should enable researchers to enter the field of spectrochemical research, where innovation is through the use and development of new sources and the application of new types of spectrometers, and also to face challenges from emerging fields of application, which is as straightforward today as it was even in the time of Bunsen and Kirchhoff. This work should appeal both to chemists and physicists, the cooperation of whom is instrumental for progress to be made in this field of analytical chemistry as well as to users from different areas of science, including the life sciences, material sciences, environmental sciences, geochemistry, chemical process technology, etc. This present work could also be viewed as a resume of the theoretical background, which manufacturers of instrumentation for atomic absorption spectrometry, arc, spark and glow discharge emission spectrometry as well as ICP emission spectrometry and plasma mass spectrometry with ICPs or glow discharges and laser based techniques can recommend to their interested users to make the most efficient use of these analytical methods in their respective fields of application. Also research associates entering the field of atomic spectroscopy with flames and plasmas should find the necessary basics and references to further literature in this book.

The work also describes a number of achievements of over thirty years of research performed at the University of Gent (Belgium), the Institute for Spectrochemistry and Applied Spectroscopy (ISAS), Dortmund, the University of Dortmund and the University of Leipzig, which have been made possible through many interactions and collaborations with experts in the field, whom I thank thoroughly. A great deal of knowledge gained from my teachers and in interaction with prominent senior researchers in the field worldwide and especially at the Council for Scientific and Industrial Research and the University of Stellenbosch (South-Africa), the Universitaire Instelling Antwerpen (UIA) (Belgium) and Indiana University, Bloomington (IN, USA) as well as results obtained while collaborating with colleagues and with students made this book possible, for which all of them are gratefully acknowledged and thanked.

Leipzig, June 2001   José A. C. Broekaert

Introduction

Atomic spectroscopy is the oldest instrumental elemental analysis principle, the origins of which go back to the work of Bunsen and Kirchhoff in the mid-19th century [1]. Their work showed how the optical radiation emitted from flames is characteristic of the elements present in the flame gases or introduced into the burning flame by various means. It had also already been observed that the intensities of the element-specific features in the spectra, namely the atomic spectral lines, changed with the amount of elemental species present. Thus the basis for both qualitative and quantitative analysis with atomic emission spectrometry was discovered. These discoveries were made possible by the availability of dispersing media such as prisms, which allowed the radiation to be spectrally resolved and the line spectra of the elements to be produced.

Around the same time it was found that radiation of the same wavelength as that of the emitted lines is absorbed by a cold vapor of the particular element. This discovery was along the same lines as the earlier discovery made by Fraunhofer, who found that in the spectra of solar radiation line-shaped dark gaps occurred. They were attributed to the absorption of radiation by species in the cooler regions around the sun. These observations are the basis for atomic absorption spectrometry, as it is used today. Flames proved to be suitable sources for determinations in liquids, and in the work of Bunsen and Kirchhoff estimations were already being made on the smallest of elemental amounts that would still produce an emission or absorption signal when brought into a flame. From this there was already a link appearing between atomic spectroscopy and the determination of very small amounts of elements as being a basis for trace analysis.

With industrial developments arose a large need for the direct chemical analysis of solids. This resulted from expansion of production processes, where raw materials are subjected to large-scale processes for the production of bulk materials, from which products of increased value, complying to very strict specifications, are manufactured. The search for appropriate raw materials became the basis for mining, which was then developed on a large scale. Geological prospecting with the inevitable analyses of large amounts of samples for many elements, often down to low concentration levels as in the case of the noble metals, took place. Also trading of raw materials developed, which intensified the need for highly accurate characterization of ores and minerals, a development that today is becoming more stringent. Accordingly, analytical methodology that allows widely diverse materials to be characterized for many elements became necessary. Because of economic implications this information must frequently be obtained rapidly, which necessitates so-called multielement methods for the direct analysis of solids.

Not only is there a need for the characterization of raw bulk materials but also the requirement for process controled industrial production introduced new demands. This was particularly the case in the metals industry, where production of steel became dependent on the speed with which the composition of the molten steel during converter processes could be controlled. After World War II this task was efficiently dealt with by atomic spectrometry, where the development and knowledge gained about suitable electrical discharges for this task fostered the growth of atomic spectrometry. Indeed, arcs and sparks were soon shown to be of use for analyte ablation and excitation of solid materials. The arc thus became a standard tool for the semi-quantitative analysis of powdered samples whereas spark emission spectrometry became a decisive technique for the direct analysis of metal samples. Other reduced pressure discharges, as known from atomic physics, had been shown to be powerful radiation sources and the same developments could be observed as reliable laser sources become available. Both were found to offer special advantages particularly for materials characterization.

The need for environmentally friendly production methods introduced new challenges for process control and fostered the development of atomic spectrometric methods with respect to the reliable determination of elements and their species in both solids, liquids and gaseous samples. The limitations stemming from the restrictive temperatures of flames led to the development of high temperature plasma sources for atomic emission spectrometry. Thus, as a result of the successful development of high-frequency inductively coupled plasmas and microwave plasmas these sources are now used for routine work in practically all large analytical laboratories. Accordingly, atomic emission spectrometry has developed into a successful method for multielement analyses of liquids and solids as well as for determinations in gas flows. This is due to the variety of sources that are available but also to the development of spectrometer design. The way started with the spectroscope, then came the spectrographs with photographic detection and the strongest development since photoelectric multichannel spectrometers and flexible sequential spectrometers, has recently been with array detectors becoming available.

In time, the use of flames as atom reservoirs for atomic absorption spectrometry was also transformed into an analytical methodology, as a result of the work of Walsh [2]. Flame atomic absorption spectrometry became a standard tool of the routine analytical laboratory. Because of the work of L'vov and of Massmann, the graphite furnace became popular as an atom reservoir for atomic absorption and gave rise to the widespread use of furnace atomic absorption spectrometry, as offered by many manufacturers and used in analytical laboratories, especially for extreme trace analysis. However, in atomic absorption spectrometry, which is essentially a single-element method, developments due to the multitude of atomic reservoirs and also of primary sources available, is far from the end of its development. Lasers will be shown to give new impetus to atomic absorption work and also to make atomic fluorescence feasable as an extreme trace analysis method. They will also give rise to new types of optical atomic spectrometry such as laser enhanced ionization spectrometry.

The sources investigated and that are being used successfully for optical atomic spectrometry are also powerful sources for elemental mass spectrometry. This led to the development of classical spark source mass spectrometry moving through to what are today important plasma mass spectrometric methods, such as glow discharge and inductively coupled plasma mass spectrometry. The development of elemental mass spectrometry started with the work of Aston on the elemental mass separator. Here the ions of different elements are separated on the basis of their deflection in electrical and magnetic fields. This development lead to high-resolution but expensive instrumentation, with which highly sensitive determinations could be performed, as e.g. are necessary for high-purity materials that are required for the electronics industry. Towards the end of the 1970s, however, so-called quadrupole mass filters developed to such a high standard that they could replace conventional mass spectrometers for a number of tasks. The use of mass spectrometry instead of optical emission spectrometry enabled considerable gains in the power of detection to be made for the plasma sources developed around that time. The development of this field is still proceeding fully, both with respect to the ion sources and the types of mass spectrometers used as well as with respect to detector technology. All the items mentioned will finally make an impact on the analytical performance of plasma mass spectrometry. Certainly the latter is considerably more advantageous than optical methods as a result of the possibility of detecting the various isotopes of particular elements.

The different atomic spectrometric methods with flames and plasmas have to be judged by comparing their analytical figures of merit with those of other methods for elemental analysis, a point which has to be seen through a critical eye with respect to the analytical problems to be solved. The scope for plasma spectrometry is still growing considerably, as it is no longer elemental determinations but the determination of the elements as present in different compounds that is becoming important, for problems associated with the design of new working materials, challenges in life sciences as well as for environmental and bio-medical risk assesment. This gives the area of interfacing atomic spectrometry with separation sciences a strong impetus, which needs to be treated in depth, both from the development of types of interfaces as well as from the point of view of reshaping existing and developing new sources of suitable size and cost–performance ratios.

1
Basic Principles

1.1
Atomic structure

The basic processes in optical atomic spectrometry involve the outer electrons of the atomic species and therefore its possibilities and limitations can be well understood from the theory of atomic structure itself. On the other hand, the availability of optical spectra was decisive in the development of the theory of atomic structure and even for the discovery of a series of elements. With the study of the relationship between the wavelengths of the chemical elements in the mid-19th century a fundament was obtained for the relationship between the atomic structure and the optical line emission spectra of the elements.

In 1885 Balmer published that for a series of atomic lines of hydrogen a relationship between the wavelengths could be found and described as:

$$\lambda = k \cdot n^2 / (n^2 - 4) \tag{1}$$

where $n = 2, 3, 4, \ldots$ for the lines $H_\alpha, H_\beta, H_\gamma$ etc.

Eq. (1) can also be written in wavenumbers as:

$$v' = 1/\lambda = R(1/2^2 - 1/n^2) \tag{2}$$

where v' is the wavenumber (in cm^{-1}) and R the Rydberg constant (109 677 cm^{-1}). The wavenumbers of all so-called series in the spectrum of hydrogen are given by:

$$v' = 1/\lambda = R(1/n_1^2 - 1/n_2^2) \tag{3}$$

where n_2 is a series of numbers $> n_1$ and with $n_1 = 1, 2, 3, 4, \ldots$ for the Lyman, Balmer, Paschen and Pfund series, respectively.

Rydberg applied the formula of Balmer as:

$$v' = R \cdot Z^2 (1/n_1^2 - n_2^2) \tag{4}$$

where Z is the effective charge of the atomic nucleus. This formula then also allows calculation of the wavelengths for other elements. The wavenumbers of the atomic spectral lines can thus be calculated from the difference between two positive numbers, called terms, and the spectrum of an element accordingly contains a large number of spectral lines each of which is related by two spectral terms.

The significance of the spectral terms had already been reflected by Bohr's theory, where it is stated that the atom has a number of discrete energy levels related to the orbits of the electrons. These energy levels are the spectral terms. As long as an electron is in a defined orbit no electromagnetic energy is emitted but when a change in orbit occurs, another energy level is reached and the excess energy is emitted in the form of electromagnetic radiation. The wavelength is given according to Planck's law as:

$$E = h \cdot v = h \cdot c/\lambda \tag{5}$$

Here $h = 6.623 \times 10^{-27}$ erg s, v is the frequency in s^{-1}, $c = 3 \times 10^{10}$ cm/s is the velocity of light and λ is the wavelength in cm.

Accordingly:

$$v' = 1/\lambda = E/h \cdot c = E_1/(h \cdot c) - E_2/(h \cdot c)$$
$$= T_1 - T_2 \tag{6}$$

T_1 and T_2 are the Bohr energy levels and the complexity of the emission spectra can be related to the complexity of the structure of the atomic energy levels.

For an atom with a nucleus charge Z and one valence electron, the energy of this electron is given by:

$$E = -\frac{2 \cdot \pi \cdot Z^2 \cdot e^4 \cdot \mu}{n^2 h^2} \tag{7}$$

$\mu = m \cdot M/(m + M)$, with m being the mass of the electron and M the mass of the nucleus; n is the main quantum number ($n = 1, 2, 3, \ldots$) and gives the order of the energy levels. Through the movement around the atomic nucleus, the electron has an orbital impulse moment L of which the absolute value is quantitized as:

$$|L| = h/(2\pi)\sqrt{l(l+1)} \tag{8}$$

l is the orbital quantum number and has values of: $0, 1, \ldots, (n-1)$.

The elliptical orbits can take on different orientations with respect to an external electric or magnetic field and the projections on the direction of the field also are quantitized and given by:

$$L_z = h/(2\pi)m_l \tag{9}$$

L_z is the component of the orbital momentum along the field axis for a certain angle, $m_l = \pm l, \pm(l-1), \ldots, 0$ is the magnetic quantum number and for each value of l it may have $(2l + 1)$ values.

When a spectral line source is brought into a magnetic field, the spectral lines start to display hyperfine structures, which is known as the Zeeman effect. In order to explain these hyperfine structures it is accepted that the electron rotates around its axis and has a spin momentum S for which:

$$|S| = h/(2\pi)\sqrt{S(S+1)} \tag{10}$$

The spin quantum number m_s determines the angles possible between the axis of rotation and the external field as:

$$s_z = h/(2\pi)m_s \tag{11}$$

where $m_s = \pm\frac{1}{2}$.

The orbital impulse momentum and the spin momentum are vectors and determine the total impulse momentum of the electron J as:

$$J = L + S \quad \text{with } |J| = h/(2\pi)\sqrt{j(j+1)} \tag{12}$$

$j = l \pm s$ and is the total quantum number.

In the case of an external magnetic or electrical field, the total impulse momentum also has a component along the field, whose projections on the field are quantized and given by:

$$J_z = h/(2\pi) \times m_j \quad \text{with } m_j = \pm j, \pm(j-1), \dots, 0 \tag{13}$$

This corresponds with possible $2j+1$ orientations.

The atomic terms differ by their electron energies and can be characterized by the quantum numbers using the so-called term symbols:

$$n^m l_j \tag{14}$$

Here $l = 0, 1, 2, \dots$ and the corresponding terms are given the symbols s (sharp), p (principal), d (diffuse), f (fundamental), etc., originally relating to the nature of different types of spectral lines: n is the main quantum number, m is the multiplicity ($m = 2s \pm 1$) and j is the total internal quantum number. The energy levels of each element can be given in a term scheme. In such a term scheme, also indicated are which transitions between energy levels are allowed and which ones are forbidden. This is reflected by the selection rules. According to these, only those transitions are allowed for which Δn has an integer value and at the same time $\Delta l = \pm 1$, $\Delta j = 0$ or ± 1 and $\Delta s = 0$. The terms of an atom with one valence electron can easily be found, e.g., for Na ($1s^2 2s^2 2p^6 3s^1$), in the ground level: $3^2 S_{1/2}$ [$l = 0$ (s), $m = 2.1/2 + 1 = 2$ ($s = 1/2$) and $j = 1/2$ ($j = |l \pm s|$)]. When the 3s electron goes to the 3p level, the term symbol for the excited level is: $3^2 P_{1/2, 3/2}$ [$l = 1$ (p), $m = 2.1/2 + 1 = 2$ as $s = 1/2$ and $j = 1/2, 3/2$]. The terms have a multiplicity of 2 and accordingly the lines have a doublet structure.

Fig. 1. Atomic energy level diagram for the sodium atom. (Reprinted with permission from Ref. [3].)

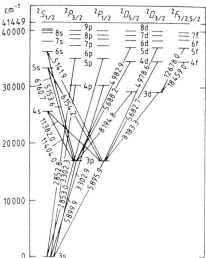

The term schemes of the elements are well documented in the work of Grotrian [3]. For the case of the Na atom the term scheme is represented in Fig. 1.

When atoms have more than one valence electron, the term schemes become more complex as a coupling between the impulse and orbital momentums of the individual electrons occurs. According to Russell and Saunders $(L - S)$ a coupling applies, where the orbital moments of all electrons have to be coupled to a total orbital momentum, as with the spin momentum. This coupling applies for elements with Z below 20, where it is accepted that the spin–orbital interactions are much lower than the spin–spin and the orbital–orbital interactions. The fact that none of the electrons in an atom can have the same set of quantum numbers is known as the Pauli rule. The total quantum number L is obtained as $L = \Sigma l$, $S = \Sigma s$ and $J = L - S, \ldots, L + S$. The term symbol accordingly becomes:

$$^{M}L_{J} \tag{15}$$

For the case of magnesium $(1s^2 2s^2 2p^6 3s^2)$, the ground level is 3^1S_0 (as there are two 3s electrons which must have antiparallel spins $L = 0$ as $l_1 = 0$ and $l_2 = 0$, $S = 0$ as $s_1 = 1/2$ and $s_2 = -1/2$, and $J = 0$ as both L and $S = 0$). The excited level $(1s^2 2s^2 2p^6 3s 3p)$ is characterized by the terms: 3^1P_1 ($L = 1$ as $l_1 = 0$ and $l_2 = 1$, $S = 0$ as $s_1 = 1/2$ and $s_2 = -1/2$, and $J = |L \pm 1| = 1$) but also 3^3P_2, 3^3P_1 and 3^3P_0 (as for the spins $s_1 = 1/2$ and $s_2 = 1/2$, $S = 1$, and further $J = 0, 1, 2$ parallel). Here singlet ($m = 1$) and triplet ($m = 3$) terms are present in the term scheme. Also a jj coupling is possible, when the interaction between spin and orbital momentum of the individual electrons is decisive.

With a number of electrons the coupling becomes more complex and leads to a high number of terms and accordingly line-rich atomic spectra. Also not only neutral atoms but also ions with different levels of ionization have term schemes, making the optical spectra very line rich. Indeed, for 90 elements between 200 and 400 nm more than 200 000 atomic lines have been listed, and many others are missing from the tables.

From Planck's law, as given by Eq. (5), the relationship between the optical atomic spectra of the elements and energy level transitions of the valence electrons can be understood. Indeed, the wavelength corresponding to a transition over an energy difference of 1 eV according to Planck's law corresponds to a wavelength of: $1 \text{ eV} = 1.6 \times 10^{-12} \text{ erg} = 6.62 \times 10^{-27} \text{ erg s} \times 3 \times 10^{10} \text{ cm/s} \times 1/\lambda$ (cm) or 1240 nm. Accordingly, the optical wavelength range of 200–800 nm corresponds to energies of 2–7 eV, this being the range involved in transitions of the valence electrons.

1.2
Plasmas

Partially ionized gases are usually denoted as plasmas [4]. They contain molecules, radicals and atoms but also ions and free electrons and result from the coupling of energy with matter in the gaseous state. As has been previously stated for atoms, radicals, molecules and ions also present in the plasma can be in their ground states and in excited states and radiation can be emitted or absorbed when transitions from one state to another occur. The wavelength of the radiation can be obtained from Planck's law whereas the intensities of the discrete lines depend on the number densities of the species and the states involved.

Transfer of energy for the different species in a plasma results from the non-radiative as well as from the radiative processes taking place. Non-radiative processes involve collisions and radiative processes involve emission, absorption and fluorescence of radiation. The efficiency of collision processes is described by the cross section $\sigma(v)$. This reflects the loss in impulse a particle with mass m and velocity v undergoes when it collides with a particle with mass M. It can be given by:

$$\sigma(v) = 2\pi \int_0^\pi p(v, \theta)(1 - \cos\theta) \sin\theta \, d\theta \tag{16}$$

This expression shows that apart from loss of momentum a change in direction may also result from collisions. The mean collision cross section is denoted as: $\langle \sigma(v) \rangle$. A collision frequency is described as $\langle \sigma(v) \cdot v \rangle$ and a mean collision frequency as $\langle \sigma(v) \cdot v \rangle / \langle v \rangle$.

Apart from the cross section, however, the velocity distribution for a given species is important for describing the energy transfer in a plasma.

In the case of a Maxwell distribution the velocity distribution is given by:

$$dn/n = 2/(\sqrt{\pi}) \cdot \sqrt{u'} \cdot e^{-u'} \cdot du' \tag{17}$$

In the case of a so-called Druyvenstein distribution:

$$dn/n = 1.039 \cdot \sqrt{u'} \cdot \exp(-0.548 \cdot u'^2)\, du' \tag{18}$$

$u' = E/kT$, where E is the mean energy of the particles and T the absolute temperature.

For a plasma contained in a closed system and which is in so-called thermal equilibrium, the population of the excited levels for every type of particle is given by Boltzmann's law:

$$n_q/n_0 = g_q/g_0 \cdot \exp(-E_q/kT) \tag{19}$$

n_q is the number density of particles in the excited state, n_0 is the number density of particles in the ground state, g_q and g_0 are the statistical weights of the corresponding levels, E_q is the excitation energy of the state q, k is Boltzmann's constant (1.38×10^{-16} erg K) and T is the absolute temperature. In Eq. (19) a relationship is formulated between the temperature and the atom number densities in a single excited state and in the ground state, respectively. As the latter is not constant, the Boltzmann equation can be better formulated as a function of the total number of particles n distributed over all states. Then

$$n_q/n = [g_q \cdot \exp(-E_q/kT)]/[\Sigma_m g_m \cdot \exp(-E_m/kT)] \tag{20}$$

as $n = \Sigma_m n_m$. The sum $Z_m = \Sigma_m g_m \cdot \exp(-E_m/kT)$ is the partition function. This is a function of the temperature and the coefficients of this function for a large number of neutral and ionized species are listed in the literature (see e.g. Ref. [5]). When E_q is expressed in eV, Eq. (20) can be written as:

$$\log n_{aq} = \log n_a + \log n_q - (5040)/T \cdot V_q - \log Z \tag{21}$$

1.3
Emission and absorption of radiation

In a steady-state plasma the number of particles leaving an energy level per unit of time equals the number returning to this level [6]. In order to characterize such an equilibrium, all processes which can lead to excitation as well as to de-excitation have to be considered. The most important energy exchange processes in a plasma are as follows.

- (1a) Collisions where atoms are excited to a higher level by collision with energetic neutrals (collisions of the first kind).
- (1b) Collisions where excited neutrals loose energy through collisions without emission of radiation (collisions of the second kind).
- (2a) Excitation by collision with electrons.

- (2b) De-excitation where energy is transferred to electrons.
- (3a) Excitation of atoms by the absorption of radiation.
- (3b) De-excitation of atoms by spontaneous or stimulated emission.

When n is the number density of the first type of particles and N the one of a second species that is present in excess $(n \ll N)$, the following equilibria can be considered:

$$\alpha \cdot N \cdot n_0 = \beta \cdot N \cdot n_q \tag{22}$$

$$\alpha_e \cdot n_e \cdot n_0 = \beta_e \cdot n_e \cdot n_q \tag{23}$$

$$B' \cdot \rho_v \cdot n_0 = (A + B \cdot \rho_v) \cdot n_q \tag{24}$$

n_e is the electron number density, A, B and B' are the Einstein transition probabilities for spontaneous emission, stimulated emission and absorption and $\alpha_e, \alpha, \beta_e$ and β are functions of the cross sections for the respective processes as well as of the velocity distribution of the particles involved. ρ_v is the radiation density (frequency v).

When the system is in so-called thermodynamic equilibrium, the neutrals and the electrons have the same Maxwell velocity distribution and at a temperature T we have:

$$n_q/n_0 = \alpha/\beta = \alpha_e/\beta_e = B'/(A/\rho_v + B) = g_q/g_0 \cdot \exp(-E_q/kT) \tag{25}$$

Thus each process is in equilibrium with the inverse process and the Boltzmann distribution of each state is maintained by collisions of the first and the second kind, including the ones with electrons, and there are no losses of energy through the emission of radiation or any absorption of radiation from an external source.

In a real radiation source this perfect equilibrium cannot exist and there are losses of energy as a result of the emission and absorption of radiation, which also have to be considered. However, as long as both only slightly affect the energy balance, the system is in so-called local thermal equilibrium and:

$$\alpha \cdot N \cdot n_0 + \alpha_e \cdot n_e \cdot n_0 + B' \cdot \rho_v \cdot n_0$$
$$= \beta \cdot N \cdot n_q + \beta_e \cdot n_e \cdot n_q + (A + B \cdot \rho_v) \cdot n_q \tag{26}$$

from which n_q/n_0 can be calculated as:

$$n_q/n_0 = (\alpha \cdot N + \alpha_e \cdot n_e + B' \cdot \rho_v)/[\beta \cdot N + \beta_e \cdot n_e + (A + B \cdot \rho_v)] \tag{27}$$

The population of the excited states is determined by the excitation processes in the radiation source, as reflected by the coefficients in Eq. (26).

In the case of a dc arc for instance $\alpha \cdot N \gg \alpha_e \cdot n_e + B' \cdot \rho_v$ and $\beta \cdot N \gg \beta_e \cdot n_e + (A + B \cdot \rho_v)$. This leads to:

$$n_q/n_0 = \alpha/\beta = g_q/g_0 \cdot \exp(-E_q/kT) \tag{28}$$

As the radiation density is low, it can be accepted that the dc arc plasma is in thermal equilibrium. The excited states are produced predominantly and decay through collisions with neutrals.

The simplification which leads to Eq. (26) does not apply to discharges under reduced pressure, where collisions with electrons are very important as are the radiation processes. Moreover, the velocity distributions are described by the Druyvenstein equation. These sources are not in thermal equilibrium.

Excited states are prone to decay because of their high energy and therefore mainly have short lifetimes. The decay can occur by collisions with surrounding particles (molecules, atoms, electrons or ions) or by emission of electromagnetic radiation. In the latter case the wavelength is given by Planck's law. When the levels q and p are involved, the number of spontaneous transitions per unit of time is given by:

$$-\mathrm{d}N_q/\mathrm{d}t = A_{qp} \cdot N_q \tag{29}$$

where A_{qp} is the Einstein coefficient for spontaneous emission (in s^{-1}). When considering an optically thin sytem with atoms in the excited state q, which decay spontaneously to a level p under the emission of radiation, the number of transitions per unit time at each moment is proportional to the number of atoms in the state q. When several transitions can start from level q ($q \rightarrow p_1, q \rightarrow p_2, \ldots, q \rightarrow p_n$) Eq. (29) becomes:

$$-\mathrm{d}N_q/\mathrm{d}t = N_q \cdot \Sigma_p A_{qp} = N_q \cdot v_q \tag{30}$$

Here v_q is the inverse value of the mean lifetime of the excited state q. For levels in which a decay by an allowed radiative transition can take place the lifetime is of the order of 10^{-8} s. When no radiative transitions are allowed we have metastable levels (e.g. Ar 11.5 and 11.7 eV), which only can decay by collisions. Therefore, such levels in the case of low pressure discharges may have very long lifetimes (up to 10^{-1} s).

In the case of the absorption of electromagnetic radiation with a frequency v_{qp} and a radiation density ρ_v, the number density of N_q increases as:

$$\mathrm{d}N_q/\mathrm{d}t = B_{qp} \cdot N_q \cdot v_p \tag{31}$$

For the case of stimulated emission, atoms in the excited state q only decay when they receive radiation of the wavelength λ_{qp} and

$$-\mathrm{d}N_q\,\mathrm{d}t = B_{qp} \cdot N_q \cdot \rho_v \tag{32}$$

For the case of thermal equilibrium:

$$g_q \cdot B_{qp} = g_p \cdot B_{pq} \tag{33}$$

and:

$$A = (8\pi h v^3/c^3) \times B_{qp} = (8\pi h v^3/c^3) \times (g_q/g_p) \times B_{qp} \tag{34}$$

where g_p and g_q are the degeneratives of the respective levels with $g = 2J + 1$.

The intensity (I_{qp}) of an emitted spectral line is proportional to the number density of atoms in the state q:

$$I_{qp} = A_{qp} \cdot n_{aq} \cdot h \cdot v_{qp} \tag{35}$$

or after substitution of n_{aq}, or n_q for atomic species, according to Eq. (20):

$$I_{qp} = A_{qp} \cdot h \cdot v_{qp} \cdot n_a \cdot (g_q/Z_a) \cdot \exp(-E_q/kT) \tag{36}$$

When multiplying with $d/(4\pi)$, where d is the depth of the source (in cm), one obtains the absolute intensity. T is the excitation temperature, which can be determined from the intensity ratio for two lines (a and b) of the same ionization stage of an element as:

$$T = [5040(V_a - V_b)]/\{\log[(gA)_a/(gA)_b] - \log(\lambda_a/\lambda_b) - \log(I_a/I_b)\} \tag{37}$$

In order to determine the excitation temperature with a high precision, the thermometric species should have a high degree of ionization, otherwise the temperature or the geometry of the discharge change when the substance is brought into the source. Furthermore, the difference between V_a and V_b must be large. Indeed, the error of the determination can be obtained by differentiating Eq. (37):

$$dT/T = T/[5040(V_a - V_b)] \times 0.434 \times dI/I \tag{38}$$

$(gA)_a/(gA)_b$ must be large when $(V_a - V_b)$ is large and I_a/I_b should not be particularly small or large. In addition the transition probabilities must be accurately known. Indeed, the error of the determination of the temperature strongly depends on the accuracy of $(gA)_a/(gA)_b$, as by differentiating Eq. (37) with respect to $\log[(gA)_a/(gA)_b]$ one obtains:

$$dT = T^2/[5040(V_a - V_b)] \times d[(gA)_a/(gA)_b] \tag{39}$$

Often the line pair Zn 307.206/Zn 307.59 is used, for which: $V_a = 8.08$ eV and $V_b = 4.01$ eV and $(gA)_a/(gA)_b = 380$ and:

$$T = 20510/[2.58 + \log[I_{307.6 \text{ nm}}/I_{307.2 \text{ nm}}] \tag{40}$$

This line pair is very suitable because the ionization of zinc is low as a result of its relatively high ionization energy, the wavelengths are close to each other, which minimizes errors introduced by changes in the spectral response of the detector, and the ratio of the *gA* values is well known.

The excitation temperature can also be determined from the slope of the plot $\ln[I_{qp}/(g_q \cdot A_{qp} \cdot v_{qp})]$ or $\ln[I_{qp} \cdot \lambda/(gA_{qp})]$ versus E_q, which is $-1/k \cdot T$ as:

$$\ln[I_{qp}/(g_q \cdot A_{qp} \cdot v_{qp})] = \ln(h \cdot n/Z) - E_q/(k \cdot T) \tag{41}$$

The λ/gA values for a large number of elements (such as argon, helium, iron) and their lines have been compiled [6]. The determination of excitation temperatures in spatially inhomogeneous plasmas has been treated extensively by Boumans [7] and is described later (see Ref. [8]).

• Example

For the case of a 2 kW inductively coupled plasma the four iron lines Fe I 381.58 nm, Fe I 383.04 nm, Fe I 382.44 nm and Fe I 382.58 nm have relative intensities of 5, 10, 2.3 and 7.4 a.u., respectively. When using the transition probability products *gA* of 66, 36, 1.26 and 26 (see Ref. [9]) as well as the excitations energies of 38175, 33096, 26140 and 33507 cm^{-1}, respectively, it can be calculated that the excitation temperature should be 5000 K.

Limitations to the spectroscopic measurement of the temperatures from line intensities lie in possible deviations from ideal thermodynamic behavior in real radiation sources, but also in the poor accuracy of transition probabilities. They can be calculated from quantum mechanics, and have been determined and compiled by Corliss and Bozman at NIST [10] from measurements using a copper dc arc. These tables contain line energy levels, transition probabilities and the so-called oscillator strengths for ca. 25000 lines between 200 and 900 nm for 112 spectra of 70 elements. Between the oscillator strength f_{qp} (being 0.01–0.1 for non-resonance and nearer to 1 for resonance lines) there is the relationship [11]:

$$f_{qp} = (g_q/g_p) \times A_{qp} \times [(m \cdot c^3)/(8 \cdot \pi^2 \cdot e^2 \cdot v^2)] \tag{42}$$

and

$$f_{qp} = 1.499 \times 10^{-16} \times \lambda^2 \times g_q/g_p \times A_{qp} \tag{43}$$

As g is known to be $2J + 1$, oscillator strengths can be converted into transition probabilities and vice versa.

According to the classical dispersion theory, the relationship between the absorption and the number density of the absorbing atoms is given by:

$$\int K_v \cdot dv = (\pi e^2)/mc \cdot N_v \cdot f \tag{44}$$

K_v is the absorption coefficient at a frequeny v, m is the mass and e the charge of the electron, c is the velocity of light, N_v is the density of atoms with a line at a frequency between v and $v + dv$ and is almost equal to N; f is the oscillator strength. This relationship applies purely to monochromatic radiation. As the widths of the absorption lines in most atom reservoirs are of the order 1–5 pm, the use of a primary source which emits very narrow lines would be advantageous. Indeed, when using a continuous source one would need a spectral apparatus with a practical resolving power of at least 500 000 to reach the theoretically achievable values of K_v and this certainly would lead to detector noise limitations as a result of the low irradiances. Therefore, it is more advantageous to use sources which emit relatively few narrow lines and to use a low-resolution monochromator which just isolates the spectral lines in the spectra.

The relationship between the absorption A and the concentration of the absorbing atoms in an atom reservoir is given by the Beer–Lambert law. When I_0 is the intensity of the incident radiation, l the length of the atom reservoir and I the intensity of the sorting radiation, the change in intensity dI resulting from the absorption within the absorption path length dl is given by:

$$-dI = k \cdot I_0 \cdot c \cdot dl \tag{45}$$

or

$$\int_{I_0}^{I} (dI)/I = k \cdot c \cdot \int_{0}^{l} dl \tag{46}$$

$A = \log(I_0/I) = \log(1/T)$, in which A is the absorbance and T the transmission. Accordingly, Eq. (46) becomes:

$$A = k \cdot c \cdot l \tag{47}$$

The absorbances are additive. The Lambert–Beer law, however, is only valid within a restricted concentration range. This is due to the fact that not all radiation reaching the detector has been absorbed to the same extent by the analyte atoms. Normally the calibration curve at high concentrations levels off asymptotically to the signal for the non-absorbed radiation. The latter consists of contributions from non-absorbed lines of the cathode material in the hollow cathode lamp or of lines of the filler gas within the spectral bandwidth of the monochromator. The calibration curve is only linear in the concentration range where the ratio of the widths of the emission to the absorption line is $< 1/5$. Also incomplete dissociation of analyte molecules leads to a curvature towards the concentration axis, and this incomplete dissociation then becomes limiting at high concentrations. A decrease in the ionization at higher analyte concentrations, however, may cause curvature away from the concentration axis. For all these reasons, in atomic absorption spectrometric measurements deviations from linearity are common and the linear dynamic range is much smaller than in atomic emission or atomic fluorescence.

Line broadening

Atomic spectral lines have a physical width as a result of several broadening mechanisms [12].

The natural width of a spectral line is due to the finite lifetime of an excited state (τ). The corresponding halfwidth in terms of the frequency is given by:

$$\Delta \nu_N = 1/(2\pi\tau) \tag{48}$$

For lines corresponding to transitions that are allowed according to the selection rules, lifetimes are of the order of 10^{-8} s and, accordingly, for most spectral lines this results in a natural broadening with a halfwidth of the order of 10^{-2} pm.

The Doppler spectral width results from the movement of the emitting atoms and their velocity component in the viewing direction. The respective halfwidth is given by:

$$\Delta \nu_D = [2 \cdot \sqrt{(\ln 2)}/c] \cdot \nu_0 \cdot \sqrt{[(2 \cdot R \cdot T)/M]} \tag{49}$$

where c is the velocity of light, ν_0 is the frequency of the line maximum, R is the gas constant and M the atomic mass. The Doppler broadening thus strongly depends on the temperature. Accordingly, it is also often denoted as temperature broadening and reflects the kinetic energy of the radiating species (atoms, ions or molecules). The relevant temperature is denoted as the gas temperature or Doppler temperature. The measurement of the Doppler broadening thus allows the determination of the gas temperatures in spectroscopic sources (see line profiles). For the light elements, the Doppler broadening is larger than it is for analytical lines with shorter wavelengths. For the Ca 422.6 nm line in the case of a hollow cathode discharge at a few mbar pressure, the Doppler broadening at 300 K for instance is 0.8 pm whereas at 2000 K it is 2 pm [13].

The Lorentzian broadening or pressure broadening results from the interaction between the emitting atoms of the element considered and atoms of other elements. The halfwidth is given by:

$$\Delta \nu_L = (2/\pi) \cdot \sigma_L^2 \cdot N \sqrt{[2 \cdot \pi \cdot R \cdot T \cdot (1/M_1 + 1/M_2)]} \tag{50}$$

M_1 and M_2 are the atomic masses, N is the concentration of the foreign atoms and σ_L is their cross section. The pressure broadening is low in the case of discharges under reduced pressure. For the case of the Ca 422.6 nm line this type of broadening at a temperature of 300 K and a pressure of 9 mbar is only 0.02 pm. At atmospheric pressure, however, this type of line broadening is the predominant one.

Furthermore, isotopic structure and hyperfine structure, and also resonance broadening, resulting from the interaction between radiating and non-radiating atoms of the same species, and Stark broadening, resulting from interaction with electrical fields, contribute to the physical widths of the spectral lines.

The natural broadening and the Lorentzian broadening have a Lorentzian distribution, which is given by:

$$I(v) = I_0 / \{1 + [2(v - v_0)/\Delta v]^2\} \tag{51}$$

The Doppler broadening has a Gaußian distribution, described by:

$$I(v) = I_0 \cdot \exp{-\{[2(v - v_0)/2\Delta v_D] \cdot \sqrt{(\ln 2)}\}} \tag{52}$$

The combination of both types of profile functions (which is normally due to the predominance of both pressure and temperature broadening) results in a so-called Voigt profile, which can be described by:

$$V(\alpha, \omega) = \alpha/\pi \int_{-\infty}^{+\infty} \{[\exp(-y^2 \cdot dy)]/[\alpha^2 + (\omega - y)]^2\} \tag{53}$$

$$\omega = 2(v - v_0)/\Delta v_D \cdot \sqrt{(\ln 2)} \quad \text{and} \quad \alpha = \Delta v_D \cdot \sqrt{(\ln 2)}$$

when the contribution of the natural width is neglected. From the physical line widths of the spectral lines, which in most cases are between 1 and 20 pm, and the information on the profile functions discussed above, the contributions of the different processes can be calculated by deconvolution methods. However, the experimentally measured full-widths at half maximum (FWHM) must be corrected for the bandwidth of the spectral apparatus used. In the determination of absorption profiles with the aid of very narrow bandwidth tunable laser sources, the spectral bandwidth of the probing beam can often be neglected. Equation (53) can e.g. be solved by a mathematical progression as:

$$V(\alpha, \omega) = \alpha/[\text{constant} \times (\alpha^2 + \omega^2)]\{1 + [3(\alpha^2 - \omega^2)/2(\alpha^2 + \omega^2)^2]$$

$$+ [3(\alpha^4 + 10\alpha^2\omega^2 + 5\omega^4)/4(\alpha^2 + \omega^2)^4] + \cdots\} \tag{54}$$

From this solution α and ω and accordingly Δv_L and Δv_D can be determined and from the latter the gas kinetic temperature. This is a measure of the kinetic energy of atoms and ions in the plasma.

Owing to the line broadening mechanisms, the physical widths of spectral lines in most radiation sources used in optical atomic spectrometry are between 1 and 20 pm. This applies both for atomic emission and atomic absorption line profiles. In reality the spectral bandwidth of dispersive spectrometers is much larger than the physical widths of the atomic spectral lines.

The profiles of spectral lines as are obtained in plasma sources at atmospheric pressure are illustrated by the high-resolution records of a number of rare earth spectral lines obtained in the case of an inductively coupled plasma source (ICP) (Fig. 2) [14].

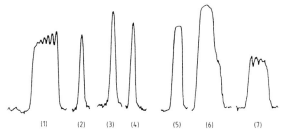

Fig. 2. Line profiles of some rare earth element atomic emission lines in inductively coupled plasma atomic emission spectrometry (photographic measurements obtained with a high-resolution grating spectrograph): theoretical resolving power: 460 000. (1): Ho II 345.6 nm, (2): Er II 369.2 nm, (3): Yb II 369.4 nm, (4): Y II 371.0 nm, (5): Tm II 376.1 nm, (6): Eu II 381.9 nm and (7): La II 398.8 nm. (Reprinted with permission from Ref. [14].)

Self-absorption

The radiation emitted in a radiation source is absorbed by ground state atoms of the same species. This phenomenon is known as self-absorption (for an explanation, see e.g., Ref. [15]). As the chance that an absorbed photon is re-emitted is <1, this causes the observed radiation to be smaller than the emitted radiation. When I_0 is the intensity emitted at the wavelength of the line maximum and $P_E(v)$ is the profile function, the intensity distribution for a line emitted by a radiation source over the profile of a line equals $I_0 \cdot P_E(v)$ and the intensity observed after the radiation has passed through a layer with a number density of absorbing atoms of n_A is:

$$I(v) = I_0 \cdot P_E(v) \cdot \exp[-p \cdot P_A(v)/P_A(v_0)] \tag{55}$$

v_0 is the frequency at the line center, $P_A(v)$ is the absorption profile function, $P_A(v_0)$ is its value at the line center and p is an absorption parameter:

$$p \approx B \cdot P_A(v) \cdot n_A \tag{56}$$

p increases with the absorption transition probability and thus is larger for resonance lines which stem from transitions ending at the ground level. It is also larger in sources with a high analyte number density n_A. As the absorption is maximum in the center of the line, self-absorption always leads to flatter line profiles. Self-absorption increases with the analyte number densities in the source and with the number densities of emitting and absorbing analyte atoms the intensity of a line goes to that given by Planck's law for black body radiation. When a minimum occurs in the absorption profile, then the line undergoes so-called self-reversal, when the absorption parameter is then >1. Self-reversal only occurs when there is a strong temperature gradient in the radiation source and when the analyte number densities both in the hot as well as in the cooler zones of the plasma are high.

Continuum radiation

Apart from the atomic and ion lines of the species present in a plasma source an emission spectrum has a continuum on which the emission lines are superimposed. This extends over the whole spectrum. It is due to the interactions between free electrons ("Bremsstrahlung") and to the interaction of free and bound electrons ("recombination continuum"). The former is particularly important in the UV spectral region, whereas the latter is important at longer wavelengths. The spectral intensity distribution for the continuum radiation is given by:

$$I(v) \cdot dv = K \times n_e \times n_r \times r^2/(T_e)^{1/2} \times \exp[(-hv)/(kT_e)] \times dv (\text{free–free})$$
$$+ K' \times 1/(j^3) \times n_e \times n_Z \times (Z^4)/(T_e^{3/2})$$
$$\times \exp[-(U_j - hv)/kT_e](\text{free–bound}) \tag{57}$$

K and K' are constants, n_e is the electron number density, n_r is the number density of the ions with a charge of r times the elementary charge, T_e is the electron temperature, and is a measure of the kinetic energy of the electrons in the plasma. v is the frequency, h and k are the Planck and Boltzmann constants, respectively, U_j is the ionization energy from the term with quantum number j, n_Z is the number density of the atoms with atomic number Z.

The intensity per unit of wavelength (radiant density B_λ) is obtained by multiplying with $4\pi c/\lambda^2$. Accordingly, for a hydrogen plasma the intensity of the "Bremsstrahlung" is given by:

$$B_\lambda = (2.04 \times 10^{-32} \times n_e^2)/(\lambda^2 \times T_e^{1/2}) \exp[-(1.44 \times 10^8)/(\lambda T_e)] \tag{58}$$

At complete ionization of the hydrogen (e.g. when added to a plasma with another gas as the main constituent) $n_e = p/(2 \times k \times T_e)$ has a maximum at a wavelength of $\lambda = (7.2 \times 10^7)/T_e$ or at a fixed wavelength, the maximum intensity is found at a temperature $T_e = (5.76 \times 10^7)/\lambda$. Thus, the electron temperature can be determined from the wavelength dependence of the continuum intensity. As T_e is the electron temperature, absolute measurements of the background continuum emission in a plasma, e.g. for the case of hydrogen, allow determination of the electron temperature in a plasma, irrespective of whether it is in local thermal equilibrium or not. Similar methods also make use of the recombination continuum and of the ratio of the "Bremsstrahlung" and the recombination continuum.

1.4
Ionization

Provided that sufficient energy is transferred to a plasma, atoms can also be ionized. This depends on the temperature of the plasma and also on the ionization energy of the elements. As these ions have discrete energy levels between which transitions are possible, just as atoms do, ionic spectra also will be important when considering the emission of radiation by a plasma. The ionization of atoms (a) of the element j into ions (i) is an equilibrium:

$$n_{aj} \rightleftharpoons n_{ij} + n_e \tag{59}$$

and the equilibrium constant $S_{nj}(T)$, known as the Saha constant, is given by:

$$S_{nj}(T) = (n_{ij} \cdot n_e)/n_{aj} \tag{60}$$

The degree of ionization α_j for an element j is given by:

$$\alpha_j = n_{ij}/n_j = n_{ij}/(n_{aj} + n_{ij}) \tag{61}$$

n_{aj} and n_{ij} are the concentrations of the atoms and the ions and can be expressed as a function of the total number of atoms n_j by:

$$n_{aj} = (1 - \alpha_j) \cdot n_j \quad \text{and} \quad n_{ij} = \alpha_j \cdot n_j \tag{62}$$

Accordingly, using the notation given in Ref. [7] the intensity of an atom line can be written as:

$$I_{qp} = A_{qp} \cdot h \cdot v_{qp} \cdot g_q/Z_{aj} \cdot (1 - \alpha_j) \cdot n_j \cdot \exp(-E_q/kT) \tag{63}$$

and the intensity of an ion line is given by:

$$I_{qp}^+ = A_{qp}^+ \cdot h \cdot v_{qp}^+ \cdot g_q^+/Z_{ij} \cdot \alpha_j \cdot n_j \cdot \exp(-E_q^+/kT) \tag{64}$$

These expressions for the intensities contain three factors that depend on the temperature, namely the degree of ionization, the Boltzmann factors and the partition functions. In particular, α_j can be written as a function of the electron number density and the Saha function as:

$$[\alpha_j/(1 - \alpha_j)] \cdot n_e = S_{nj}(T) \tag{65}$$

However, the latter is also given by the well-known Saha equation. With the aid of wave mechanics and through differentiation of the Boltzmann equation, the Saha function in terms of the partial pressures can also be expressed as:

$$S_{pj}(T) = (p_{ij} \cdot p_e)/(p_{aj}) = [(2\pi m)^{3/2} \cdot (kT)^{5/2}]/h^3$$
$$\cdot 2Z_{ij}/Z_{aj} \cdot [\exp(-E_{ij}/kT)] \tag{66}$$

The factor of 2 is the statistical weight of the free electron (2 spin orientations),

$k = 1.38 \cdot 10^{-16}$ erg/K, $m = 9.11 \times 10^{-28}$ g, $h = 6.67 \times 10^{-27}$ erg s and 1 eV = 1.6×10^{-12} erg. This leads to the expression:

$$(P_{ij} \cdot p_e)/p_{aj} = S_{pj}(T) = 6.58 \times 10^{-7} \times T^{3/2}$$
$$\times Z_{ij}/Z_{aj} \times [\exp_{10}(-5040 V_{ij}/T)] \tag{67}$$

or in the logarithmic form:

$$\log S_{pj} = 5/2 \log T - 5040/T \cdot V_{ij} + \log(Z_{ij})/(Z_{aj}) - 6.18 \tag{68}$$

where V_{ij} is the ionization energy (in eV).

The Saha equation is only valid for a plasma which is in local thermal equilibrium, where the temperature in the equation is then the ionization temperature. When this condition is not fullfilled, the equilibrium between the different states of ionization is given by the so-called Corona equation [16].

Accordingly, the degree of ionization in a plasma can be determined from the intensity relationship between an atom and an ion line of the same element as:

$$\log[(\alpha_j)/(1 - \alpha_j)] = \log(I_{qp}^+/I_{qp}) - \log[(g_q^+ \cdot A_{qp}^+ \cdot v_{qp}^+)/(g_q \cdot A_{qp} \cdot v_{qp})]$$
$$+ (5040/T)(V^+ - V_q) + \log(Z_{ij}/Z_{aj}) \tag{69}$$

This method can again only be applied for a plasma in local thermal equilibrium, the temperature of which is known. The partition functions Z_{aj} and Z_{ij} for the atom and ion species, respectively, are again a function of the temperature and the coefficients of these functions have been calculated for many elements [5]. Furthermore, the accuracy of the gA values and of the temperatures is important for the accuracy of the determination of the degree of ionization. One often uses the line pairs Mg II 279.6 nm/Mg I 278.0 nm and Mg II 279.6 nm/Mg I 285.2 nm for determinations of the degree of ionization of an element in a plasma.

Once α_j is known, the electron pressure can also be determined. Indeed, from

$$\log[\alpha_j/(1 - \alpha_j)] = \log[S_{pj}(T)/p_e] \tag{70}$$

one can calculate

$$\log p_e = -\log[\alpha_j/(1 - \alpha_j)] + \log S_{pj}(T) \tag{71}$$

When taking into account Eqs. (68) and (69), this results in:

$$\log p_e = -\log(I_{qp}^+)/(I_{qp}) + \log(g_q^+ \cdot A_{qp}^+ \cdot v_{qp}^+)/(g_q \cdot A_{qp} \cdot v_{qp})$$
$$- (5040/T)(V_{ij} + V_q^+ - V_q) + 5/2(\log T) - 6.18 \tag{72}$$

By differentiating Eq. (72) with respect to temperature it can be found that the error in the determination of the electron pressure as a result of errors in the determination of the temperature is given by:

$$d(\log p_e) = [(0.434 \times 5/2) + (5040/T)(V_{ij} + V_q^+ - V_q)] \times (dT/T) \tag{73}$$

Eq. (72) also shows that the intensity ratio of the atom and ion lines of an element will change considerably with the electron pressure in the plasma. Elements with a low ionization energy such as Na will thus have a strong influence on the intensity ratios of the atom and ion lines of other elements. This is analytically very important as it is the cause of the so-called ionization interferences, found in classical dc arc emission spectrometry but also in atomic absorption and plasma optical emission as well as in mass spectrometry.

When the plasma is not in local thermal equilibrium (LTE), the electron number densities cannot be determined on the basis of the Saha equation. Irrespective of the plasma being in local thermal equilibrium or not, the electron number density can be derived directly from the Stark broadening of the H_β line or of a suitable argon line. This contribution to broadening is a result of the electrical field of the quasi-static ions on one side and the mobile electrons on the other side. As described in Ref. [17] it can be written as:

$$\delta\lambda = 2[1 + 1.75\alpha(1 - 0.75\rho)]\omega \tag{74}$$

where ρ is the ratio of the distance between the ions (ρ_m) and the Debye path length (ρ_D), ω is the broadening due to the interaction of the electrons ($\approx n_e$) and α is the contribution of the interaction with the quasi-static ions ($\approx n_e^{1/4}$).

$$\rho_m = (4\pi \times n_e/3)^{-1/3} \tag{75}$$

$$\rho_D = [(k \times T)/(4\pi \times e^2 \times n_e)]^{1/2} \tag{76}$$

$\delta\lambda$ can thus be calculated as a function of n_e. Accordingly, from the widths of the Ar I 549.59 or the Ar I 565.07 nm lines, which are due mainly to Stark broadening, n_e can be determined directly and is independent of the existence of LTE. Thus the temperature can also be determined when combined with measurement of the intensities of an atom and an ion line of the same element. Indeed,

$$\log n_i/n_a = -\log n_e + 3/2 \log T - (5040/T) \cdot V_{ij} + \log(Z_{ij}/Z_{aj}) + 15.684 \tag{77}$$

which can be combined with Eq. (72).

Because with the "two-line method" using lines of the same ionization level for the determination of temperatures, it is difficult to fulfil all conditions necessary to obtain highly accurate values [see Eqs. (38) and (39)], a method was developed that enables the plasma temperature to be determined from intensities of lines belonging to different ionization levels. When I_i is the intensity of an ion line and I the intensity of an atom line (in general both lines have to belong to two adjacent ionization levels), one can write:

$$I_i/I = 2(A_i g_i \lambda_i / A g \lambda) \times [(2\pi m k T)^{3/2}/(h^3)] \times (1/n_e)$$
$$\times (T^{3/2}) \times \exp[-(E_i - E + E_i - \Delta E_i)/(kT)] \tag{78}$$

ΔE_i is a correction for the ionization energy of the lowest level. The plasma temperatures can also be determined from the measurements of absolute line intensities. A survey of all methods used and discussed in the various chapters is given in Refs. [7, 8, 12].

Norm temperatures

From Eqs. (63) and (64), which give the intensity of a line, and from the Saha equation [Eq. (68)], it can be understood that for each spectral line emitted by a plasma source there is a temperature where its emission intensity is maximum. This is the so-called norm temperature. In a first approximation [18], it can be written as:

$$T_n = (0.95 V_{ij} \times 10^3)/[1 - 0.33 \cdot \alpha + 0.37 \cdot \log(V_{ij}/10) - 0.14 \log P_e^*] \tag{79}$$

where V_{ij} is the ionization energy, $\alpha = V_a/V_{ij}$ and V_e is the excitation energy. P_e^* is the electron pressure (in atm; 1 atm $\approx$ 101 kPa) and is a function of T and n_e:

$$P_e^* = 1.263 \times 10^{-12} n_e \times T \tag{80}$$

In the case where we have to consider the norm temperature for a line of an element which is only present as an impurity in a plasma, e.g., one formed in a noble gas, the dilution in the plasma (a) also has to be considered. For a system with more components P_e^* is given by:

$$P_e^* = [(2i + 1)/(2i + 3)] \cdot (g_i/g_{i+1}) \cdot [4a/(1 + a)^2] \cdot P \tag{81}$$

g_i and g_{i+1} are the statistical weights of the ions with charge i and $i + 1$, respectively. Accordingly, as a result of the dilution the change in the norm temperature (T_n) at a dilution of a will be given by:

$$\Delta T_n/T_n = 0.14 \log[4a/(1 + a)^2] < 1 \tag{82}$$

At a dilution of 0.1 the change in norm temperature will thus be −7.2%. In a source such as the inductively coupled plasma the analyte dilution can be very high [of the order of 10^8 (1 mL/min of a 1–10 µg/mL solution for an element with a mass of 40, which is nebulized with an efficiency of 1% into an argon flow of

10 L/min). In an ICP the norm temperatures for lanthanum atom lines such as La I 418.7 nm ($V_{ij} = 5.61$ eV and $V_a = 2.96$ eV), will thus change from ca. 5000 K for a pure lanthanum plasma to 2830 K as a result of the large dilution. However, for atom lines of elements with relatively low ionization energies, although the change in the norm temperatures as a result of the analyte dilution is high, it is much less than for ion lines. For the La II 412.3 nm line (second ionization energy: 11.43 eV and V_a: 3.82 eV) T_n is 9040 K.

From what is known about the norm temperatures, it becomes clear which types of lines will be optimally excited in a plasma of a given temperature, electron pressure and gas composition, and the norm temperatures thus give important indications for line selection in a source of a given temperature. Atom lines often have their norm temperatures below 4000 K, especially when the analyte dilution in the plasma is high, whereas ion lines often reach 10 000 K. Both types of lines are often denoted as "soft" and "hard" lines, respectively.

1.5
Dissociation

The dissociation of molecular plasma gases or analyte molecules which are brought into the radiation source is an equilibrium reaction. Accordingly, thermally stable radicals in particular or molecules are always present in a radiation source. They emit molecular bands which occur along with the atomic and ionic line spectra in the emission spectrum. Radicals and molecules may also give rise to the formation of cluster ions, the signals of which will be present in the mass spectra. The main species stemming from the plasma gases are: CN, NH, NO, OH and N_2 (or N_2^+). From the analytes, predominantly thermally stable oxides remain (e.g., AlO^+, TiO^+, YO^+, etc.). A thorough treatment of molecular spectra is available in many classical textbooks (see e.g., Refs. [19, 20]).

Molecules or radicals have different electronic energy levels ($^1\Sigma, {}^2\Sigma, {}^2\Pi, \ldots$), which have a vibrational fine structure ($v = 0, 1, 2, 3, \ldots$) and the latter again have a rotational hyperfine structure ($J = 0, 1, 2, 3, \ldots$). The total energy of a state is then given by:

$$E_i = E_{el} + E_{vibr} + E_{rot} \tag{83}$$

E_{el} is of the order of 1–10 eV, the energy difference between two vibrational levels of the same electronic state is of the order of 0.25 eV and for the case of two rotational levels of a vibrational band the energy difference is of the order of only 0.005 eV. Through a transition between two rotational levels a rotational line is emitted. When the rotational levels considered belong to the same electronic level, the wavelength of the radiation emitted will be in the infrared region. When they belong to different electronic levels, their wavelengths will be in the UV or in the visible region. Transitions are characterized by the three quantum numbers of the states involved, namely: n', v', j' and n'', v'', j''. All lines which originate from transitions between rotational levels belonging to different vibrational levels of two electronic states form the band: n', $v' \rightarrow n''$, v''. For these band spectra the selection rule is $\Delta j = j' - j'' = \pm 1, 0$. Transitions for which $j'' = j' + 1$ then give rise to the P-branch, $j'' = j' - 1$ to the R-branch and $j' = j''$ to the Q-branch of the band. The line corresponding with $j' = j'' = 0$ is the zero line of the band. When $v' = v'' = 0$, it is also the zero line of the system. The difference between the wavenumber of a rotation line (in cm^{-1}) and the wavenumber of the zero line in the case of the P and the R branch is a function of the rotation quantum number j and the rotational constant B_v for which:

$$E_j/(hc) = B_v \cdot j(j+1) \tag{84}$$

The functional relationship is quadratic and is known as the Fortrat parabola.

For the CN radical and the N_2^+ molecular ion, the transitions giving rise to band emission between 370 and 400 nm, together with the rotational line pattern are represented in Fig. 3 [21]. For the violet system of the CN band, there is no Q-branch and the lowest j in the R-branch is $j = 1$.

Molecular and radical band spectra thus consist of electronic series, which in their turn consist of various vibrational bands, which again consist of rotational lines, many often only partially resolved. As in the case of atomic spectral lines, the intensity of a rotation line can be written as:

$$I_{nm} = N_m \cdot A_{nm} \cdot h \cdot \nu_{nm} \cdot 1/2\pi \tag{85}$$

where N_m is the population of the excited level and ν_{nm} the frequency of the emitted radiation. The transition probability for dipole radiation is:

$$A_{nm} = (64 \cdot \pi^4 \cdot \nu_{nm}^3)/3k \times 1/(g_m) \times \Sigma |R_{n_i m_k}|^2 \tag{86}$$

i and k are degenerate levels of the upper (m) and the lower state (n). $R_{n_i m_k}$ is a matrix element of the electrical dipole moment and g_m is the statistical weight of the upper state. N_m is given by the Boltzmann equation:

$$N_m = N \cdot (g_m)/Z(T) \cdot \exp(-E_r/kT) \tag{87}$$

where E_r is the rotational energy of the excited electronic and vibrational level and is given by:

$$E_r = h \cdot c \cdot B_{v'} \cdot J' \cdot (J' + 1) \tag{88}$$

$B_{v'}$ is the rotational constant and J' is the rotational quantum number of the upper state (m). For a $^2\Sigma_g$–$^2\Sigma_u$ transition the term $\Sigma |R_{n_i m_k}|^2 = J' + J'' + 1$ where J' and J'' are the rotational quantum numbers of the upper and the lower states, respectively. Accordingly:

$$I_{nm} = (16 \cdot \pi^3 \cdot c \cdot N \cdot v_{nm}^4)/3Z(T) \cdot (J' + J'' + 1)$$
$$\cdot \exp(-h \cdot c \cdot B_{v'} \cdot J'(J' + 1)/kT) \tag{89}$$

or

$$\ln[I_{nm}/(J' + J'' + 1)] = \ln[16 \cdot \pi^3 \cdot c \cdot N \cdot v_{nm}^4]/[3Z(T)]$$
$$- [h \cdot c \cdot B_{v'} \cdot J'(J' + 1)]/kT \tag{90}$$

By plotting $\ln[I_{nm}/(J' + J' + 1)]$ versus $J'(J' + 1)$ for a series of rotational lines a so-called rotational temperature can be determined. It characterizes the kinetic energy of the molecules and radicals, by which the band spectra are emitted. It is also a good approximation of the temperature reflecting the kinetic energy of the neutrals and ions in a plasma. For the case of a hollow cathode discharge the Boltzmann plot and the temperatures as measured from CN and N_2^+ band hyperfine structures are given in Fig. 3.

Spectral lines of molecular bands emitted by molecules and radicals present in a plasma often interfere with the atomic spectral lines in atomic emission spectrometry. However, in atomic absorption spectrometry the absorption by molecular bands stemming from undissociated molecules in the atom reservoir also lead to systematic errors and require correction. Furthermore, in mass spectrometry molecular fragments give rise to signals, which can also interfere with the signals from the ionized analyte atom. Therefore, it is important to study the dissociation of molecular species in the high-temperature sources used as radiation sources, atom reservoirs or ion sources. In different plasma sources a series of band-emitting species stem from the working gas. In this respect, for instance, N_2, N_2^+, CN, OH and NH band emission has to be mentioned. However, undissociated sample and analyte species are also present in the plasma. In particular, thermally stable molecules such as AlO, LaO, BaO, AlF, CaF_2 and MgO may be present in atomic spectrometric sources. It is important to understand their dissociation as a function of the plasma temperature and the plasma composition. This dependence can be described by a dissociation equation, which is similar to the Saha equation:

$$K_n = [(2\pi/h^2) \cdot (m_X \cdot m_Y/m_{XY}) \cdot (kT)^{3/2}] \cdot (Z_X \cdot Z_Y/Z_{XY}) \cdot [\exp(-E_d/kT)] \tag{91}$$

where:

$$K_n = (n_X \cdot n_Y)/n_{XY} \tag{92}$$

Z represents the partition functions for the different atomic and molecular species, M_X, M_Y and M_{XY} are the respective masses and E_d is the dissociation energy.

This can be rewritten as:

a)

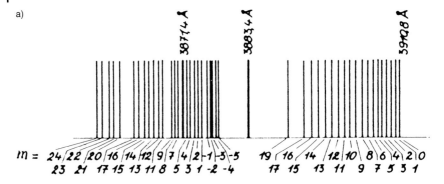

b)

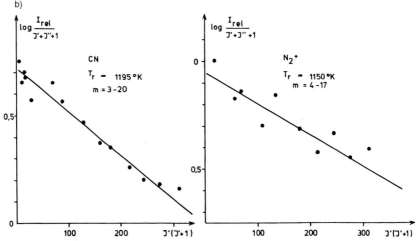

Fig. 3. Determination of rotational temperatures in a transitional hollow cathode. Rotational lines used are of the R-branch of $^2\Sigma-^2\Sigma$ (0,0) CN 388.3 nm and $^2\Sigma-^2\Sigma$ (0,0) N_2^+ 391.4 nm band (positive $m = J'$ values belong to the R-branch) (a); Boltzmann plots for a graphite hollow cathode ($\phi_i = 4.76$ mm) operated at 80 mA and 330 Pa argon (b). (Reprinted with permission from Ref. [21].)

$$\log K_n = 20.432 + 3/2 \log(M_X \cdot M_Y/M_{XY}) + \log(Z_X \cdot Z_Y/Z_{XY})$$
$$- \log g + 1/2 \log T + \log B + \log(1 - 10^{-0.625\omega/T})$$
$$- (5040/T) \times V_d \tag{93}$$

For most diatomic molecules B and ω (in cm^{-1}) have been listed in tables (see e.g. Ref. [19]). V_d then is given in eV and for a diatomic molecule:

$$Z = [kT/(h \cdot c \cdot B)] \times [g \times \exp(-E_d/kT)]/[1 - \exp(h \cdot c \cdot \omega/kT)] \tag{94}$$

with $g = 1$ ($^1\Sigma$), 2 ($^2\Sigma, ^1\pi, ^2\pi_{1/2}, ^2\pi_{3/2}, \ldots$), 3 ($^3\Sigma, \ldots$).

Thus for a metal oxide (XY), from $n_{XY}/n_X = N_Y/K_n$ the ratio of the number densities for the metal oxide and the metal atoms (n_X) as well as the degree of dissociation can be calculated when the plasma temperature, the partial pressure of oxygen in the plasma (p_Y) and the dissociation constant are known. For refractory oxides of relevance in dc arc analysis, these data are listed, for example, in Ref. [7].

1.6
Sources for atomic spectrometry

In atomic spectrometry the sample material is brought into a high-temperature source (plasma, flame, etc.) with the aid of a sampling device. The sample, which may be a liquid, a gas mixture or a solid must be transformed into a vapor or an aerosol. This involves sample nebulization or various volatilization processes (e.g., by thermal evaporation or sputtering). It is advantageous to supply as much energy as possible for this process. The volatilization processes, the principle of which is to lead to a physical or chemical equilibrium, will possibly result in complete atomization, irrespective of the state of aggregate, of the eventual solid state structure or of the chemical composition of the sample. This is very important both for obtaining the highest sensitivity as well as for keeping the matrix interferences involved in the analyses at the lowest possible level. The effectiviness of the volatilization processes involved, the plasma temperatures describing the kinetic energy distributions of the various plasma components as well as their number densities will all influence the atomization of the sample in the source.

The rotational temperatures are relevant to all processes in which molecules, radicals and their dissociation products are involved. They can be obtained from the intensity distribution for the rotational lines in the rotation–vibration spectra, as described by Eqs. (83–90). The molecules OH, CN etc. have often been used to measure temperature (see e.g. Refs. [21–23]).

The gas temperature is determined by the kinetic energy of the neutral atoms and the ions. It can be determined from the Doppler broadening of the spectral lines, as described by Eq. (49). However, to achieve this contributions of Doppler broadening and temperature broadening have to be separated, which involves the use of complicated deconvolution procedures as e.g. shown for the case of glow discharges in Ref. [24].

Whereas the rotational and the gas temperature are particularly relevant to the evaporation processes in the plasma, the electron temperature, being a measure of the kinetic energy of the electrons, is relevant to the study of excitation and ionization by collisions with electrons. This is an important process for generation of the analyte signal both in optical atomic emission and in mass spectrometry. The electron temperature can be determined from the intensity of the recombination continuum or of the "Bremsstrahlung", as described by Eq. (57).

The excitation temperature describes the population of the excited levels of atoms and ions. Therefore it is important in studies on the dependence of analyte line intensities on various plasma conditions in analytical emission spectrometry.

Tab. 1. Temperatures (K) for sources in atomic spectrometry.

Source	$T_{rotational}$	$T_{excitation}$	$T_{electron}$	$T_{ionization}$	
Arc discharges	5000	5000	5500	5000	LTE
Spark	—	20 000	20 000	2000	LTE (transient)
Inductively coupled plasma	4800	5000	6000	6000	$\approx$LTE
Microwave plasmas	2000	4000	6000	6000	departures from LTE
Glow discharges	600	20 000	30 000	30 000	non-LTE

It can be determined from the intensity ratio of two atomic emission lines of the same element and ionization state [see Eq. (42)] or from plots of the appropriate function for various atomic emission lines versus their excitation energies.

The ionization temperature is relevant for all phenomena involving equilibria between analyte atoms, ions and free electrons in plasmas. In the case of thermal equilibrium, it occurs in the Saha equation [Eqs. (66, 68)] and can be determined from the intensity ratio of an ion and an atom line of the same element. In all other cases ionization temperatures can be determined from the n_e value obtained from Stark broadening [see Eqs. (74, 77)].

The different temperatures for the most important sources in atomic spectrometry are listed in Table 1.

In a plasma, which is at least in local thermal equilibrium, all the temperatures discussed are equal. In addition, this implies that the velocity distribution of all types of particles in the plasma (molecules, atoms, ions and electrons) at any energy level can be described by the Maxwell equation [Eq. (17)]. For all species the population of the different levels is then given by the Boltzmann equation [Eq. (19)]. Furthermore, the ionization of atoms, molecules and radicals can be described by the Saha equation [Eqs. (66, 68)] and the related equations for the chemical equilibrium. Finally, the radiation density in the source conforms to Planck's law and both the exchange of kinetic energy between the particles as well as the electromagnetic radiation exchange are in equilibrium with each other. The real plasma sources used in atomic spectrometry are at best in so-called local thermal equilibrium. However, contrary to the case of thermal equilibrium all processes between the particles do not involve emission or absorption of electromagnetic radiation, as the plasma cannot be considered as a completely closed system. Moreover, real plasma sources are extremely inhomogeneous with respect to temperature and species number density distributions. Accordingly, the above mentioned equilibria only occur within small volume elements of the sources, where gradients can be neglected.

Many plasmas, however, have a cylindrical symmetry and can be observed laterally. The observer then integrates the information provided for many volume elements along the observation direction and:

$$I(x) = 2 \int_0^\infty I(x, y) \cdot dy \qquad (95)$$

where x and y are the coordinates of the volume element and integration of an

Fig. 4. Abel inversion procedure used for the determination of radial distributions in plasma sources. x: lateral position, y: direction of observation, r: radial distance.

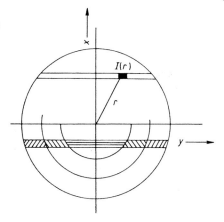

intensity I is performed along the x-axis. When considering $I(r)$, with r the radial distance away from the plasma center (Fig. 4), this becomes:

$$I(x) = 2 \int_x^\infty I(r) \cdot r \cdot dr / [\sqrt{(r^2 - x^2)}] \tag{96}$$

From these integral values, however, the values at a well-defined radial distance away from the plasma center can be calculated. This necessitates side-on observation along several parallel axes, which are equidistant with respect to each other and an Abel-inversion. Indeed, when $I'(x)$ is the first derivative of the function $I(x)$, describing the variation of a measured value as a function of the lateral position (x) during side-on observation (Fig. 4), the radial values at a distance r are given by:

$$I(r) = -(1/\pi) \int_r^\infty [I'(x) \cdot dx] / \sqrt{(r^2 - x^2)} \tag{97}$$

This integral can only be solved numerically. One has to write each laterally measured intensity as a sum of the different volume elements with a given coefficient denoting their contribution. For certain numbers of lateral observation positions $(3, 10, \ldots)$ these Abel coefficients are listed in the literature [25]. By inversion of the corresponding matrix the radial values are then obtained. This allows determination of the radial distributions of emissivities, absorbances, temperatures or species number densities in plasmas. By repeating this procedure at different heights in a plasma one can perform plasma tomography. Similar results, which do not require the assumption of cylindrical symmetry in the source, can now be obtained by imaging spectrometry using two-dimensional detectors (for a discussion, see e.g. Ref. [26]).

In real plasmas departures from thermal equilibrium often occur. For the extreme case, as encountered in plasmas under reduced pressure, the emission or the

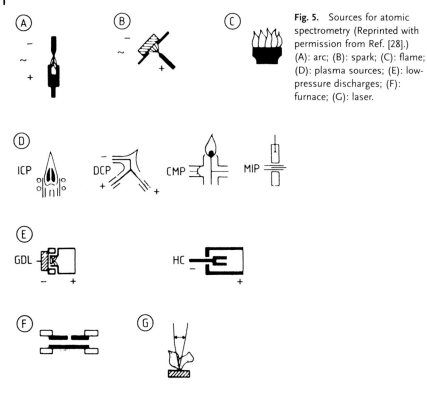

Fig. 5. Sources for atomic spectrometry (Reprinted with permission from Ref. [28].) (A): arc; (B): spark; (C): flame; (D): plasma sources; (E): low-pressure discharges; (F): furnace; (G): laser.

absorption of radiation becomes so important that there is no longer a clear relationship between the mean kinetic energies of the species and the excitation temperatures. The latter then lose the physical meaning of a temperature. The absence of local thermal equilibrium in these so-called non-LTE plasmas relates to the existence of high field gradients or ac fields, by which only the light electrons and not the heavy atoms and ions can follow the field changes and take up the dissipated energies fully. Accordingly, the mean kinetic energy of the electrons and thus also the electron temperature will be much higher than the gas kinetic temperature. For a plasma, which is not in local thermal equilibrium, the equilibrium between species at different ionization levels at different temperatures will be given by the so-called Corona equation, as is, for example, derived for the sun by astrophysicists [27].

Sources for atomic spectrometry include flames, arcs, sparks, low-pressure discharges, lasers as well as dc, high-frequency and microwave plasma discharges at reduced and atmospheric pressure (Fig. 5) [28]. They can be characterized as listed in Table 2. Flames are in thermal equilibrium. Their temperatures, however, at the highest are 2800 K. As this is far below the norm temperature of most elemental lines, flames only have limited importance for atomic emission spectrometry, but they are excellent atom reservoirs for atomic absorption and atomic fluorescence spectrometry as well as for laser enhanced ionization work. Arcs and sparks are

Tab. 2. Use of sources in analytical atomic spectrometry.

Source	Optical spectrometry			Mass spectrometry
	emission	absorption	fluorescence	
Chemical flame	+	+	+	
Arc	+			+
Spark	+			+
Electrically-heated furnace	+	+	+	+
Arc jet plasma	+			
Inductively coupled plasma	+		+	+
Microwave plasma	+	+		+
Glow discharge	+	+	+	+
Laser plasma	+	+	+	+
Exploding wire/foil	+			

well known as sources for atomic emission spectrometry. The high temperatures obtained in spark sources anticipate that ion lines in particular will be excited, the norm temperatures of which are often beyond 10 000 K, whereas in arc sources atom lines will be predominant. In plasma sources under reduced pressure the gas kinetic temperatures are low. Accordingly, their atomization capacity will be limited. When these sources are used as both atomic emission sources and as primary sources in atomic absorption the line widths will be very narrow. Moreover, particularly when gases with high ionization energies such as helium are used, lines with high excitation energies, such as the lines of the halogens, can also be excited. The discharges under reduced pressure are valuable ion sources for mass spectrometry as analyte ionization takes place, and because of the low pressure the coupling with a mass spectrometer operated at 10^{-5} mbar becomes easier. So-called plasma jet and plasma sources at atmospheric pressure are of particular use for the emission spectrometric analysis of solutions. Their gas kinetic temperatures are high enough so as to achieve a complete dissociation of thermally stable oxides and both atom lines as well as ion lines occur in the spectra. As reflected by the fairly high ionization temperatures, they are powerful ion sources for mass spectrometry and plasma mass spectrometry is now one of the most sensitive methods of atomic spectrometry. Lasers are very suitable devices for material ablation in the case of solids. Owing to their high analyte number densities, the plasmas are subjected to high self-absorption thus it is more appropriate to use them as sources for material volatilization only and to lead the ablated material into a second source.

1.7
Analytical atomic spectrometry

Analytical atomic spectrometry nowadays includes the use of flames and plasma discharges for optical and mass spectrometry. The sources are then used directly as

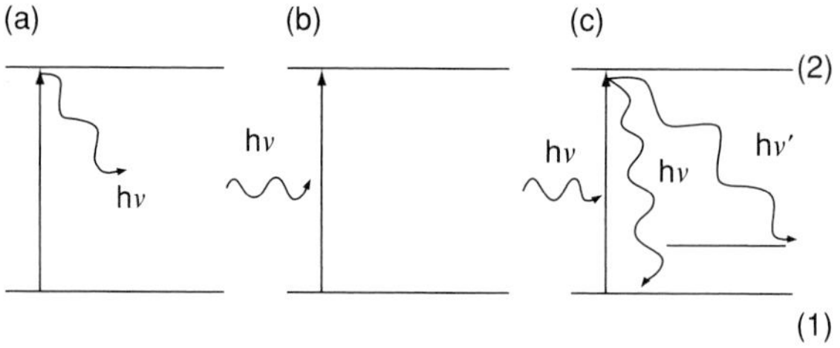

(a)   (b)   (c)

Fig. 6. Term schemes for (a): atomic emission, (b): absorption and (c): fluorescence spectrometry (resonant: $h\nu$ and non-resonant: $h\nu'$).

emission sources or atom reservoirs for atomic absorption or atomic fluorescence or they are used for ion production. In optical atomic spectrometry, atomic emission, absorption and fluorescence all have their specific possibilities and analytical features. The type of information obtained is clear from the transitions involved (Fig. 6).

In atomic emission, thermal or electrical energy is used to bring the analyte species into an excited state, from which they return to their ground state through emission of radiation characteristic of all species that are present and that were sufficiently excited. Thus from the principle of atomic emission spectrometry it is a clearly multielement method. The number of elements that can be determined simultaneously is only limited by the availability of sufficiently sensitive interference-free spectral lines. In the case of atomic emission spectrometry the selectivity is achieved by the isolation of the spectral lines with the aid of the exit slit of the spectrometer. This puts high demands on the optical quality of the spectral apparatus used and certainly requires a spectrometer with high spectral resolution. Moreover, the lines are superimposed on a spectral background, which is partly structured as the result of the presence of radicals and molecular ions emitting band spectra. Broad wings arising from neighboring spectral lines, e.g. of matrix constituents, may also occur and finally there will be a continuum resulting from the interaction of free and bound electrons (see Section 1.3, continuum radiation). The intensity of the spectral background on which the line to be measured is superimposed, may thus differ considerably from one sample to another. It must be subtracted from the total line intensities, as only the radiation emitted by the analyte atoms is relevant for the calibration. This can be done by estimating the spectral background intensity "under" the analytical line for the background intensities at the wavelength of the analytical line using a blank sample. It is often safer to estimate it from the spectral background intensities close to the analytical line (e.g. on each side) in the spectrum of the sample itself. The latter is certainly the case when the line is positioned on a wing of a band or a broad matrix spectral line, providing different spectral background intensities on either side of the analytical line. The means to correct for the spectral background must be available in every atomic emission spectrometer used for trace analysis.

In atomic absorption spectrometry we need a primary source delivering monochromatic radiation of which the wavelength agrees with that of a resonance line of the element to be determined. The spectral width must be narrow with respect to the absorption profile of the analyte line. From this point of view atomic absorption is a single-elemental method, of which the dynamic range is usually much lower than in atomic emission spectrometry. It is, at a first approximation, a zero-background method, when neglecting absorption due to radiation scattering and molecular absorption. As the final spectral selectivity is realized by the primary source, the spectral apparatus only has to enable line isolation in a spectrum of the element to be determined, where the spectral lines are very narrow. Accordingly, the demands on the spectral resolving power of the spectrometer are much lower than in atomic emission spectrometry.

In atomic fluorescence, the excitation can be performed both with white as well as with monochromatic sources, which consequently affects the fluorescence intensities obtainable and the freedom from stray radiation limitations. The latter are particularly low with monochromatic primary sources and when using fluorescence lines with wavelengths differing from that of the exciting radiation. Generally, in atomic fluorescence the linear dynamic range is higher than in atomic absorption and spectral interference as well as background interferences are just as low.

In the case of atomic absorption and atomic fluorescence the selectivity is thus already partly realized by the radiation source delivering the primary radiation, which in most cases is a line source (hollow cathode lamp, laser, etc.). Therefore, the spectral bandpass of the monochromator is not as critical as it is in atomic emission work. This is especially true for laser based methods, where in some cases of atomic fluorescence a filter is sufficient, or for laser induced ionization spectrometry where no spectral isolation is required at all.

For glow discharges and inductively coupled high-frequency plasmas ion generation takes place in the plasmas. In the first case mass spectrometry can be performed directly on solids and in the second case on liquids or solids after sample dissolution. In the various atomic spectrometric methods, real samples have to be delivered in the appropriate form to the plasma source. Therefore, in the treatment of the respective methods extensive attention will be given to the techniques for sample introduction.

In an atomic spectrometric source both the atomic vapor production as well as signal generation processes take place. The first processes require high energy so as to achieve complete atomization as already discussed, whereas the signal generation processes in many cases would profit from a discrete excitation which makes, use for example, of the selective excitation of the terms involved. Therefore, in a number of cases the use of so-called tandem sources, where the analyte vapor generation and the signal generation take place in a different source (for a discussion, see Ref. [29]), may offer advantages with respect to the power of detection as well as freedom from interferences.

2
Spectrometric Instrumentation

Atomic spectrometric methods of analysis essentially make use of equipment for spectral dispersion so as to isolate the signals of the elements to be determined and to make the full selectivity of the methodology available. In optical atomic spectrometry, this involves the use of dispersive as well as of non-dispersive spectrometers. The radiation from the spectrochemical radiation sources or the radiation which has passed through the atom reservoir is then imaged into an optical spectrometer. In the case of atomic spectrometry, when using a plasma as an ion source, mass spectrometric equipment is required so as to separate the ions of the different analytes according to their mass to charge ratio. In both cases suitable data acquisition and data treatment systems need to be provided with the instruments as well.

When designing instruments for atomic spectrometry the central aim is to realize fully the figures of merit of the methods. They include the power of detection and its relationship to the precision, the freedom from spectral interferences causing systematic errors and the price/performance ratio, these being the driving forces in the improvement of spectrochemical methods (Fig. 7).

2.1
Figures of merit of an analytical method

In practice atomic spectrometric methods are relative methods and need to be calibrated.

Calibration

The calibration function describes the functional relationship between the analytical signals, whether they be absorbances, absolute as well as relative radiation intensities or ion currents. The simplest form of the calibration function can be written as:

$$Y = a \cdot c + b \tag{98}$$

In the calibration this functional relationship is established from m measurements

Fig. 7. Qualities of an analytical method.

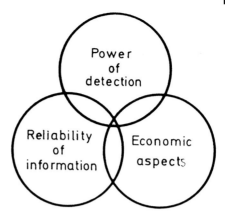

of the signals Y_i obtained for samples with a known concentration c_i. The coefficients a and b are then determined by a least square linear regression procedure, which ensures that the deviations between the measured signals and the signals calculated from the regression equation are minimal. This is the case for

$$a = (m\Sigma c_i \cdot Y_i - \Sigma c_i \cdot \Sigma Y_i)/[m\Sigma c_i^2 - (\Sigma c_i)^2] \qquad (99)$$

and

$$b = (\Sigma Y_i - a\Sigma c_i)/m \qquad (100)$$

Often the inverse function of the calibration function is used, namely:

$$c = a' \cdot Y + b' \qquad (101)$$

This is known as the analytical evaluation function. Calibration curves are often not linear, which necessitates use of a polynome of a higher degree to describe the calibration function. The latter may have the form:

$$Y = a_0 + a_1 \cdot c + a_2 \cdot c^2 + \cdots \qquad (102)$$

$a_0, a_1, a_2, \ldots$ are determined by multivariate regression. Normally, a polynome of the 2nd degree is sufficient to describe the calibration function over a large concentration range. A segmented calibration curve can also be used when the calibration function is not linear.

Precision

When performing a series of measurements of analytical signals as they are obtained for a well-defined concentration of the analyte, there is a statistical uncertainty that stems from fluctuations in the analytical system. This gives rise to a statistical error. This determines the precision of an analytical method. The precision achievable is an important figure of merit of an analytical procedure and the statistical evaluation of analytical data has been treated extensively in classical textbooks (see e.g. Refs. [30–32]). The precision of a measurement procedure is expressed in terms of the standard deviation as:

$$\sigma = \sqrt{[\Sigma(Y_m - Y_i)^2/(n-1)]} \tag{103}$$

provided the distribution of the individual values can be described by a Gaussian function and, accordingly, is a normal one. $Y_m = \Sigma(Y_i/n)$ is the mean value, Y_i is an individual measurement value, n the number of individual measurements and s is an estimate for the standard deviation.

The standard deviations for values resulting from different series of measurements are the result of a propagation of errors. Thus the coefficients of variation (σ^2) for a and b in the linear regression [Eqs. (99–100)] can be calculated as:

$$s(a)^2 = m \cdot \sigma_Y^2/[m\Sigma c_i^2 - (\Sigma c_i)^2] \tag{104}$$

$$s(b)^2 = \sigma_Y^2 \cdot \Sigma c_i^2/[m\Sigma c_i^2 - (\Sigma c_i)^2] \tag{105}$$

and the confidence levels for a and b are:

$$\Delta a = \pm t(P, f) \times s(b) \quad \text{and} \quad \Delta b = \pm t(P, f) \times s(b) \tag{106}$$

where t is the value of the Student distribution (see e.g. Ref. [33]).

For the precision of an analytical method not only the repeatability of single measurements but also the errors in the calibration procedure should be included. This is a complex problem, because according to the nature of the signal noise, the precision obtainable may vary considerably with the concentration level. In the case of a calibration the scattering of the measured Y_i values around the calibration curve is given by the standard deviation of the regression:

$$s(Y)^2 = [\Sigma_1^m(Y_i - y_i)^2]/(m-2) \tag{107}$$

y_i are the signal values obtained for a standard sample with concentration c from the regression equation. The latter is calculated by a least square procedure from the pairs (Y_i, c_i), where Y_i are the measured values for c_i; m is the total number of measurements taken with the standard samples and $F = m - 2$ is the number of degrees of freedom and:

$$\Sigma(Y_i - y_i)^2 = \Sigma Y_i^2 - b\Sigma Y_i - a\Sigma x_i Y_i \tag{108}$$

Accordingly, the standard deviation for the concentration of the analytical sample (c_X), which includes the error on the calibration, in the case of a calibration with samples, for which it is accepted that the errors in their concentration are low as compared with the scattering on a limited number of signals, can be calculated through a propagation of the errors as:

$$s_r(c) = \ln 10 \times s(\log c) = \ln 10 \times b \times s(Y) \tag{109}$$

b is the slope of the calibration curve. The magnitude of the concentrations of the analytical sample with respect to those of the calibration samples used also has to be considered, which can be included in Eq. (109) as:

$$s(c_X) = t \cdot c_X \cdot \ln 10 \cdot b \cdot s(Y)$$
$$\cdot \sqrt{[1/m + 1/n_X + (\log c_X - \log c^*)^2 / \Sigma(\log c_i - \log c^*)^2]} \tag{110}$$

t is the value from the Student table for a given probability and number of degrees of freedom, c_X is the concentration determined for the unknown sample, b is the slope of the calibration curve (sensitivity), $s(Y)$ is the standard deviation of the regression, m is the number of replicates for all calibration samples, n_X is the number of replicates for the unknown sample and $\log c^*$ is the mean for all calibration samples used [32].

Both Eqs. (109) and (110) are only valid for a limited concentration range, which is known as the linear dynamic range. This range is limited at the upper end by physical phenomena such as the detector overflow and at the lower end by the limit of determination. This limit is typical for a given analytical procedure and is the lowest concentration at which a determination can still be performed with a certain precision. When using calibration by standard additions, the sample matrix is present both in the unknown as well as in the calibration samples. Therefore, one can effectively correct for all interferences stemming from changes in the sensitivity resulting from influences of matrix constituents on the signal generation. Provided one makes a single addition of an amount of analyte, which is free of errors, the statistical error for the analysis can be derived by a propagation of errors. When X is a function of n variables, for each of which N replicate measurements are made, the estimate of the standard deviation $s(X)$ can be calculated from the individual standard deviations for the n variables [$s(k)$ with k: $1, 2, 3, \ldots$] and the estimates X_k as:

$$s^2(X) = \sum_{k=1}^{n} (\partial X/\partial X_k \cdot s_k)^2 \tag{111}$$

For:

$$Y_1 = b \cdot c \tag{112}$$

and after an addition c_A for which:

$$Y_2 = b(c + c_A) \tag{113}$$

$$c = [c_A/(Y_2 - Y_1)] \cdot Y_1 \tag{114}$$

A propagation of error according to Eq. (111) delivers:

$$s^2(c) = \{\partial/\partial Y_1[c_A \cdot Y_1/(Y_2 - Y_1)]\}^2 \cdot s(Y_1)^2$$
$$+ \{\partial/\partial Y_2[c_A \cdot Y_1/(Y_2 - Y_1)]\}^2 \cdot s(Y_2)^2 \tag{115}$$

or

$$s^2(c) = c_A^2\{[1 \cdot (Y_2 - Y_1) + Y_1]/(Y_2 - Y_1)^2\} \cdot s(Y_1)^2$$
$$+ \{[Y_1 \cdot 1]/(Y_2 - Y_1)^2\}^2 \cdot s(Y_2)^2 \tag{116}$$

and as $c_A = c \cdot (Y_2 - Y_1)/Y_1$:

$$s^2(c) = [(c/Y_1) \cdot Y_2/(Y_2 - Y_1) \cdot s(Y_1)]^2 + [c \cdot 1/(Y_2 - Y_1) \cdot s(Y_2)]^2 \tag{117}$$

or:

$$s_r^2(c) = Y_2^2/(Y_2 - Y_1)^2 [s_r^2(Y_1) - s_r^2(Y_2)] \tag{118}$$

Normally, there is also an uncertainty in the concentrations of the calibration samples used both in the case of calibration with synthetic calibration samples as well as in calibration by standard additions. In both cases these errors can also be included in the calculation of the analytical error [34].

When weighing an amount of sample M there is a weighing error σ_M and when dissolving this amount of sample in a volume V, there is an additional uncertainty σ_V in the volume to be taken into account. The resulting relative standard deviation can be calculated as:

$$\sigma_{r,c} = \sqrt{[(\sigma_M/M)^2 + (\sigma_V/V)^2]} \tag{119}$$

and when further diluting to a volume V': $c' = c \cdot (V/V')$ and the standard deviation for the sample is:

$$\sigma_{r,c'} = \sqrt{[(\sigma_c/c)^2 + (\sigma_V/V)^2 + (\sigma_{V'}/V')^2]} \tag{120}$$

When calibrating with one synthetic calibration sample (c_S) giving a signal Y_s, the concentration of the unknown sample c_X can be calculated as:

$$c_X = Y_X/Y_S \cdot c_S \tag{121}$$

and:

$$\sigma^2(c_X) = (\partial c_X/\partial Y_X)^2 \cdot \sigma^2(I_X) + (\partial c_X/\partial Y_S)^2 \cdot \sigma^2(Y_S) + (\partial c_X/\partial c_S)^2 \cdot \sigma^2(c_S) \quad (122)$$

with:

$$\partial c_X/\partial Y_X = \partial/\partial Y_X[(Y_X/Y_S) \cdot c_S] = C_S/Y_S \quad (123)$$

$$\partial c_X/\partial Y_S = \partial/\partial Y_S[(Y_X/Y_S) \cdot c_S] = -(Y_X \cdot c_S)/Y_S^2 \quad (124)$$

$$\partial c_X/\partial c_S = \partial/\partial c_S[(Y_X/Y_S) \cdot c_S] = Y_X/Y_S \quad (125)$$

and:

$$\sigma^2(c_X) = [c_S/Y_S]^2 \cdot \sigma^2(Y_X) + [-(Y_X \cdot c_S)/Y_s^2]^2 \cdot \sigma^2(Y_S) + [Y_X/Y_S]^2 \cdot \sigma^2(c_S) \quad (126)$$

Also in the case of calibration by standard additions, the uncertainty in the amount added in the case of one standard addition can be taken into account. When

$$c_X = [Y_x/(Y_{X+A} - Y_X)] \cdot c_A \quad (127)$$

$$s^2(c_X) = (\partial c_X/\partial Y_X)^2 \cdot \sigma^2(Y_X) + (\partial c_X/\partial Y_{X+A})^2 \cdot \sigma^2(Y_{X+A})$$

$$+ (\partial c_X/\partial c_A)^2 \cdot \sigma^2(c_A) \quad (128)$$

with:

$$\partial c_X/\partial Y_X = \partial/\partial Y_X[Y_x/(Y_{X+A} - Y_X) \cdot c_A] = [Y_{X+A}/(Y_{X+A} - Y_X)^2] \cdot c_A \quad (129)$$

$$\partial c_X/\partial Y_{X+A} = \partial/\partial Y_{X+A}[Y_x/(Y_{X+A} - Y_X) \cdot c_A] = -[Y_X/(Y_{X+A} - Y_X)^2] \cdot c_A \quad (130)$$

$$\partial c_X/\partial c_A = \partial/\partial c_A[Y_x/(Y_{X+A} - Y_X) \cdot c_A] = Y_X/(Y_{X+A} - Y_X) \quad (131)$$

and:

$$\sigma^2(c_X) = [Y_{X+A}/(Y_{X+A} - Y_X)^2 \cdot c_A]^2 \cdot \sigma^2(Y_X) + [-Y_X/(Y_{X+A} - Y_X)^2 \cdot c_A]^2$$

$$\cdot \sigma^2(Y_{X+A}) + [Y_X/(Y_{X+A} - Y_X)]^2 \cdot \sigma^2(c_A) \quad (132)$$

Noise

The precision which can be obtained with a certain measurement system depends on the noise in the system. The latter may have different causes and accordingly, different types of noise can be distinguished as follows (for a detailed treatment, see Ref. [35]).

- "Fundamental noise" or "random noise": this type of noise is statistically distributed and its amplitude as a function of the frequency can be written as a sum of many sinusoidal functions. This type of noise is related to the corpuscular nature of matter or to the quantization of radiation, respectively, and cannot be completely eliminated.
- "Non-fundamental noise" also called "flicker noise" or "excess noise". Here the sign or the magnitude can correlate with well-defined phenomena.
- "Interference noise", resulting in contributions at well-defined frequencies and mostly stemming from components in the system.

The last two types can often be eliminated by appropriate filtering.

In a so-called noise spectrum (Fig. 8) the noise amplitude is plotted as a function of the frequency [see Ref. [36]]. Here one distinguishes between "white noise",

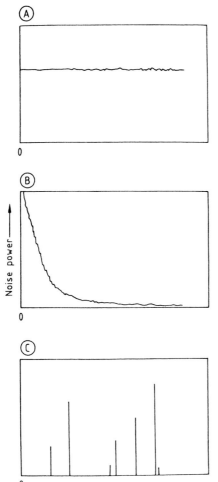

Fig. 8. Noise power spectra commonly found in chemical instrumentation. (A): "white" noise; (B): flicker $(1/f)$ noise and (C): interference noise. (Reprinted with permission from Ref. [36].)

Fig. 9. Noise power spectra for ICP-OES using a Fassel-type torch. (a): 0–500 Hz and (b): 0–24 Hz; sample: 10 000 µg/mL Al solution; power: 1.8 kW, outer gas: 18 L/min Ar; nebulizer pressure: 3 bar Ar; sample uptake rate: 1 mL/min. (Reprinted with permission from Ref. [37].)

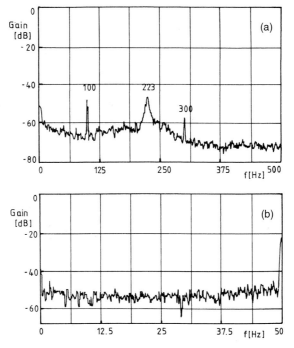

which spreads over all frequencies and is almost always "fundamental" in origin, whereas for the "$1/f$ noise" the amplitude decreases with the frequency and it has non-fundamental causes. Also discrete noise bands with well-defined causes may occur. They may stem from the source and be caused by the gas flow dynamics or contributions from the vacuum line etc.

In the case of an inductively coupled plasma for example (Fig. 9) [37], there is a $1/f$ noise contribution stemming from large droplets which vigorously and in some way unpredictably evaporate in the plasma. Furthermore, there is a white noise contribution spread out over all frequencies stemming from the nebulization. There is also a noise contribution in the form of a broad noise band at ca. 300 Hz. This stems from the pulsing of the plasma which is related to the gas flow dynamics. These pulsations can be visualized by high-speed photography [38]. It was also found, for example, that in the case of slurry nebulization, the contribution of the $1/f$ noise may increase. This can be shown by the $1/f$ noise increase obtained when using a direct powder introduction (Fig. 10). The frequency noise component depending on gas flow was found to depend both on different gas flows as well as on the power level. This is understandable, as both parameters influence the size and hence the expansion and shrinkage of the plasma and their frequencies.

In the case of a toroidal microwave discharge obtained in a microwave plasma torch, the frequency dependent noise stemming from the gas flow dynamics could also be found at a frequency of 150–250 Hz (Fig. 11) [39]. When using hydride

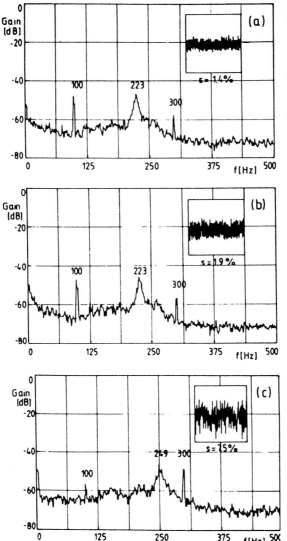

Fig. 10. Noise power spectra and precision in ICP-OES for the introduction of (a): a 10 000 µg/mL Al solution, (b): a 10 000 µg/mL Al suspension and (c): the dry powder (Al$_2$O$_3$, Sumitomo, Japan). Fassel-type torch with outer gas flow: 18.1 L/min Ar; (a), power: 1.75 kW, nebulizer pressure: 3 bar Ar, sample uptake rate: 1 mL/min; (b), power: 1.7 kW, nebulizer pressure: 3 bar Ar, sample uptake rate: 1 mL/min; (c), power: 1.8 kW, aerosol gas flow: 1.3 L/min. (Reprinted with permission from Ref. [37].)

generation only a gaseous mixture enters the MIP, which might be the reason why the 1/f noise contribution is then almost absent. For glow discharge atomic spectrometry, the sputtering processes and analyte introduction and interaction with the plasma are so smooth that both 1/f noise and white noise are virtually absent and the noise spectra are confined to some frequency dependent noise components (Fig. 12) [40], which either stem from the power line (50 or 60 Hz) or may also be introduced by the vacuum line (10–20 Hz). The latter component can be identified, as it decreases when throttling the main vacuum line.

In plasma mass spectrometry noise power spectra were also found to reflect the

(a)

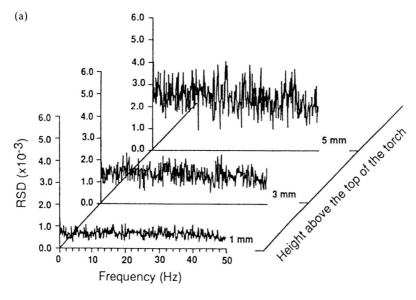

(b)

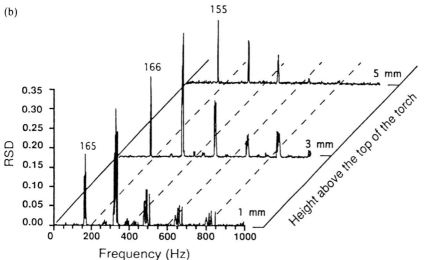

Fig. 11. Noise power spectra of the He I 501.56 nm line collected at 1, 3 and 5 mm above the top of the torch for a 220 W He MPT plasma as source for OES. Gas flow of the He support gas: 2.2 L/min; gas flow of He carrier: 0.2 L/min. (a): low-frequency noise spectra (0–50 Hz), (b): high-frequency noise spectra (0–1000 Hz). (Reprinted with permission from Ref. [39].)

gas flows dynamics (Fig. 13) [41]. However, in ICP-MS they also were found to be severely influenced by the cooling of the spray chamber. This is understandable as this will influence the droplet evaporation during aerosol transport as well as the concentration of vapor present in the aerosol. In particular, it was found that the level of the white noise follows the spray chamber temperature.

a)

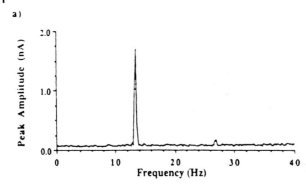

b)

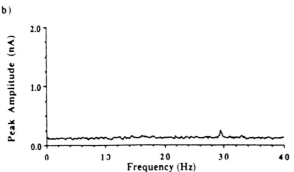

Fig. 12. Amplitude noise spectra for a matrix line in glow discharge source (GDS) atomic emission spectrometry. Steel standard sample 218A (Research Institute CKD, Czech Republic); i: 50 mA; argon pressure: 600 Pa; burning voltage: 900 V; 0.35 m McPherson monochromator; line: Fe I 371.9 nm. (a): Without needle valve between the vacumm pump and the GDS, (b): with needle valve between the pump and the GDS. (Reprinted with permission from Ref. [40].)

Accordingly, noise spectra are a strong diagnostic tool to trace the sources of noise, and to study the limitations stemming from different sources (source stability, atom reservoir stability, detector used, etc.). For instance, it will be important to see if the noise of the detector is predominant, as this type of noise can be described by Poisson statistics where:

$$\sigma^2 = n \tag{133}$$

n is the number of events per second and σ its standard deviation. Alternatively, it might be that the background noise of the source is much more important or that "flicker noise" or "frequency-dependant noise" are predominant. In the last case overtones will often occur.

Signal-to-noise ratios

The signal-to-noise ratios, which can be measured, depend on all aspects that influence the signal magnitude and also on the magnitude and the nature of the

5 °C

15 °C

25 °C

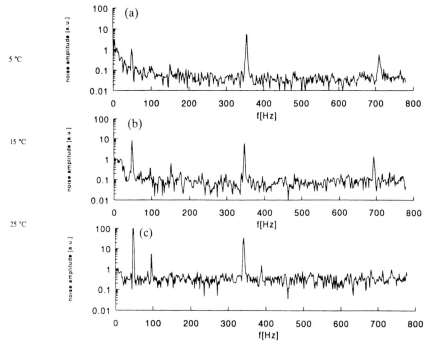

Fig. 13. Noise power spectra in ICP-MS for ArO^+ (56 dalton). Meinhard nebulizer, power: 1350 W, outer gas: 14 L/min, intermediate gas: 1.4 L/min, nebulizer gas: 1.0 L/min, sample uptake rate: 1.05 mL/min, doubly distilled water. (Reprinted with permission from Ref. [41].)

noise. Depending on the nature of the noise several measures can be taken so as to improve signal-to-noise ratios in a system (for a discussion in the case of atomic fluorescence spectrometry, see Ref. [42]).

Provided both the source (primary source and/or atom reservoir, nebulizer, etc.) as well as the detector contribute to the noise, the signal-to-noise ratio (S/N) in the case of a Poisson distribution for all types of noise is then given by:

$$S/N = S/\sqrt{(\sigma_{DC}^2 + \sigma_S^2)} \tag{134}$$

σ_{DC} is the noise of the dark current of the detector (shot noise) and σ_S is the noise of the source (radiation or ion source or in the case of absorption measurements the primary source). In the case of absorption measurements, e.g., three cases can be considered.

- A continuous source and a continuously measuring detector: here the signal is proportional to the mean radiant density of the source B_{ave}^s and the measurement time t_0, the shot noise N_s (being σ_S^2) is proportional to the signal B_{ave}^s and t_0, and the background noise of the atom reservoir is proportional to the background signal B_{ave}^b and:

$$S/N = K \cdot B_{ave}^s / \sqrt{(B_{ave}^s + B_{ave}^b)} \tag{135}$$

- A pulsed source and a continuously measuring detector: here the signal is again proportional to the mean radiant density of the source, but when the pulse width is t_p, the pulse frequency is f and the pulse heigth is B_p^s, it can be written as $B_p^s \cdot f \cdot t_p$ and the background noise is still given by B_{ave}^b. Accordingly:

$$(S/N)' = K \cdot B_p^s \cdot f \cdot t_P / \sqrt{(B_P^S \cdot f \cdot t_P + B_{ave}^b)} \tag{136}$$

When the background noise is predominant, the improvement in signal-to-noise ratio is $B_p^s \cdot f \cdot t_P / B_{ave}^b$, whereas the improvement in the case where the shot noise dominates is given by $(B_p^S \cdot f \cdot t_P / B_{ave}^s)^{1/2}$

- A pulsed source and a synchronously pulsed detector: here the signal again is proportional to B_p^s, t_P and f, however, the background level also depends on t_P and f and:

$$(S/N)'' = K \cdot B_P^S \cdot f \cdot t_P / \sqrt{[f \cdot t_P \cdot (B_P^S + B_{ave}^b)]} \tag{137}$$

or

$$(S/N)'' = K \cdot B_P^S \cdot (f \cdot t_P)^{1/2} / \sqrt{(B_P^S + B_{ave}^b)} \tag{138}$$

and the improvement as compared with S/N or $(S/N)'$ can be considerable when the background noise predominates ($B_{ave}^b > B_{ave}^S$ and $B_{ave}^b > B_P^S$). Then the improvement is $f \cdot t_P$. As $f \cdot t_P$ can be 10^{-2} to 10^{-4} and with pulsed lasers and so-called boxcar integrators large improvements of even 10^{-8} in signal-to-noise ratios are possible.

Power of detection

According to Kaiser [43], the limit of detection of an analytical procedure is the concentration at which the analytical signal can still be distinguished from a noise level with a specific degree of uncertainty. In the case of a 99.86% uncertainty and provided the signal fluctuations of the limiting noise source can be described by a normal distribution, the lowest detectable net signal Y_L is three times the relevant standard deviation:

$$Y_L = 3\sigma^* \tag{139}$$

In atomic spectrometry the net signal is determined from the difference between a

brutto signal, including analyte and background contributions, and a background
signal, a propagation of error has to be applied as:

$$\sigma^* = \sqrt{\left(\sigma_{Signal}^2 + \sigma_{background}^2\right)} \tag{140}$$

and provided the total signal and the background have almost the same absolute
standard deviation, a factor of only $\sqrt{2}$ must be introduced.

For photoelectric measurements $\sigma(I)$, the standard deviation of the measured
signals contains several contributions and:

$$\sigma^2(I) = \sigma_P^2 + \sigma_D^2 + \sigma_f^2 + \sigma_A^2 \tag{141}$$

Here, σ_P represents the noise of the photoelectrons. When the photon flux is n,
$\sigma_P \approx \sqrt{(n)} \cdot \sigma_D$ is the dark current noise of the photomultiplier and is proportional
to the dark current itself. σ_f is the flicker noise of the source and is proportional to
the signal and σ_A is the amplifier noise resulting from electronic components. The
last contribution can usually be neglected, whereas σ_f is low for very stable sources
(e.g., glow discharges) or can be compensated for by simultaneous line and background measurements. As $\sigma_D \approx I_D$ one should use detectors with low dark current,
then the photon noise of the source limits the power of detection.

In many cases it is not the background signal from the source or the measurement system but blank contributions that limit the power of detection, the limiting
standard deviation is often the standard deviation of the blank measurements and
this value must be included in Eq. (139) [44]. From the calibration function the
detection limit then is obtained as:

$$c_L = a' \cdot (3\sqrt{2} \cdot \sigma) \tag{142}$$

The detection limit thus is closely related to the signal-to-background ratio and the
signal-to-noise ratio. It is the concentration for which the signal-to-background ratio
equals $3\sqrt{2}$ times the relative standard deviation of the background or at which the
signal-to-noise ratio is $3\sqrt{2}$. The signal-to-noise ratio itself is related to the types of
noise occurring in the analytical system. From the knowledge of the limiting noise
sources well-established measures can be taken during signal acquisition to improve
the signal-to-noise, as discussed before, and accordingly, also the power of detection of a system.

Provided the fluctuations of the relevant background or blank and the fluctuations near this signal level are not identical, the detection limit cannot be calculated
using the $3\sqrt{2}\sigma$ criterion but must be defined considering the error of the first and
the error of the second kind [45] (Fig. 14). Here, the error of the first kind stems
from the scattering of the blank or background values around a mean value,
whereas the error of the second kind follows the confidence intervals of the calibration curve, which are a function of the concentration. In particular, when the
background signal stems mainly from the detector dark current whereas the fluc-

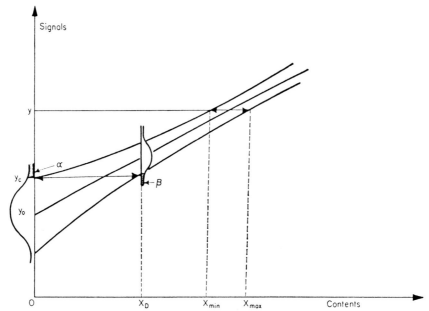

Fig. 14. Calculation of the detection limit (x_D) taking into account the error of the first (α) and the error of the second (β) kind. (Reprinted with permission from Ref. [45].)

tuations of the signals for the calibration samples are related to the source and the signal generation, the error of the first and the error of the second kind may differ.

A similar difficulty may arise when the distribution of the background measurement values and/or the standard signals is not normal. This can be tested by the χ^2 and the "Skewedness and excess" (S & E) test [46].

In the χ^2 test [47] the test value is:

$$\chi^2 = \sum_{i=1}^{k}[(\chi_i - n \cdot \theta_i)^2/(n \cdot \theta_i)] \tag{143}$$

with χ_i the number of measurements in a class i normalized to 1 as $\Sigma\chi_i = 1$ and θ_i the theoretical frequency in a class i (in tables for a normal distribution) again normalized to 1. The distribution must thus be investigated to establish whether the n measurements comply with a normal distribution with mean $\mu = m$ and standard deviation $\sigma = s$ calculated from the measurements. The theoretical normal distribution for μ and σ can be calculated from:

$$P(u) = [1/\sqrt{(2 \cdot \pi \cdot \sigma)}] \cdot \exp\{-[(x - \mu)/\sqrt{2} \cdot \sigma]\} \cdot dx \tag{144}$$

Which theoretical frequency is found for the different classes is thus calculated. Every term $[(\chi_i - n \cdot \theta_i)^2]/(n \cdot \theta_i)$ is quadratic in a normally distributed value. The sum of k squares of normally distributed values is χ^2 distributed with k degrees of freedom (k is the number of classes). As μ and σ are calculated from the data and $\Sigma\theta_i = 1$, three degrees of freedom have to be subtracted. The resulting test value can be compared with tabulated values for a given confidence (e.g., 95%) and number of degrees of freedom and this can be done for the values or for a given function of them. When the latter is a logarithmic one for example, it can be controlled if the distribution of the data is normal or logarithmically normal.

Whereas the χ^2 test only gives information on the distribution of measured values over different classes the S & E test gives further information on whether the extreme values are not only systematically too high or too low and whether or not there is an excess of values around the mean. For a number of measured values one determines the parameter Q [46]:

$$Q = [\Sigma(x_i - x^*)^3 \cdot f_i]/(\sigma^3 \cdot N) \tag{145}$$

Here Q is a measure for the skewedness, x^* is the mean, x_i is a measured value, f_i is the number of measured values of a magnitude x_i, σ is the standard deviation and N the total number of measurements. One can further determine:

$$\varepsilon = [\Sigma(x_i - x^*)^4 \cdot f_i]/(\sigma^4 \cdot N) - 3 \tag{146}$$

For a normal distribution both values should simultaneously be zero. With a finite number of N:

$$S = \{\sqrt{[N(N-1)]/(N-2)}\} \cdot Q \tag{147}$$

and

$$E = (N-1)/[(N-2)(N-3)] \cdot [(N+1)\varepsilon + 6] \tag{148}$$

In the case of a normal distribution the values of S and E should also have normal distributions with relative standard deviations:

$$\sigma_S = \pm\sqrt{[6N(N-1)/(N-2)(N+1)(N+3)]} \tag{149}$$

and:

$$\sigma_E = \pm\sqrt{[24N(N-1)^2/(N-3)(N-2)(N+3)(N+5)]} \tag{150}$$

If 66% of the experimental S and E values lie within the $\pm\sigma_S$ or $\pm\sigma_E$ limits it can be controlled. If the values scatter according to a normal distribution.

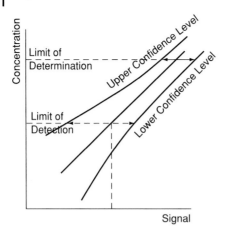

Fig. 15. Limit of determination and limit of detection.

Limit of determination

The limit of determination indirectly relates to the limit of detection. It is the concentration level from where a determination can be performed with a preset precision. The definition can be understood from the confidence lines at each side of the calibration curves (Fig. 15), which diverge both at lower concentrations, as a result of sample inhomogeneities or noise magnitude, as well as at large concentrations, as a result of deviations from linearity in the calibration or source instabilities.

When for a sample with a given concentration a number of signal measurements n are obtained with a standard deviation σ_S, the probability P of finding the value between the limits $\mu_s \pm ux\sigma_S$ will be the ratio of the integral under the whole population from $-\infty < x < +\infty$ to the integral within the limits $\mu - u \cdot \sigma_S < x < \mu + u \cdot \sigma_S$. On the contrary, for a single value, P will be the probability within which limits it deviates from the true value μ_s as a result of statistical fluctuations. For the mean μ_S of n measurements:

$$\mu_S - u(P) \cdot \sigma_S/(\sqrt{n}) < x < \mu_S + u(P) \cdot \sigma_S/(\sqrt{n}) \tag{151}$$

The analytical value then differs with a probability P by less than $\pm u(P) \cdot \sigma_S/(\sqrt{n})$ from the value μ and this is the confidence interval. It is the interval belonging to the mean value μ in which with a preset probability one can assume a certain analytical error. This interval changes with the concentration and this determines the confidence limits.

The limit of determination is often pragmatically defined as being found at a concentration being 5–6 times the detection limit. Indeed, it is at this concentration level that the full precision of the analytical method is likely to be realized.

Limit of guarantee of purity

The guarantee of purity (c_G) is the highest concentration that can be present in a sample, without being able to obtain an analytical signal, which can be differentiated with a given probability from the limiting noise level. When excluding errors from sample heterogeneity, c_G is given by:

$$c_G = 2c_L \tag{152}$$

This analytical figure of merit is of special importance especially in the control of high-purity substances and materials for microelectronics, and also in food analysis and other disciplines.

In atomic spectrometry, the analytical signals measured often include contributions from non-spectrally resolved features stemming from constituents other than the analyte (e.g. matrix constituents). These contributions are known as spectral interferences. They can be corrected for by subtracting their contributions to the signal, which can be calculated from the magnitude of the interference and the concentration of the interferent. A special type of spectral interference is that which influences the background signal on which the analyte signals are superimposed. For this type of interference a number of corrections are known. The degree of freedom from interferences is an important figure of merit for an analytical method.

2.2
Optical spectrometers

In optical atomic spectrometry the radiation emitted by the radiation source or the radiation which comes from the primary source and has passed through the atom reservoir has to be lead into a spectrometer. In order to make optimum use of the source, the radiation should be lead as complete as possible into the spectrometer. The amount of radiation passing through an optical system is expressed by its optical conductance. Its geometrical value is given by:

$$G_0 = \int_A \int_B (\cos \alpha_1 \cdot \cos \alpha_2 \cdot dA \cdot dB)/a_{12}^2 \tag{153}$$

$$\approx (A \cdot B)/a_{12}^2 \tag{154}$$

dA and dB are surface elements of the entrance and the exit apertures, a_{12} is the distance between them. α_1 and α_2 are the angles between the normals of the aperture planes and the radiation. When n is the refractive index of the medium, the optical conductance is given by:

$$G = G_0 \cdot n^2 \tag{155}$$

The radiant flux through an optical system is given by:

$$\phi = \tau \cdot B \cdot G \tag{156}$$

τ is the transmittance determined by reflection or absorption losses at the different optical elements and B is the radiant density of the source (in W/m^2 sr). For an optimal optical illumination of a spectrometer, the dispersive element, which serves to provide the spectrum, should be fully illuminated so as to obtain full resolution. However, no radiation should bypass the dispersive element, as this would cause stray radiation. Furthermore the optical conductance at every point of the optical system should be maximum.

2.2.1
Optical systems

Illumination of the spectrometer

The type of illumination system with which these conditions can be fullfilled, will depend on the dimensions of the source and the detector, the homogeneity of the source, the need to fill the detector homogeneously with radiation, the distance between the source and the entrance aperture of the spectrometer and on the focal length of the spectrometer.

In conventional systems lenses as well as imaging mirrors are used. In the case of lenses, the lens material is important. Here glass lenses can only be used at wavelengths above 330 nm. In the case of quartz, radiation with wavelengths down to 165 nm can still be transmitted. However, as a result of the absorption of short-wavelength radiation by air, evacuation or purging of the illumination system and the spectrometer with nitrogen or argon is required when lines with wavelengths below 190 nm are measured. At wavelengths below 160 nm, evacuation and the use of MgF_2 or LiF optics is required.

With lenses mainly three illumination systems are of use.

Imaging on the entrance collimator. A lens is placed immediately in front of the entrance slit and should image the relevant part of the radiation source on the entrance collimator (Fig. 16A). This has the advantage that the entrance slit is homogeneously illuminated, however, stray-radiation may easily occur inside the spectrometer. The distance between the source and the entrance slit (a) is given by the magnification required, as:

$$x/W = a/f_k \tag{157}$$

x is the width (or diameter) of the source, W is the width of the entrance collimator and f_k its focal length. The *f*-number of the lens than is given by:

$$1/f = 1/a + 1/f_k \tag{158}$$

• Example
When a grating spectrometer with a focal length of 1.2 m and collimator dimensions of 55×55 mm^2 must be illuminated for a wavelength of 200 nm with a radiation source of 7 mm in diameter the enlargement for $55\sqrt{2} = 77.8$ mm is

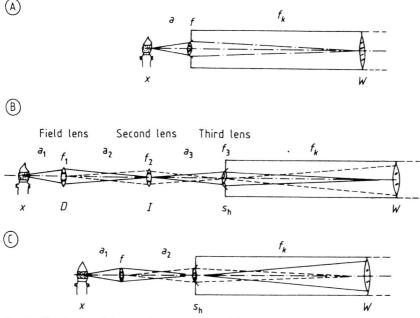

Fig. 16. Illumination of the optical spectrometer with lenses.
(A): Imaging on the entrance collimator, (B): illumination with
intermediate image, (C): imaging on the entrance slit.

$\beta = 77.9/7$ and the distance d between the entrance slit of the spectrometer and the source should be $\beta = f_k/d$ or $d = 108$ mm, with f_k the focal length. According to Eq. (158):

$$1/f = 1/1200 + 1/108$$

and therefore a lens with a focal length of 99 mm at 200 nm is required. It must be considered that the focal length of a lens depends on the wavelength as:

$$f(\lambda_1)/f(\lambda_2) = n_1/n_2 \tag{159}$$

n_1 and n_2 are the refractive indices of the lens material at the respective wavelengths. For quartz, e.g., a factor of 0.833 has to be applied when passing from the Na 583 nm D-line to 200 nm.

Illumination with an intermediate image.   Here a field lens is used to produce an intermediate image on a diaphragm. To illuminate the collimator mirror fully, the appropriate zone can be selected with the aid of a lens placed immediately in front of the exit slit. A third lens is used to illuminate the entrance slit homogeneously (Fig. 16B). The magnification is then divided over all 3 lenses, thus chromatic aberrations are minimized, but the set-up is highly inflexible. The distances between the three lenses must be chosen so as to achieve the respective magnifications. Accordingly:

$$x/I = a_1/a_2 \tag{160}$$

$$D/s_h = a_2/a_3 \tag{161}$$

and

$$I/W = a_3/f_k \tag{162}$$

I is the diameter of the intermediate image, D the diameter of the field lens, s_h the entrance slit height and a_1, a_2 and a_3 are the distances between the respective lenses. Furthermore:

$$a_1 + a_2 + a_3 = A \tag{163}$$

is usually fixed because of the construction of the system, and

$$1/f_1 = 1/a_1 + 1/a_2 \tag{164}$$

$$1/f_2 = 1/a_2 + 1/a_3 \tag{165}$$

$$1/f_3 = 1/a_3 + 1/f_k \tag{166}$$

give the f-numbers of the respective lenses. One parameter (e.g. the width of the intermediate image) can be freely selected as x, W, s_h, A and f_k are fixed.

• Example

When a radiation source with a diameter of 4 mm is to be coupled with a 1 m monochromator with an entrance collimator of 50 mm width, an entrance slit heigth of 20 mm and a total distance between the radiation source and the entrance slit of 1 m, for an intermediate image I of 10 mm Eq. (162) gives: $a_3 = 200$ mm. Eq. (166) gives f_3. According to Eq. (160): $2.5 \times a_1 = a_2$ and from Eq. (163) $3.5 \times a_1 = 800$ mm, so that: $a_1 = 230$ mm and $a_2 = 590$ mm; f_1 is given by Eq. (164) and from Eq. (161): $D = 50$ mm.

Image on the entrance slit. With the aid of one lens this is also possible. Here the structure of the source appears on the entrance collimator and on the detector. This allows spatially-resolved line intensity measurements to be made when a detector with two-dimenional resolution such as a photographic emulsion or an array detector (see Section 2.2.2) are employed. This type of imaging is often used for diagnostic studies.

Quartz fiber optics. These have been found to be very useful for radiation transmission. Small lenses are used so as to respect the opening angle of the fiber. This depends on the refractive index of the material, which because of optical transmission

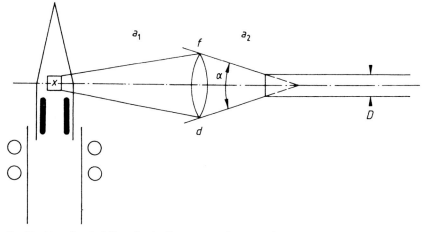

Fig. 17. Use of optical fibers for the illumination of an optical spectrometer. α: Opening angle of the fiber; d: lens diameter, D: diameter of the fiber, x: dimension of the zone in the radiation source to be selected, a_1: distance between lens and radiation source, a_2: distance between fiber entrance and lens.

reasons is usually quartz, and is often of the order of 30–40°. A typical illumination of a spectrometer with an optical fiber (Fig. 17) uses a lens (diameter d) for imaging the source on the fiber. Fibers or fiber beams with a diameter of $D = 600$ μm are often used. Then the magnification (x/D) as well as the entrance angle $tg\alpha = (d/2)/a_2$ and the lens formula determine the f number of the lens and the diameter d. At the exit of the fiber, a lens is used so as to allow the radiation to enter the spectral apparatus, without causing stray radiation.

With quartz fibers it is easier to lead the optical emission into the spectral apparatus, however, it should be mentioned that the transmittance decreases seriously below 220 nm. This may give rise to detector noise limitations for analytical lines at lower wavelengths.

As an alternative to lenses and optical fibers so-called light pipes which use total reflection of radiation may also be used.

Spectrometers

Spectral apparatus are used to produce the spectrum or to isolate narrow spectral ranges, without further spectrum deconvolution. In dispersive spectral apparatus, the spectrum is produced with a prism or a diffraction grating. In non-dispersive spectral apparatus, spectral areas are isolated from a radiation beam, without any further spatial deconvolution, by reflection, interference or absorption in an interferometer or a filter monochromator, respectively. The filter is now only of use in flame emission spectrometry.

A dispersive spectral apparatus contains an entrance collimator, a dispersive element and an exit collimator (for an example discussion, see Refs. [48, 49]).

With an *entrance collimator* a quasi parallel beam is produced from the radiation coming through the entrance aperture, which has a width s_e and a heigth h_e. The entrance collimator has a focal length f_k and a width W. The diffraction slit width (s_0) and the diffraction slit heigth (h_0) are the half-widths given by:

$$s_0 = \lambda \cdot f / W \quad \text{and} \quad h_0 = \lambda \cdot f / h \tag{167}$$

The entrance aperture dimensions should not be made smaller than s_0 and h_0, as then diffraction will limit the resolution that can be obtained. The value of f/W is a measure of the amount of radiation energy entering the spectral apparatus.

The *exit collimator* images the monochromatized radiation leaving the dispersive element on the exit slit. Here we have a series of monochromatic images of the entrance slit. In the case of one exit slit in which one line after another can be isolated by turning the dispersive element, we have a monochromator. In a polychromator the dispersive element is fixed and there are many exit slits placed at locations where monochromatic images of the entrance slit for the lines of interest are obtained. They are often on a curved surface with a radius of curvature R (Rowland circle). Here simultaneous measurements of several lines and accordingly simultaneous multielement determinations are possible. In the case of a spectrographic camera, the lines are focussed in a plane or on a slightly curved surface, where a detector with two-dimensional resolution can be placed. With such a detector (photoplate, diode array detector, etc.), part of the spectrum over a certain wavelength range as well as the intensities of the signals, eventually at several locations in the source, can be recorded simultaneously. The energy per unit of surface on the detector is given by the irradiance:

$$E = \phi \cdot \cos \alpha / A \tag{168}$$

α is the angle between the surface of the radiation detector (A) and the incident radiation. Virtually the only dispersive elements now used are diffraction gratings. Prisms are used only as predispersers. Distinction can be made between plane and concave gratings, of which the latter have imaging qualities. One also has to distinguish between mechanically ruled and holographically ruled gratings. As a result of the profile of the grooves, they have a fairly uniform radiant output over a large spectral area. Mechanically ruled gratings always have a so-called blaze angle and accordingly a blaze wavelength where the radiant energy delivered is at a maximum. In modern spectrometers reflection gratings are usually used. As a result of interference, a parallel beam of white radiation incident at an angle ϕ_1 with the grating normal is dispersed and at an angle ϕ_2, radiation of wavelength λ is diffracted (Fig. 18), according to the Bragg equation:

$$\sin \phi_1 + \sin \phi_2 = m\lambda / a \tag{169}$$

m is the order, ϕ_2 is the angle the diffraction beam of wavelength λ makes with the grating normal and $a = 1/n_G$ is the grating constant, where n_G is the number of

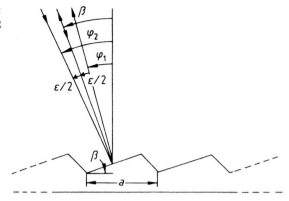

Fig. 18. Spectral dispersion at a diffraction grating. a: Grating constant, β: Blaze angle, ϕ_1: angle of incident radiation, ϕ_2: angle of diffracted radiation with wavelength λ.

grooves per mm. When B is the width of the grating, the total number of grooves, $N = B \cdot n_G$, determines the theoretical resolving power R_0 as:

$$R_0 = B \cdot n_G \cdot m \tag{170}$$

The angular dispersion can be obtained by differentiation of Eq. (169) with respect to λ:

$$d\phi_2/d\lambda = m/(a \cos \phi_2) \tag{171}$$

The angular dispersion and the theoretical resolving power R_0 are related as: $d\phi_2/d\lambda = R_0/D_s'$, where D_s' is the width of the monochromatic beam where it exits the dispersive element. The reciprocal linear dispersion is given as:

$$\begin{aligned} dx/d\lambda &= (dx/d\phi_2)(d\phi_2/d\lambda) \\ &= (f_k/\cos \theta')(m/a \cdot \cos \phi_2) \end{aligned} \tag{172}$$

where θ' is the angle between the plane of the detector and the direction of the sorting beam. The spectral slit width of the spectrometer is given by:

$$\Delta\lambda = (d\lambda/dx) \cdot s_e \tag{173}$$

where s_e is the entrance slit width. The form and the depth of the grooves determine the intensity distribution in the spectrum. The angle for the maximum intensity is the so-called "blaze" angle β. The blaze wavelength for the order m can be calculated as:

$$\sin \beta = \lambda_B \cdot m/[2a \cos(\varepsilon/2)] \tag{174}$$

where $(\phi_1 - \phi_2) = \varepsilon$.

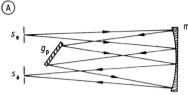

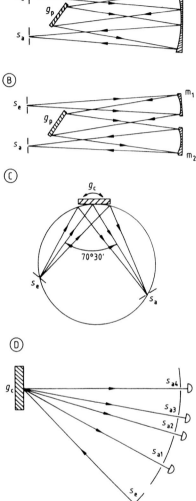

Fig. 19. Important optical mountings for optical spectrometers with a plane (A: Ebert, B: Czerny–Turner) and a concave (C: Seya–Namioka, D: Paschen–Runge) grating. m: Mirror, s_e: entrance slit, s_a: exit slit, g_p: plane grating, g_c: concave grating.

In a stigmatic spectral apparatus the height and the width of the slit are imaged in the same plane. In an autocollimation spectral apparatus the rays before and after the spectral dispersion pass through the same optical elements. With a polychromator all optical elements are usually fixed, whereas in monochromators the desired wavelength is normally brought into the exit slit by turning the grating around its axis.

With a *plane grating* several mountings (Fig. 19) can be used. In the Czerny–Turner mounting, two spherical mirrors with slightly different focal lengths are positioned at sligthly different angles. In this way spherical aberration can be adequately corrected for. This mounting is used for high-luminosity monochromators with fairly short focal lengths and highly dispersive gratings (more than 1800 grooves/mm). With the Ebert mounting, there is only one mirror serving as the entrance and the exit collimator. Accordingly, aberrations occur and the use of curved slits is necessary, so as to be able to realize the highest resolution and optical conductance. The Fastie–Ebert mounting is used for large spectrographs, where the entrance collimator and the exit collimator plane lie above each other. The spectrum is focussed in a plane and the f/W value can be as low as 1/30, which is very low and may require the use of a cylindrical lens to increase the irradiance on the detector.

Photoplates, films and photo multipliers are used as detectors. Normally, gratings are used at low orders ($m < 4$) and they have a small grating constant ($1/3600$ mm $< a < 1/300$ mm). The different orders can be separated by using special photomultipliers. For instance, with a solar-blind photomultiplier only radiation with a wavelength below 330 nm can be detected. This allows separation of the 1st order radiation at 400 nm from the 2nd order radiation at 200 nm. This can, for example, be applied in polychromators to double the practical resolution.

Apart from gratings used at low orders, the so-called echelle gratings can be used. Their groove density is low (a up to $1/100$ mm) but therefore order numbers of up to 70 can be used [50]. Here, the orders overlap and must be separated by using a second dispersive element (e.g. a prism) either with its axis parallel to the one of the echelle grating or in so-called crossed-dispersion mode. In the latter case the spectrum occurs as a number of dots with a height equalling that of the entrance slit. The optical conductance is given by:

$$\phi = (B_\lambda \cdot t \cdot s \cdot h \cdot W \cdot H \cdot \cos \beta)/f^2 \tag{175}$$

where B_λ is the spectral radiance, t is the transmittance of the optics, W is the width and H the height of the grating. Echelle spectrometers often use an Ebert mounting and allow a high practical resolution to be obtained at low focal length (f). Therefore, they are thermally very stable and are used both with monochromators as well as polychromators. As two-dimensionally resolving array detector technology [vidicons, charge coupled devices (CCD) now also with image intensifiers] with high spatial resolution and high sensitivity as well as signal-to-noise ratios, have become available, such Echelle spectrometers have become very useful. They allow the whole spectrum to be collected on one detector of limited dimensions (e.g. 1 in × 1 in) and enable flexibility in selecting any analytical wavelength. The practical resolution then is finally limited by the pixel dimensions (e.g. 1000 pixels in one direction) and blooming as well as cross talk. In modern instrumentation a separate spectrometer is often built-in to zoom into remote spectral areas, as required, for example, for the alkali sensitive lines (Fig. 20).

Mountings with a *concave grating* are also used. Here the radius of the grating determines the so-called Rowland circle. In the direction of dispersion the spectral lines are focussed on the Rowland circle and are monochromatic images of the

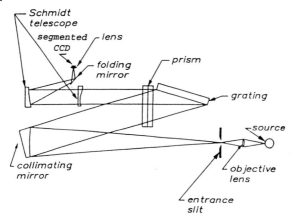

Fig. 20. Echelle mounting in optical set-up for CCD-ICP-OES (Optima) (PerkinElmer Inc.).

entrance slit. In the Paschen–Runge mounting, the grating, the entrance slit and all the exit slits are fixed on the Rowland circle. This mounting is most often used in large simultaneous polychromators with photoelectric detection. The number of detectors used here is often more than 30 and they are usually used in several orders. In this mounting, and when using conventional photomultipliers behind an exit slit, background acquisition is possible in a sequential way by slewing the entrance slit. This can be done very reproducibly by positioning the entrance slit in a stepping-motor driven unit drawn by a spring against a plate. This may provide for dynamic background correction as well as for sequential atomic emission spectrometry when working with time-stable sources.

As the price of individual CCDs has come down, the whole relevant Rowland circle can be covered with CCDs. By positioning them below and above the plane of the exiting radiation and providing mirrors placed at 45°, coverage of the whole spectrum, without missing any wavelength areas, becomes possible and a fully simultaneous spectrometer with data acquisition possibilities for analytical lines and background measurements at any wavelength can be realized (Fig. 21). Such

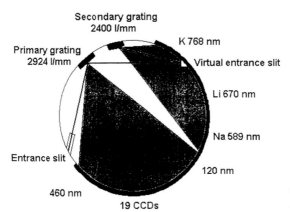

Fig. 21. Optical set-up for CCD-ICP-OES (CIROS) (Spectro Analytical Instruments GmbH).

instruments are now commercially available, so the capacity of a photographically recording spectrograph for obtaining simultaneous information has now been realized in atomic spectrometry with however, the ease, precision and time-resolution of a photoelectrical system.

In the Eagle mounting, the grating can be turned around its axis and moved along the direction of the radiation at the same time. The center and the focal plane change when switching to other wavelengths. This mounting is mechanically easy to construct, the astigmatism is low and the system can be easily housed in a narrow tank. It is often used in vacuum monochromators. The directions of the entering and the dispersed radiation beam in the Seya–Namioka mounting are constant and the angle between them is 70° 30′. The entrance and exit slit as well as the grating axis are fixed and the wavelength selection at the exit slit is performed by rotating the grating. At large angles a serious defocussing occurs and the aberrations are larger than in the case of the Eagle mounting. The Seya–Namioka system is also often used in vacuum monochromators.

With most sequential spectrometers switching from one line to another by turning the grating is usually fairly quick. From Eq. (169) it can be calculated that, as the spectral lines have widths of 1–3 pm, the angle selection must be performed very accurately for wavelength selection. Indeed, for a grating with $a = 1/2400$ mm and a line width of 1 pm the grating must be positioned with an accuracy of 10^{-4} degrees.

Therefore, one can scan the profile of the line or realize random access by using angle encoders. Computer-controlled stepper motors are used for turning the grating. Their angular resolution is often above 1000 steps per turn. In the case of a so-called sinebar drive, the number of steps performed by the motor is directly proportional to the wavelength displacement. Here a ball-bearing runs along while being pressed against a block moving on the spindle driven by the stepper motor. A further system uses a Paschen–Runge spectrometer with equidistant exit slits and a detector moving along the focal plane. Fine adjustment of the lines is then done by computer-controlled displacement of the entrance slit.

2.2.2
Radiation detectors

Photographic emulsions and photoelectric detection devices can be used as detectors for electromagnetic radiation between 150 and 800 nm. Among the photoelectric devices, photomultipliers are the most important but new solid state devices have become a useful alternative.

Photographic emulsions

These were frequently used for analytical spectrography. They allow the whole spectrum to be recorded simultaneously and, accordingly, for multielement survey analysis they have a high information capacity. However, their processing is long, the precision obtainable is low and they do not allow on-line data processing. Accordingly, the quantitative treatment of photographically recorded spectra remained only of limited importance for analytical use and the interest in photographic detection is now limited to research where documentation of the whole spectrum is required. Therefore, it will be treated only briefly here.

When a photographic emulsion is exposed to radiation a blackening is produced, which is a function of the radiant energy accumulated during the exposure time. It is given by:

$$S = \log(1/\tau) = -\log \tau = \log \phi_0/\phi \tag{176}$$

τ is the transmission, ϕ_0 is the flux through a non-irradiated part of the emulsion and ϕ is the flux obtained for an exposed part of the emulsion, when a white light beam is sent through the emulsion in a so-called densitometer. The emulsion characteristic for a selected illumination time gives the relationship between the logarithm of the intensity ($Y = \log I$) and the blackening and has an S-shape (Fig. 22).

Originally, only the linear part of the characteristic was used for quantitative work. However, soon transformations were described, which also allowed this characteristic to be linearized at low blackenings. The *P*-transformation has been

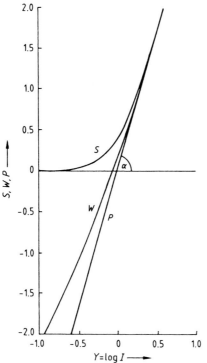

Fig. 22. Emulsion characteristic (*S*) and transformation functions (*P*, *W*).

described by Kaiser [51] as:

$$P = \gamma(Y - Y_0) = S - \kappa \cdot D \tag{177}$$

where $D = \log[1/(1 - \tau)]$.

The Seidel function is given by:

$$W = \log(10^S - 1) \tag{178}$$

and thus:

$$P = (1 - \kappa)S + \kappa W \tag{179}$$

Y_0 is the inertia of the emulsion and the constants κ and γ describe the properties of the emulsion but they also depend on the densitometer. Y_0 in most cases is not known as ΔP or ΔY values are usually used. $\gamma = \tan \alpha$ and is the contrast of the emulsion. The emulsion can be calibrated with the aid of a step filter which is placed in front of the spectrometer. When the ratio of the intensities passing through two sectors of the filter is given by:

$$\Delta Y_m = \log(I_{m(1)}/I_{m(2)}) \tag{180}$$

the transformation equation becomes:

$$\Delta S = \gamma \cdot \Delta Y_m + \kappa \cdot \Delta D \tag{181}$$

where ΔD is the independent variable and ΔS the function. A linear regression for a number of line pairs gives κ as slope and $\gamma \cdot \Delta Y_m$ as intersection with the x-axis [52]. $Y = \log I$ values for a given blackening then can be calculated as:

$$Y = \kappa/\gamma - \log(1 - \tau) - 1/\gamma \cdot \log \tau + Y_0 \tag{182}$$

and:

$$\Delta Y = -\kappa/\gamma \cdot \Delta D + 1/\gamma \cdot \Delta S \tag{183}$$

Errors on the intensity measurement stem from the graininess of the emulsion. According to Siedentopf [53], the standard deviation for measurements of small blackenings is given by:

$$\sigma(S) = \sqrt{[0.5 \cdot (a/F) \cdot S]} \tag{184}$$

a is the surface of a grain and F the surface of the densitometer slit (e.g. 10 μm × 1 mm).

• Example:

For the Kodak SA 1 emulsion at a photometrically measured surface of 10 μm × 1 mm and $S \approx 0.2$ the standard deviation $\sigma_S = 0.003$, from which it can be calculated according to Eq. (184) that the grain diameter is about 1 μm.

From $\sigma(S)$ one can calculate the relative standard deviation of the intensities $\sigma_r(I)$ according to the rule of the propagation of errors as:

$$\sigma_r(I) = \ln 10 \cdot \sigma(S) \cdot 1/\gamma\{1 + [\kappa/(10^S - 1)]\} \tag{185}$$

Typical values for γ are 1–2.5 and for κ 0.5–2. Non-sensitized emulsions can be used in the wavelength range 220–450 nm. At longer wavelengths green- or red-sensitized emulsions can be used. They have a large γ and accordingly a small dynamic range. At UV or VUV wavelengths the gelatine layer absorbs, so emulsions with less gelatine have to be used, which are fairly insensitive, or the radiation must be transformed to longer wavelengths with the aid of a scintillator, such as sodium salicylate.

Whereas in optical atomic spectrometry photographic emulsions are now rarely used, they are made use of in mass spectrometry with spark sources in particular. Here they are very useful for survey semi-quantitative analyses, and they still are of great importance for routine high-purity metals analysis and semiconductor material analysis.

For the precise and quick measurement of radiant intensities but also of ion currents photoelectric techniques are used almost exclusively. The measurements can easily be automated. The detectors used are photomultipliers, electron multipliers, photodiode array detectors, camera systems and other solid state detectors.

Photomultipliers

These are most commonly used for precise measurements of radiant intensities. The photomultiplier has a photocathode and a series of so-called dynodes. Its properties and their theoretical basis have been well treated in Ref. [54]. The radiation releases photoelectrons from the photocathode as a result of the Compton effect. After acceleration these photoelectrons impact on the dynodes so that a large number of secondary electrons are produced, which after successive action on a number of dynodes leads to a measurable photocurrent at the anode (I_a). For a photon flux N_ϕ passing through the exit slit of the spectrometer and impacting on the photocathode, the flux of photoelectrons produced (N_e) is:

$$N_e = N_\phi \cdot Q.E.(\lambda) \tag{186}$$

$Q.E.$ is the quantum efficiency (up to 20%). The cathode current I_c is given by:

$$I_c = N_e \cdot e \ (e = 1.9 \times 10^{-19} \text{ A s}) \tag{187}$$

and:

$$I_a = I_c \cdot \theta(V) \tag{188}$$

$\theta(V)$ is the amplification factor (up to 10^6). The dark current I_d is the I_a measured when no radiation enters the photomultiplier (about 1 nA). Q.E., $\theta(V)$ and I_d are characteristics of a certain type of photomultiplier. The photocurrents are lead to a measurement system. Firstly, I_a is fed to a preamplifier containing a high-ohmic resistor. The voltage produced is fed to an integrator. Each measurement channel mostly contains a preamplifier and an integrator. By selection of the resistors and capacitors, a high dynamic measurement range can be realized. When I_a accumulates in the capacitor with capacitance C (in Faradays) during an integration time t the potential obtained is:

$$U = I_a \cdot t/C \tag{189}$$

This voltage is digitized in an A/D convertor. The dynamic range of measurement is then finally limited at the lower side by the dark current of the photomultiplier and at the high side by saturation at the last dynode. The linear dynamic range of a photomultiplier is usually more than 5 decades of intensity.

In the case of a photomultiplier the relative standard deviation of the measured intensities $[\sigma_r(I)]$ is determined by the noise of the photoelectrons and the dark current electrons. The quantum efficiency and the amplification at the dynodes are also important. Approximately:

$$\sigma_r(I) \approx 1.3(N_{e,d})^{-1/2} \tag{190}$$

where $N_{e,d}$ is the total number of photoelectrons and dark current electrons in the measurement time t. The dark current should be so low that the lowest intensity signal to be measured (e.g. the spectral background intensity) produces a photoelectron flux near which the dark current is negligible. Then the noise of the photoelectrons is predominant.

Photomultipliers for the spectral range between 160 and 500 nm usually have an S20 photocathode, where $Q.E.(\lambda) \approx 5$–25%, the amplification factors are 3–5 at dynode voltages between 50 and 100 V and there are 9–15 dynodes mounted in head-on or side-on geometry. For selected types the dark current may be below 100 photoelectrons/s $(I_a < 10^{-10}$ A). Red-sensitive photomultipliers have a bialkali photocathode. Their dark current is higher, but can be decreased by cooling. For wavelengths below 160 nm photomultipliers with MgF_2 or LiF windows are used. So-called "solar-blind" photomultipliers are only sensitive for short-wavelength radiation (e.g. below 330 nm).

• Example:
With an ICP atomic emission spectrometer, for the spectral background intensity at 250 nm a signal of 104 digits is obtained with a 10 V A/D convertor and 12 bit resolution (4096), thus representing a voltage of 0.254 V. When a capacitor with 1.11×10^{-8} F and a measurement time of 10 s are used, this corresponds to a

photomultiplier anode current of 2.8×10^{-10} A. For the EMI 9789 QB photomultiplier with a voltage of 780 V, the amplification is 1.11×10^5, which means that the photocathode current is 2.5×10^{-15} A or 1.5×10^4 photoelectrons/s. As the quantum efficiency is 17% this represents a photon flux at the photomultiplier of 8.7×10^4 photons/s. When the dark current of the photomultiplier is 0.1 nA, this means that the flux of the dark current electrons is below 1000 electrons/s and that even with the spectral background at this wavelength, the noise of the photoelectrons is still dominant.

Solid-state detectors

In optical atomic spectrometry solid-state detectors have become more and more important. They are multichannel detectors and include vidicon, SIT-vidicon, photodiode array detectors and image dissector tubes [55]. Furthermore, charge-coupled devices (CCD) and charge injection devices (CID) [56] have now been introduced. Photodiode arrays (PDA) may consist of matrices of up to 512 and even 1024 individual diodes (e.g. Reticon) which individually have widths of 10 μm and heights of up to 20 mm. The charge induced by the photoelectric effects gives rise to photocurrents which are sequentially fed into a preamplifier and a multichannel analyzer. Therefore, they are rapid sequential devices. They have the advantage that memory effects due to the incidence of high radiant densities on individual diodes (lag) as well as cross talk between individual diodes (blooming) are low. Their sensitivity in the UV and the VUV spectral range is low as compared with photomultipliers. Nevertheless, at present they are of interest for atomic emission spectrometry in particular as they can be coupled with microchannel plates giving an additional amplification. This plate can be considered as being built up from a semiconductor plate in which parallel channels of low diameter at a high density have been provided. When a high potential is applied between both sides of the plate each tube acts as a single electron multiplier (Fig. 23). The signal-to-noise ratios

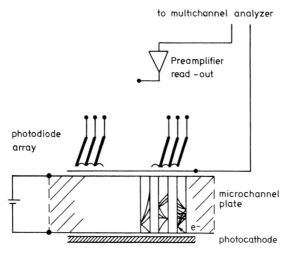

Fig. 23. Principle of the microchannel-plate/diode-array camera.

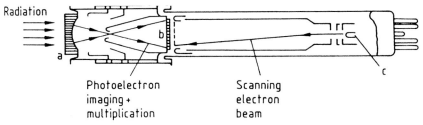

Radiation

Photoelectron
imaging +
multiplication

Scanning
electron
beam

Fig. 24. Early SIT vidicon system. (Courtesy of SSR Instruments, Inc.)

of photodiode arrays can be considerably improved by cooling them with the aid of a Peltier element or liquid nitrogen. Commercial instrumentation is now available (e.g. EG&G OMA 4; Spectroscopy Instruments Inc.) and has replaced silicon intensified target (SIT) vidicon detectors for most applications. In vidicons the charge is integrated into the individual pixels and the simultaneously collected information is read out sequentially with the aid of an electron beam. When using an electrostatic beam intensifier (Fig. 24), in a number of cases these devices are quite sensitive and also permit two-dimensionally resolved measurements. However, they suffer from blooming. This means that in pixels near places receiving a high radiance a charge decay and thus a signal is also induced. Furthermore, lag is present, through which an irradiated pixel has some hysteresis effects, deteriorating subsequent signal acquisition. Photodiode array detectors have been used successfully, for example, in a segmented echelle spectrometer (Plasmarray, Leco Co) [57]. Here, spectral segments around the analytical lines are sorted by primary masks subsequent to spectral dispersion with an echelle grating and are detected after a second dispersion on a diode array. Accordingly, more than ten analytical lines and their spectral background contributions can be measured simultaneously. An example of this can be shown in the case of a glow discharge lamp (see Fig. 25) [58], however, it should be mentioned that restrictions arise due to detector noise limitations at low wavelengths.

Image dissector tubes (Fig. 26) make use of an entrance aperture behind the photocathode, by which the photoelectrons stemming from different locations of the photocathode can be scanned and measured after amplification in the dynode train, as in a conventional photomultiplier. Although used in combination with an echelle spectrometer with crossed dispersion for flexible rapid sequential analyses [59], these systems have not had any commercial breakthrough. This might be due to the limited cathode dimensions but also to stability problems.

Charge transfer devices (CTD) [60] are solid-state multichannel detectors and integrate signal information as light strikes them, just as a photographic film does. An individual detector in a CTD array consists of several conducting electrodes overlaying an insulating layer that forms a series of metal oxide semiconductor (MOS) capacitors. The insulator separates the electrodes from a doped silica region used for photogenerated charge storage. In this n-doped silicon, the major current carrier is the electron, and the minor carrier is the hole. When the electrodes are negatively charged with respect to the silicon, a charge inversion region is created

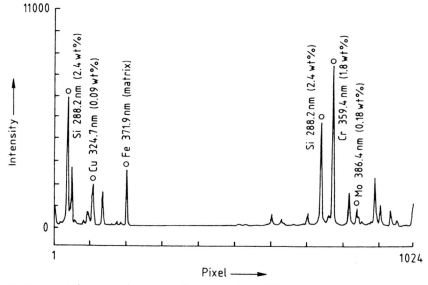

Fig. 25. Spectral-segmented spectrum of a glow discharge [58]. Grimm-type glow discharge lamp with floating restrictor (8 mm diameter): 50 mA, 3.5 torr, 2 kV [42]; Plasmaarray spectrometer (LECO); steel sample: 217 A (Research Institute, CKD, Prague, Czech Republic).

under the electrodes. This region is an energetically favorable location for mobile holes to reside in. The promotion of an electron into the semiconductor conduction band, such as by the absorption of a photon, creates a mobile hole that migrates to and is collected in the inversion region. The amount of charge generated in a CTD detector is measured either by moving the charge from the detector element, where it is collected, to a charge-sensing amplifier, or by moving it within the detector element and measuring the voltage change induced by this movement. These two modes of charge sensing are employed by the charge-coupled device (CCD) and the charge injection device (CID), respectively. In the CCD, the charge from each detector element is shifted in sequence to an amplifier located at the end of a linear array or the corner of a two-dimensional array of detector elements. Charge is transformed to the on-chip amplifier by sequentially passing the charge packets

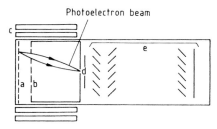

Fig. 26. Image dissector tube. (a): Photocathode, (b): accelerating electrode (mesh), (c): deflection coils, (d): aperture, (e): electron multiplier.

Fig. 27. Charge-coupled device (CCD) detector. Layout of a typical three-phase CCD. Photogenerated charge is shifted in parallel to the serial register. The charge in the serial register is then shifted to the amplifier. (Reprinted with permission from Ref. [63].)

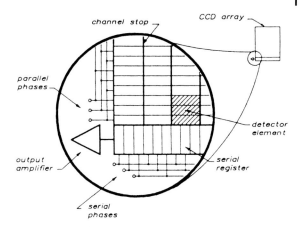

through a row of detectors. The organization of a two-dimensional, three-phase CCD array is illustrated in Fig. 27. Columns are read in parallel. With each reading, all of the charge in the imaging array is shifted towards the serial register by one row, whereas charge from the row adjacent to the serial register is transferred into the serial register. Once this done, the charge packets are shifted sequentially to the on-chip amplifier. Modern CCDs are capable of thousands of transfers with almost immeasurable charge transfer losses. By their ability to transfer the photogenerated signal from the photoactive element to an on-chip amplifier, a very high signal-to-noise ratio can be obtained, as compared with other detectors (see Fig. 28) [60].

Whereas in photodiode arrays the limiting noise is read noise, this is not the case here with CTDs. With CIDs the charge information can be read out non-destructively further increasing the signal-to-noise ratios. With CCDs binning can be applied, which is the process of summing the charge contained in multiple elements on the detector before sensing the total charge [61]. CTDs which can be used from the soft x-ray region to the IR are now commercially available and the signal-to-noise ratios achievable have become as high as with photomultipliers. The dynamic range has risen up to 7 orders of magnitude in the case of a CID [62].

Fig. 28. Calculated signal-to-noise ratios for different detectors at a photon flux of 10 photons/s for observation times ranging from 1 to 100 s at a wavelength of 600 nm. (Reprinted with permission from Ref. [60].)

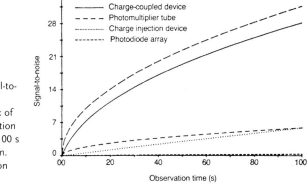

Tab. 3. Evaluation of some new radiation detectors for atomic emission spectrometry.

Detector	Dimensions	Spectral range	Sensitivity versus photomultiplier	Ref.
Vidicon	12.5×12.5 mm^2 bidimens	with scintillator to 200 nm	poorer particularly at <350 nm dynamics <10^2 lag/blooming	55
Diode array	up to 25 mm up to 1024 monodimens	particularly for >350 nm	poor in UV dynamics 10^3	55, 64
Diode array MCP	12.5 mm monodimens	200–800 nm	similar to PM dynamics >10^3	55
Dissector tube	up to 60 mm bidimens	200–800 nm	similar to PM dynamics >10^3	59
CCD	up to 13×2 mm^2 bidimens	200–800 nm	similar to PM dynamics >10^3	56

These two-dimensional detectors [63] are ideally suited for coupling with an echelle spectrometer, which is state of the art in modern spectrometers for ICP atomic emission spectrometry as well as for atomic absorption spectrometers. As for CCDs the sensitivity is high and along with the signal-to-noise ratios achievable, they have become real alternatives to photomultipliers for optical atomic spectrometry (Table 3) and will replace them more and more.

2.2.3
Non-dispersive spectrometers

Filter monochromators are now used almost only for flame photometry. They make use of interference filters, which may have a fairly low spectral bandpass (less than a few nm). However, it is also possible to use such filters for dynamic measurements of line and background intensities, and for transient signals, as occur in gas chromatography. The use of oscillating filters has been described, where the wavelength bandpass is slightly shifted by inclining them towards the radiation beam [65].

Non-dispersive multiplex spectrometers include Hadamard transformation spectrometers and Fourier transform spectrometers and are particularly useful for the case of very stable sources. In both cases the information, such as intensities at various wavelengths, is coded by a multiplex system, so that it can be recorded with a conventional detector. A suitable transformation is then used to reconstruct the wavelength dependence of the information. In Hadamard spectrometry use is made of a codation of the spectrum produced by recombining the information with the aid of a slit mask which is moved along the spectrum [66].

Fourier transform spectrometry [67] makes use of a Michelson interferometer (Fig. 29) to produce the interferogram. With the aid of a beam splitter the radiation is split into two parts each of which is directed to a mirror. When shifting the

Fig. 29. Principle of Fourier transform emission spectrometry.

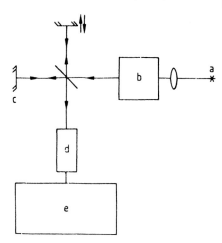

mirror in one of the side arms, the interference for each wavelength is given by:

$$I(x) = B(\sigma)[1 + \cos(2\pi\sigma x)] \tag{191}$$

x is the optical path difference, σ is the wavenumber of the radiation (cm^{-1}), I is the intensity measured with the detector, and B is the radiation density of the source. The frequency of the signal at the detector for a radiation with a frequency v is a function of the velocity v by which the optical path difference is changed:

$$f = \sigma \cdot v \tag{192}$$

Accordingly rapid changes in optical frequencies are transformed to lower frequencies, where they can be measured with the normal detector systems. For a wavelength of 400 nm or a frequency of 25000 s^{-1} and a scanning velocity of 0.08 cm/s the frequency becomes 2 kHz.

Polychromatic radiation from a source accordingly gives an interferogram where the intensity of each point is the sum of all values resulting from Eq. (191). The central part contains the information at low resolution and the ends contain information at high resolution. A detector with a high dynamic range is thus required and this especially applies for line-rich atomic emission spectra with lines of widely differing intensities. Furthermore, the source must be very stable during the recording of the interferogram. The resolution obtained depends on the recording time, the spectral bandwidth and the number of repetitive scans. By applying a Fourier transform of the signal for each point of the interferogram, one can write:

$$I(x) = \int_{-\infty}^{+\infty} B(\sigma)\, d\sigma + \int_{-\infty}^{+\infty} B(\sigma)\cos(2\pi\sigma x)\, d\sigma \tag{193}$$

$$= C + \int_{-\infty}^{+\infty} B(\sigma)\cos(2\pi\sigma x)\, d\sigma \tag{194}$$

C is a constant term and must be subtracted before the transformation and one finally obtains:

$$B(\sigma) = \int_{-\infty}^{+\infty} I(x) \cos(2\pi\sigma x)\, dx \tag{195}$$

In the transformation the physical units are inverted. When the interferogram is expressed in optical path difference units (cm), the spectrum is obtained in wave-numbers (cm^{-1}) and when the interferogram is expressed in time units (s) the spectrum is in frequency units (s^{-1}). Apart from sine and cosine functions, box-car and triangular, etc. functions are also known, for which the Fourier transformation can be calculated. When applying the Fourier transformation over the whole area $\pm\infty$, the arm of the interferometer also would have to be moved from $-\infty$ to $+\infty$. When making a displacement over a distance of $\pm L$ only, the interferogram has to be multiplied by a block function, which has the value of 1 between $+$ and $-L$ and the value 0 outside. *L* then influences the resolution that can be obtained.

By digitizing the interferogram, a quick Fourier transform becomes possible provided a powerful computer is used. In the case of line-rich spectra this is possible with a limited but highly accurate displacement of the sidearm in a number of steps. In addition, repetitive scanning is applied so as to obtain a sufficiently intense image, during which a reference laser is used to ensure that the positioning of the mirror is reproducible. The technique has been well known in infrared spectroscopy for some time and now can be used for spectra in the visible and UV regions also, as sufficiently precise mechanical displacement systems and high-capacity computers have become available. Fourier transform atomic emission spectrometry is attractive for sources with a low radiance but a high stability. Furthermore, the wavelength calibration can be very accurate and the highest resolution can be realized. This makes the technique very attractive for line wavelength and line-shape measurements for diagnostics as well as for the development of line tables. However, only in the case of low-noise sources is it possible to obtain reasonable signal-to-noise ratios.

2.3
Mass spectrometers

As in the sources used in optical atomic spectrometry a considerable ionization takes place, they are also of use as ion sources for mass spectrometry. Although an overall treatment of instrumentation for mass spectrometry is given in other textbooks [68], the most common types of mass spectrometers will be briefly outlined here. In particular, the new types of elemental mass spectrometry sources have to be considered, namely the glow discharges and the inductively and eventually the microwave plasmas. In contrast with classical high voltage spark mass spectrometry (for a review see Ref. [69]) or thermionic mass spectrometry (see e.g. Ref. [70]), the plasma sources mentioned are operated at a pressure which is considerably higher (around 0.1 mbar up to atmospheric pressure) than the pressure in the mass spectrometer itself (10^{-5} mbar). Consequently, an interface with the appropriate apertures has to be provided, along with high displacement vacuum pumps for bridging the pressure difference between the source and the spectrometer itself.

In commercial instrumentation fairly low-cost quadrupole mass spectrometers and also expensive double-focussing sector field mass spectrometers are usually used (for a survey of mass spectrometers for analytical use, see Ref. [71]) and today new types of mass spectrometers such as time-of-flight mass spectrometers are being find utilized in plasma atomic spectrometry.

2.3.1
Types of mass spectrometers

In all mass spectrometers used for atomic spectrometry a vacuum of 10^{-5} mbar or better is maintained so as to avoid further ion formation from residual gas species or collisions of analyte ions with these species, by which the background signals would be increased. To this aim, turbomolecular pumps are used at present. They are preferred over diffusion pumps as their maintenance is easier and backflow of oils does not occur. They are backed up by rotation pumps. To monitor the vacuum in the mass spectrometer a Penning gauge is generally used, whereas the vacuum in the sampling region, which is at the 0.1–1 mbar level, is often controlled with the aid of a capacitance pressure gauge.

Quadrupole instruments

The spectrometer consists mainly of 4 equidistant and parallelly mounted rods (diameter 10–12 mm) between which a dc voltage and in addition a high-frequency field (frequency: up to 1 MHz) or $U + V \cdot \cos(\omega t)$ is provided (Fig. 30). The dc voltage at the quadrupole should be slightly below the energy of the ions entering (usually below 30 eV). By varying the electric alternating field, ions with different m/z can leave the analyzer sequentially and scanning becomes possible. A mathe-

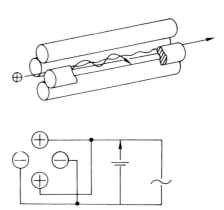

Fig. 30. Principle of a quadrupole mass spectrometer.

matical treatment of the processes taking place is possible with the aid of the Mathieu equations:

$$A = 8 \cdot e \cdot U/m \cdot r_0^2 \cdot \omega^2 \qquad (196)$$

and

$$Q = 4 \cdot e \cdot V_0/m \cdot r_0^2 \cdot \omega^2 \qquad (197)$$

with U the dc potential applied, m the mass of the ion, $\omega/2\pi$ is the frequency of the alternating current, V_0 the amplitude of the alternating current potential, r_0 the radius of the quadrupole field and e the elementary charge. Plotting A and Q on a diagram there is a zone for stable oscillation for ions within a narrow m/z range.

The ions enter the instrument through the so-called skimmer and in addition have to pass the ion lenses and eventually a mask, which prevents the UV radiation from entering the spectrometer ("beam-stop"). By changing the voltages at the ion lenses, at the beam-stop and eventually at the quadrupole rods, the transmission of the spectrometer and the mass resolution for ions of a given mass and energy can be optimized. As these parameters all are interdependent, the optimization of the mass spectrometer is complex. By changing the quadrupole field, the transmission for a certain ion changes. Accordingly, ions of a certain mass can be filtered out by manual setting of the field or the spectrum can be scanned. The scanning velocity is normally limited to up to 30 000 dalton/s as a result of the high-frequency and dc components used in the spectrometer. Thus a mass interval of 0–300 dalton can be scanned in about 30 ms. The mass resolution achievable is mainly determined by the quality of the field and in quadrupoles used for plasma mass spectrometry in the best case it is about 1 dalton. The line profiles to a first approximation are triangular, but side-wings also occur. Together with the isotopic abundances they determine the magnitude of the spectral interferences. Quadrupole mass spectrometers are rapid sequential measurement systems. Accordingly, the precision obtainable and also the figures of merit in the case of isotopic dilution techniques for calibration are limited by noise occurring in the source. Therefore, the price of quadrupole instruments is reasonable and also their transmission is high. This is important because it allows their use in plasma mass spectrometry, where the source itself can be kept at earth potential. This is a particular advantage in the case of inductively coupled plasmas.

Magnetic mass analyzers

The oldest type of mass analyzers used, going back to the early 1920s, are magnetic mass analyzers. They consist of a curved flight tube located between the poles of an electromagnet, so that the field is perpendicular to the flight direction of the ions. The ions first pass through an electrical field and then enter a magnetic field where mass selection takes place. The electrical field is used to extract the ions by a force:

$$F = z \cdot U \tag{198}$$

where z is the charge and U the voltage (up to 10 kV).

The ions then enter a magnetic field B and start to follow a curved trajectory on which they are held by the centrifugal force so that:

$$m \cdot v^2 / r = B \cdot z \cdot v \tag{199}$$

where r is the radius of the curved trajectory and v the velocity of the particle. Then:

$$v = B \cdot z \cdot r / m \tag{200}$$

As

$$z \cdot U = 1/2 \cdot m \cdot v^2 \tag{201}$$

$$v = \sqrt{(2 \cdot z \cdot U / m)} \tag{202}$$

and by combining with Eq. (200):

$$B \cdot z \cdot r / m = \sqrt{(2 \cdot z \cdot U / m)}$$

or

$$m/z = B^2 r^2 / 2U \tag{203}$$

By sweeping the field B it is possible to scan the mass spectrum (m/z) at an exit slit. As ions with different masses are deflected towards different locations in the focal plane simultaneous detection of ions with different masses is thus possible with these spectrometers, provided multichannel detection is applied.

As the energy spread of the ion beam limits the resolving power of a deflecting magnetic sector mass analyzer, energy focusing must be applied. This can be done by providing an electrostatic analyzer in the ion optics. This electrostatic analyzer consists of two curved plates with a voltage of the order of 1.5–1 kV applied between them. For a mean radius r and a gap d operated with a voltage E, a singly charged ion with a mass m and a velocity v will be focussed through the exit slit only when:

$$m \cdot v^2 / z = E \cdot r / d \tag{204}$$

with the radius of the circular path being defined by the kinetic energy as:

$$r = (2 \cdot d / E \cdot z) \times 0.5 \times (mv^2) \tag{205}$$

When bringing the magnetic sector and the electrostatic sector together, a double-focusing instrument is obtained, where ions are selected by mass only and where the ion energy of the particles does not play a role. The use of sector field mass spectrometers requires the ion source to be brought to a high positive potential relative to the mass spectrometer entrance or the whole mass spectrometer to be at a high negative voltage with respect to the ion source. The low transmission of the complex arrangement accordingly is compensated for by a higher ion extraction yield. With sector field instruments the resolution can easily be improved to below 0.1 dalton, enabling a number of spectral interferences in plasma mass spectrometry to be eliminated. The profiles of the lines may be flat-topped but can be adjusted by optimizing with the aid of the widths of the source slit and the collector slit. The scanning speed cannot be increased in an unlimited manner, as hysteresis of the magnet may hamper this. When the signals are transient and multielement detection is required, there are certainly limitations here.

Advanced types of mass spectrometers such as ion traps, ion cyclotron resonance and specifically time-of-flight mass spectrometers have also been considered for use in plasma mass spectrometry.

Time-of-flight mass spectrometer

This spectrometer is based on the detection of ions produced from a single temporally well-defined event with high time resolution. The principle is well known and has long been applied in secondary ion mass spectrometry [72] and in laser micro mass spectrometry [73].

The basic formula for mass resolution is [71]:

$$m/\Delta m = 1/2(t/\Delta t) \tag{205}$$

The velocity of the ions is given by:

$$v = 1.4 \times 10^4 \sqrt{(U/m)} \tag{206}$$

where m is the mass and U the accelerating voltage. Inserting 50 dalton for m and 5000 V for U would result in $v = 1.4 \times 10^5$ m/s. Hence for a flight tube having a length l of 1.4 m one obtains, neglecting the time required to accelerate the ions: $t = l/v = 10^{-5}$ s. With modern instruments a time spread Δt of 5 ns and even lower can be obtained. Thus according to Eq. (205) a resolution of 2000 is feasable. The resolution can be increased further by using a reflectron, doubling the flight time (Fig. 31). With this type of instrumentation high transmission can be achieved and high mass resolution can also be realized. In addition to ICP, mass spectrometry [74, 75] with the MIP [76] and glow discharges [77–79] has been developed. In the latter, packages of ions can be extracted on- or off-axis by using a pulsed field and it is even possible to deflect the abundant argon or helium filler gas ions efficiently to clean the spectrum, by applying a repeller pulse (Fig. 32). The technique has been shown to be particularly useful for transient signals, where a fair dynamic range and power of detection are achievable (see Ref. [80]). This also applies in the case of conventional ion detectors, such as dynode electron multipliers (Fig. 33).

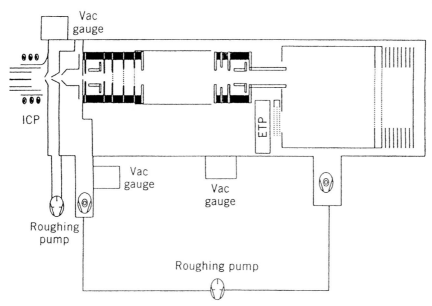

Fig. 31. On-axis ICP-time-of-flight mass spectrometer (Renaissance). (Courtesy of Leco Corp.)

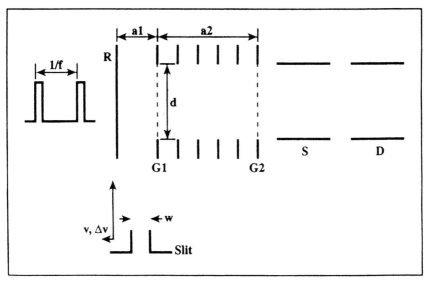

Fig. 32. Two-stage acceleration region in an ICP-TOF instrument. w: Beam width, v: beam velocity, R: repeller plate, f: repetition frequency, G1: extraction region entrance grid, d: packet length, a1: extraction region, a2: acceleration region, G2 flight-tube entrance grid, S: steering plates, D: deflection plates. (Reprinted with permission from Ref. [74].)

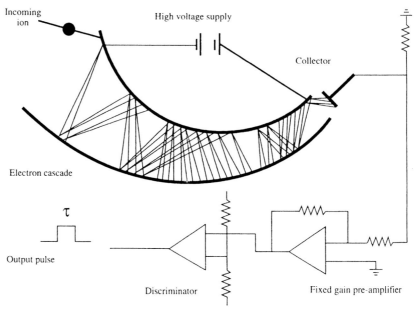

Fig. 33. Continuous dynode electron multiplier. (Reprinted with permission from Ref. [81].)

Ion trap. In an ion trap [81] we have a tridimensional quadrupole mass spectrometer (Fig. 34), in which ions may be trapped or stored. This is achieved by applying a suitable rf voltage to a specially shaped electrode. Even during storage collisional dissociation as well as excitation may be achieved. The sensitiviy is excellent and a high resolution may also be achieved (in excess of 10 000 particularly when working with long acquisition times). This means one can also work with very weak ion beams [82]. The first applications of an ion trap with an ICP have been published, both when using the ion trap behind a conventional quadrupole [83] as well as when coupling it through a suitable lens directly to the ICP [84].

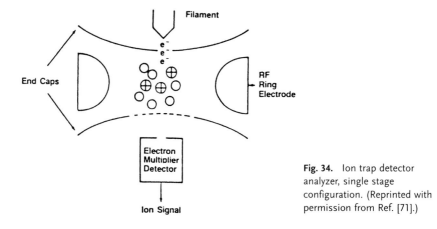

Fig. 34. Ion trap detector analyzer, single stage configuration. (Reprinted with permission from Ref. [71].)

Fig. 35. Operating principles of ion cyclotron resonance mass spectrometry. (Reprinted with permission from Ref. [81].)

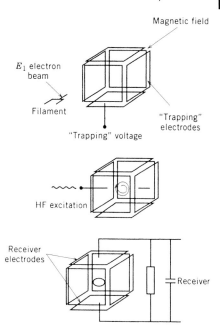

Ion cyclotron resonance analyzer

In this type of analyzer [85] the ions are trapped in circular orbits in a magnetic field. To achieve this, three pairs of mutually perpendicular plates are placed in a strong magnetic field. Externally produced ions may be injected into the cell, in which the time t for one revolution (Fig. 35) [81, 86] is given by:

$$t = (2 \cdot \pi \cdot r_m)/v \tag{207}$$

where v is the velocity and r_m the radius of the orbit. Combination with Eq. (199) delivers:

$$m/z = B/2 \cdot \pi \cdot f \tag{208}$$

where f is the frequency.

At a constant magnetic field strength B, the m/z for the orbit depends on the frequency of the cyclotron. When applying an RF pulse of this frequency to two opposite plates, the respective ions will gain energy and when pulsing with a broad pulse distribution, many ions will come into resonance. This will lead to the respective signals in the receiving electrodes being a function of the ions in the cell. As the frequency distribution is determined by a Fourier transform of the induced currents, this method is known as Fourier transform mass spectrometry. In ion cyclotron resonance mass spectrometry a very high resolution can be achieved, as frequency distributions can be determined with a very high accuracy. Further-

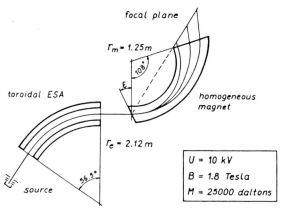

focal plane

$r_m = 1.25\,m$

108°

ε

toroidal ESA

homogeneous magnet

$r_e = 2.12\,m$

source

56.5°

| $U = 10\ kV$ |
| $B = 1.8\ Tesla$ |
| $M = 25000\ daltons$ |

Fig. 36. Marge Mattauch–Herzog system. (Reprinted with permission of Ref. [71].)

more, the method has a multiplex advantage and therefore allows for true simultaneous detection of ions over a wide mass range.

Simultaneous mass spectrometer

This can be realized by using Mattauch–Herzog geometry, as is known from spark source mass spectrometry. Here the ions are conducted through an appropriate lens system into a magnetic analyzer and separated according to mass before reaching a linear focal plane (Fig. 36). A linear photodiode array with a microchannel plate can be used for the simultaneous and time-resolved detection of ions over the whole mass range.

Multiple-collector magnetic mass analyzers

There are particularly useful when a highly accurate recording of ratios of ion currents for different isotopes is required. This is known from earlier work with thermal ionization used for isotopic analysis, and is important, for instance, in geochronology. An double-focusing instrument with a stigmatic focusing magnet and several individual collectors placed in the focal plane can be used to achieve such a system. Here the dispersion of the ions in the focal plane is larger than with the Mattauch–Herzog arrangement, allowing for the placement of several detectors. One also can use a beam-splitter in front of two quadrupole instruments [87]. As isotopic dilution becomes more and more important for tracer work as well as for highly accurate calibration, similar types of mass spectrometers may gain considerable importance.

2.3.2
Ion detection

For ion detection, several approaches are possible. From classical spark source mass spectrometry, the use of photographic plates is known, where very hard emulsions with low gelatine contents are required. The procedure for emulsion calibration is similar to the one described in optical emission spectrography, but different exposure times are normally used instead of the optical step filters. Provided an automated microdensitometer is used, mass spectrography is still a useful tool for survey analysis of solids down to the sub-µg/g level.

However, for the precise work that is possible with the plasma sources, electron multipliers and pulse counting are usually used for ion detection. In elemental mass spectrometry the magnitudes of the signal to be detected may vary between 0.1 ion/s for the case of ultratrace analysis and 10^{10} ions/s in the case of major elements, and depending on the ion source used. As the charge of an electron is 1.6×10^{-19} Coulomb, these ion densities correspond to currents of between 10^{-20} and 10^{-9} A and for a such a range, several detectors need to be used. For low signals ($<10^6$ ions/s) ion counting is usually applied whereas for higher ranges analogue detection is used.

In an ion counting device (Fig. 33) the ions entering the analyzer are directed to a suitable surface held at high potential, so that the incident ions release several electrons. The surface may consist of a continuous dynode, the curvature of which guides the electrons, which increase in number at each interaction with the dynode surface and finally onto an anode. In the case of discrete dynodes the construction of the detector is similar to that of a photomultiplier. These detectors have a dead time, as the interaction of an ion takes a finite time (usually less than 10 ns), during which the detector is blind to the next ion. One can make a correction of the measured count rate (n_{meas}) for this dead time (τ) as:

$$n_{\mathrm{corr}} = n_{\mathrm{meas}}/(1 - n_{\mathrm{meas}} \cdot \tau) \qquad (209)$$

where n_{corr} is the corrected count rate. At a count rate of 1 MHz a correction of about 1% will usually be made and this will be the upper limit for which ion counting can be used. Above this rate the uncertainties in the corrections applied become too large and moreover, damage of the electron multiplier may occur as ion counters also have a high gain.

The ion counters suffer from further problems when ions are allowed to impact directly on the dynode. Indeed, depending on the mass and ionization potential of the respective elements the number of electrons released by the impact may differ from one element to another, which introduces a mass bias and in addition, when high masses impact on a dynode, the lifetime of the dynode may decrease considerably. Therefore, the ion current is generally deflected onto a conversion dynode, which is more robust and is at a high negative potential for the ion impact, and the current then carried by electrons enters the dynode system.

For the registration of larger signals, the voltage applied at the electron multiplier can be decreased and then a current is obtained, which in the case of ion currents above 10^6 ions/s is directly proportional to the input current. A gain of 10^3 is thus generally applied instead of the 10^8 used for ion counting. Another solution is the use of a so-called Faraday collector (Fig. 37) [81]. With this, currents of down to 10^{-15} A (10^4 ions/s) can be counted, corresponding to the upper range accessible

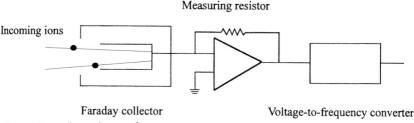

Fig. 37. Dc analogue detector for mass spectrometry. (Reprinted with permission from Ref. [81].)

with ion counting. The variable feedback-resistor allows the measuring range to be adjusted and voltage-to-frequency conversion is applied for read-out. Faraday collectors have low mass bias and are robust but slow.

So-called dual detection can also be applied, providing for both ion counting and analogue counting, and switching can even be made within a single scan. This is realized with a dual mode electron multiplier by sweeping the gain voltage.

Microchannel plate detectors can also be used. When using a phosphor, the detector can be coupled with photodiode arrays or charge coupled devices, which are known from optical atomic spectrometry, and it becomes possible to detect ions of different masses simultaneously as in the case of photoplate detection.

The signal currents are fed into a preamplifier and the sorting signals can be directly displayed. However, it is more convenient for the signal to enter a multichannel analyzer after analogue-to-digital conversion. Accordingly, accumulation of the spectrum or a part of it during a high number of scan cycles becomes possible.

2.3.3
Ion extraction

In the case of an rf spark one can work at high vacuum in the source. However, the ions formed in the analytical zone of plasmas, such as the ICP or a glow discharge, which are at atmospheric pressure or at a reduced pressure (mbar range), are extracted with the aid of a metal aperture. In the case of the ICP they enter into an interspace where the pressure is around 0.1 mbar. For glow discharges, this is achieved without any additional precautions, as these sources are operated in the mbar region. With plasma discharges at atmospheric pressure such as the ICP and MIP as ion sources, we therefore need a powerful pumping system using, for example, high-displacement rotary vacuum pumps. Also the aperture cannot be much larger than 1 mm. Its minimum dimensions are dictated by the fact that below a certain aperture diameter ion sampling from the plasma would no longer be possible due to barrier-layer effects. According to Turner et al. [81] the plasma–MS interface should fulfill the following requirements.

1. The plasma MS pressure should be sufficiently low as to minimize chemical reactions.
2. The diameter of the sampling aperture should be as large as possible to maximize the analyte signal and to minimize orifice clogging, while keeping the pumps used for this section to a reasonable size.
3. To sample the plasma in as undisturbed a state as possible, the diameter of the sampling orifice must be large compared with the Debye length in the plasma, which is a measure of the distance of the electric field of an ion as seen by the electrons and is given by:

$$\lambda_D = 6.9\sqrt{(T_e \cdot N_e)} \tag{210}$$

where N_e is the electron number density and T_e the electron temperature.
4. The gas flow through the sampler should not exceed the whole gas flow in the plasma, as only well defined zones of the plasma (namely those with the highest analyte concentrations) should be sampled.
5. Behind the sampler a concentric shock wave structure is formed, which surrounds a zone of silence and ends in a shock wave front called the Mach disk. The interface should be such that the skimmer aperture still lies in the zone of silence for an optimal extraction.

In the case of the ICP the sampler must be cooled, so as to withstand the high temperature of the ICP and corrosion from the sample aerosol. Therefore, in the early work on ICP-MS, samplers were made of nickel or copper [88]. With very corrosive acids (HF and HNO_3), which are often present after the decomposition of geological materials, samplers made of Pt were also proposed [89]. Furthermore, nebulizing a Ti solution before the analysis of corrosive samples has also been proposed so as to build up a protective TiN layer on the sampler. Owing to the large pressure difference between the plasma discharge and the first vacuum stage, the ions enter the interface with high energy. There is also a considerably high adiabatic expansion of the plasma in the interface. This jet expansion is critical with respect to the analytical performance of the system. A second aperture (skimmer) is used to sample a part of the ion beam and to bring it into the high vacuum of the mass spectrometer (10^{-5} mbar). Behind the skimmer the mutual repulsions of ions within the ion beam, known as space charge effects, also influence the ion trajectories. The magnitude of this effect depends on the total radial voltage drop (E_r), the current density (j), the total current (I) and the velocity (v) and is given by [90]:

$$\Delta\phi = \int_0^a E_r \cdot dr = \int_0^a (2 \cdot \pi \cdot j \cdot r/v)\, dr = (9 \times 10^{11})I/v \tag{211}$$

where $v = 2.5 \times 10^5$ cm/s and $I = 10^{-6}$ A, $\Delta\phi = 3.6$ V. With a field such as this, ions will move away from the beam axis and this movement is mass dependent and has to be corrected for with the aid of suitable ion optics.

The vacuum in the mass spectrometer itself is maintained by a cryopump or by turbomolecular pumps backed by a rotary pump. The aperture of the skimmer is again below 1 mm.

2.3.4
Ion optics and transmission

A cylindrically symmetric lens is used to transfer the ions sampled through the skimmer to an analyzer [91]. To reduce the noise level, a metallic disk denoted as the "photon stop" is often placed in the center of the so-called Bessel box. Thus radiation coming from the plasma sampled as it enters through the axially centered sampler and or skimmer can be eliminated. Furthermore, the voltage at the lens is optimized so as to obtain as low a background (stemming from radiation and neutrals) as possible without decreasing the sensitivity. Alternatively, the mass spectrometer and/or the detector can be placed off-axis. A series of various lenses at different voltages is usually provided so as to direct the analyte ions as efficiently as possible into the analyzer. Often 3 so-called "Einzel" lenses at negative potentials relative to the skimmer (varying from around −10 to −150 V) are used. The voltages applied determine the trajectories of the ions, which, however, also depend on their kinetic energies. The latter are element-dependent and also relate to the zones in the ion source that were sampled.

A computer program, MacSimion, [92] can be used to model the ion trajectories in ion optics after various assumptions. From these considerations it becomes clear that when dealing with several elements any optimization will also lead to compromises. Thus there will always be a difference in performance in mass spectrometry with respect to sensitivity and to the lowest matrix influences between single-element optimization and compromise conditions.

2.4
Data acquisition and treatment

Modern optical spectrometers as well as mass spectrometers are controlled by a computer, responsible for pre-setting the wavelength or ion mass and parameter selection, as well as for control of the radiation or ion source and of the safety systems. The software for the data acquisition and processing may consist of routines for the calculation of calibration functions with the aid of linear or higher regression procedures, for the calculation of the concentrations in the samples according to standard additions techniques, as well as for calibration with synthetic samples, for the calculation of statistical errors and the correction of drift. It must also be possible to display parts of the spectrum graphically or to superimpose spectra from several samples so as to recognize spectral interferences or to execute background correction. Furthermore, routines for the correction of signal enhancements and depressions as well as of spectral interferences must be present.

In most cases a linear calibration function can be used, as discussed earlier. It can be calculated by a linear regression with a number of values (c_i, x_i) where c_i is the concentration of a standard sample and x_i is the radiation intensity, absorption or ion current obtained for the element to be determined. It has the form:

$$x_i = a \cdot c_i \tag{212}$$

a is the sensitivity and the slope of the calibration curve. In order to eliminate fluctuations of the source and to obtain the highest precision it can be useful to plot the analytical signals as signal ratios. Then the calibration function obtained is of the form:

$$x_i/R = a \cdot c_i \tag{213}$$

R is the signal obtained for a reference element, which may be the matrix element, a main constituent or an element added in known concentration to all samples to be analyzed and not present in the sample. This approach is known as internal standardization. It may also happen that the calibration function is not linear, but can be described by a function of the 2nd or even of the 3rd degree. Then it has the form:

$$x_i/R = a_0 + a_1 \cdot c_i + a_2 \cdot c_i^2 \ (+ a_3 \cdot c_i^3) \tag{214}$$

The a coefficients have to be calculated using regression procedures. In the case of a calibration function of a higher degree, the calibration graph is curved. The latter also can be approximated by a polygone, where different linear calibration functions are used for well-defined concentration ranges.

Interferences or matrix effects occur when the analytical signals not only depend on the analyte element concentrations but also on the concentrations of other sample constituents. Additive and multiplicative interferences have to be distinguished. The first may result from spectral interferences. Here the interference can be corrected for by an estimation of the magnitude of the interfering signal from a scan of the spectral background in the vicinity of the analytical line. A mathematical interference correction can also be applied. Here the net analytical signal is calculated as:

$$x_i/R = (x_i/R)_{\text{measured}} - [d(x_i/R)/dc_{\text{interferent}}] \times c_{\text{interferent}} \tag{215}$$

where $c_{\text{interferent}}$ must be determined separately and $d(x_i/R)/dc_{\text{interferent}}$ is the interference coefficient for a certain sample constituent. The latter also has to be determined from a separate series of measurements. The calibration function can also be expressed as:

$$x_i/R = a_1 \cdot c_1 + a_2 \cdot c_2 \tag{216}$$

where a_1 and a_2 have to be determined by regression, which can easily be done in the case of a linear relationship.

It could also happen that certain matrix constituents produce signal enhancements or depressions. The latter are known as multiplicative interferences and may be due to influences on the sampling efficiency, on the transport of the analyte into the source and on the generation of the species delivering the analytical signals. They can be written as:

$$Y = a(c_{St}) \cdot c \tag{217}$$

where the sensitivity (a) is a function of the concentration of the interferent (c_{St}). This relationship can be linear but also of a higher order. Calibration by standard additions allows correction for all errors arising from signal depressions or enhancements insofar as the spectral background is fully compensated for.

Indeed, when calibrating by standard additions the concentration of the unknown sample can be determined graphically by extrapolating to zero the curve that is obtained by plotting the signals for samples to which known amounts of analyte have been added versus the added amounts. Therefore, standard additions can be very useful in atomic absorption spectrometry and plasma mass spectrometry, which are zero-background methods. However, in the case of atomic emission where the lines are superimposed on a spectral background, which stems from the atomic emission source itself and which is highly dependent on the sample matrix, it is more difficult. Here, calibration by standard additions can only be used when a highly accurate background correction is applied. When calibrating by standard additions, care must be taken that the curve remains within the linear dynamic range of the method. This might be problematic in the case of atomic absorption spectrometry, especially when background correction with the aid of the Zeeman effect is applied.

In atomic spectrometry the computerized treatment of spectra and spectral data have became very important, as it allows considerable method development to be facilitated as well as the optimum figures of merit to be obtained.

The display of spectra enables spectral interferences to be traced. To this aim spectral slots are scanned or taken from simultaneously recorded data at a multitude of wavelengths. Spectral scanning at high scan speed and with a high dynamic range is now possible, e.g. in optical emission spectrometry with the aid of high-precision grating drives coupled with a high-dynamics detector. Here, for example, the voltage at a photomultiplier can be rapidly changed through a feedback circuit, as is done in the IMAGE system from Instruments S.A., Jobin-Yvon. In this way it is possible to scan the whole spectral range of 200–800 nm within 2 min with a resolution of 5 pm in the case of a 0.6 m monochromator and a grating constant of 1/2400 mm. When working with a stable source, recording spectra of the sample and the matrix alone becomes easy, which can then be subtracted enabling a qualitative analysis of a sample. In addition, determination by calibration with aqueous solutions containing only the element to be determined is possible, as shown by the example of the determination of Fe in the presence of Zr by ICP-OES (Fig. 38) [93]. Matrix spectra for matrices with a multitude of components can even be composed in this way, without having to mix the chemicals but by working

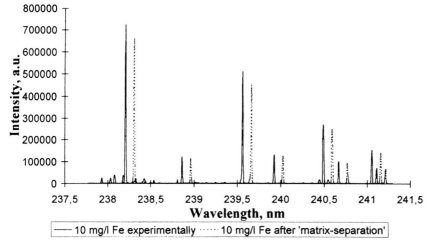

Fig. 38. Experimentally obtained ICP atomic emission spectral scan for a solution containing 10 mg/L Fe and spectral scan obtained by stripping the experimental spectral scan for a solution of 500 mg/L Zr and 10 mg/L Fe (2% Fe in Zr) from the contribution of the Zr spectral scan (shifted to higher wavelengths by 100 pm). (Reprinted with permission from Ref. [93].)

with spectra of pure substances. Also assumptions about the line profiles, which certainly become relevant for work at high resolution, can no longer be neglected. When using multichannel detectors, the quality of the spectra acquisition much improves, as now source noise appears equally at all wavelengths to the same extent and short-term drift no longer plays a role. In the case of mass spectrometry, spectrum synthesis and stripping may be less important as the spectra contain much fewer elemental lines and the formation of infering doubly charged ions or molecular species might depend more stringently on the matrix elements present.

Further requirements in graphical signal display arises from the need for background correction. Estimates of the spectral background under an analytical signal must be made from extrapolations of the spectral background intensities measured in the vicinity of the spectral line, so the availability of low-noise spectral scans in the vicinity of the analytical lines, which are free from signal drift effects, is very important. In the case of transient analytical signals, multichannel detectors with high time resolution are very helpful.

The display of signal versus time curves in real time is very important for the development of analytical procedures. In atomic absorption spectrometry with electrothermal atomization this is now indispensable and is an integral part of the development of an analytical procedure to be applied for a given analytical task. It is of further importance during the optimization of the plasma working parameters in ICP-AES and is certainly very useful for the optimization of the spectrometer with respect to drift and as a result of changes in any of the working parameters.

3
Sample Introduction Devices

In atomic spectrometry the sample must first be brought into the form of an aerosol and atomized, before excitation involving emission, absorption or fluorescence and/or ionization can take place. Accordingly, the sample introduction is a very important step in all atomic spectrometric methods. Many approaches, which can be used for very different types of samples with respect to size, state of aggregate, stability, etc., have been investigated and treated extensively in textbooks (see e.g. Refs. [94–98]).

Sample volatilization and signal generation can take place in a single source. However, sample volatilization can also take place in one source, and then to generate the signal the sample vapor or atom cloud is taken into a second source, usually with the aid of a carrier-gas flow. The use of separate sources for sample volatilization and signal generation is known as the "tandem source concept", a term which was introduced by Borer and Hieftje [99]. This concept has the advantage that the best conditions for sample volatilization, with respect to source temperature and analyte residence time, are selected for the first source and at the same time conditions can be selected in the second source for which, for example, the signal-to-background ratios and, accordingly, the power of detection are optimum or the matrix effects are lowest. This makes sense as these conditions may be totally different and using a single source for both sample volatilization and signal generation may lead to compromise conditions with losses in analytical performance. This has been demonstrated by the combination of spark ablation and plasma spectrometry for the direct analysis of electrically-conducting samples, when compared with conventional spark emission spectrometry. Both matrix effects from sample volatilization as well as from analyte excitation can be decreased as compared with spark emission spectrometry (see e.g. Ref. [100]) without losing the high power of detection of plasma atomic spectrometry. Similar effects are experienced when laser ablation is combined with the relatively cheap microwave plasma, where a matrix-independent calibration in many cases becomes feasible (see e.g. Ref. [101]).

For sample volatilization of solids or dry solution residues both thermal evaporation and cathodic sputtering may be useful. The latter is particularly important in the case of low-pressure discharges, as here the ions are highly energized when they pass through the high-energy zones of the discharge and lose only a little en-

Fig. 39. Techniques for sample
introduction in plasma
spectrometry [102].

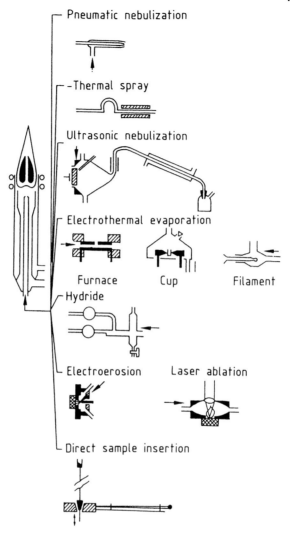

Pneumatic nebulization

-Thermal spray

Ultrasonic nebulization

Electrothermal evaporation

Furnace · Cup · Filament

Hydride

Electroerosion · Laser ablation

Direct sample insertion

ergy, because of the low number of collisions occurring in the low-pressure environment. Thermal volatilization takes place in sources where ohmic heat development is predominant, and this will be the case for sources where a large amount of electrical power is dissipated or high amounts of energy are absorbed.

In the case of solutions, various nebulization devices are often used for the production of aerosols. For plasma sources at atmospheric pressure or flames, the aerosol generation is frequently performed in a separate device. As shown schematically in Fig. 39, a number of sampling techniques for the analysis of solutions as well as for direct sampling of solids can be applied. The techniques used for plasma spectrometry [103] as well as sampling based on thermal evaporation and on cathodic sputtering, as performed in low-pressure discharges, will be treated in detail here.

3.1
Sample introduction by pneumatic nebulization

In the analysis of solutions, pneumatic nebulization and then the introduction of the aerosol into the source are already well known from the early work on flame emission spectrometry. Pneumatic nebulizers must enable a fine aerosol to be produced, which leads to a high efficiency, and the gas flow used must be low enough to transport the droplets of the aerosol through the source with a low velocity. The production of fine droplets and the use of low gas flows are essential to obtaining complete atomization in the source.

Pneumatic nebulization of liquids is based on the viscous drag forces of a gas flow passing over a liquid surface and entraining parts of the liquid, by which small independent droplets are produced. This may occur when the liquid is forced through a capillary tube and at the exit the gas then flows concentrically around the tube or perpendicularly with respect to the liquid stream. A frit can also be used, which is continuously wetted and through which the gas passes.

Nebulizers are mounted in a nebulization chamber, which has the function of separating off the larger particles and leading them into the waste. Impact surfaces are often provided, so as to fractionate the aerosol droplets further, by which the efficiency of aerosol production increases. Whereas much research has been done on nebulizers with the aim of maximizing their efficiency, much less has been done on spray chambers. They form a unit along with the nebulizer and the two can only be evaluated and optimized together. With respect to spray chambers when working with flames, impact surfaces are often provided so as to fractionate the droplets formed by the nebulizer further and to trap the large droplets. According in plasma spectrometry the nebulization chamber according to Scott et al. [104] is used in the rule. However, by providing smooth sample run-off, the noise, for example in ICP spectrometry, can be much reduced, as shown in early work by Schutyser and Janssens, who investigated many types of spray chambers [105]. In the case of ICP mass spectrometry the amount of vapor formed from the liquid needs to be limited, as analyte is thus not brought into the plasma and this only increases risk of formation of molecular ions. To this aim it is very efficient to cool the nebulization chamber down. This leads on the one hand to a decrease in the background and, accordingly, an improvement in the signal-to-noise ratios (Table 4, [41]) and on the other hand to a decrease in the white noise level, as shown by the noise power spectra (Fig. 2).

The spray chamber needs some time to fill up with the aerosol produced and some tailing of signals is seen, resultng from removal of the sample aerosol by the new incoming gas and aerosol. These build-up and wash-out times limit the speed of analysis and lead to a flattening of transient signals, when a vapor cloud passes through the spray chamber, as is the case in flow injection analysis. In order to

Tab. 4. Influence of cooling of the spray chamber on the stability of ICP-MS (30 min short-term stability) with a Meinhard nebulizer and an aqueous standard solution; analyte concentration: 20 ng/mL of each element mentioned. (Reprinted with permission from Ref. [41].)

Element	RSD* without cooling	RSD^a with cooling 10 °C
^{9}Be	3.03	2.46
^{29}Mg	2.18	2.75
^{59}Co	2.17	2.31
^{60}Ni	1.94	1.84
^{69}Ga	2.49	2.26
^{115}In	3.07	1.29
^{154}Sm	3.08	1.28
^{208}Pb	3.52	1.38
^{209}Bi	3.66	1.61
^{238}U	3.38	1.97

^{a} RSD (relative standard deviations) in % resulting from 10 replicate measurements.

minimize these dead times, small spray chambers with short build-up and rinsing times have been constructed by decreasing their volume and optimizing their form with respect to ideal aerosol droplet fractionation and very efficient surface renewal. This resulted e.g. in the cyclone spray chamber described by Wu and Hieftje [106]. They showed that with this cyclone spray chamber the wash-out times were short without sacrificing analytical precision, as shown by the noise spectra, or deterioration in the power of detection. Indeed, for a 10 µg/mL solution of Mg in ICP-OES the wash-out time, defined as the time required to reach a signal below 1% of the analyte signal, is below 10 s and the signals are of a similar magnitude as with a Scott-type chamber. Systematic investigations are just beginning on the frequency and nature of pulsations in spray chambers of varying forms, based on the calculation of the flow patterns on the one hand and measurements of droplet sizes and droplet velocities on the other [106a].

In the pneumatic nebulizers used in atomic spectrometry, the liquid flow is usually of the order of 1 to a few mL/min and the full efficiency of the nebulizer (a few %) can actually be realized at gas flows of 2 L/min. However, even with gas flows below 2 L/min, droplet diameters as low as about 10 µm and injection velocities below about 10 m/s are obtained. Pneumatic nebulization can be realized with a number of types of nebulizers. For flame emission and atomic absorption as well as for plasma spectrometry, they include concentric nebulizers, so-called cross-flow nebulizers, Babington nebulizers and fritted-disk nebulizers (Fig. 40) [95].

Concentric nebulizers

These have been known since the work of Gouy [107] in the middle of the 19th century, and in principle are they are self-aspirating. They can be made of metal, plastic or of glass. In atomic absorption spectrometry, Pt–Ir capillaries are often used so as to allow the aspiration of highly acidic solutions. In plasma atomic

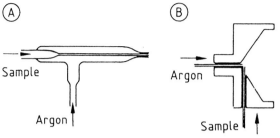

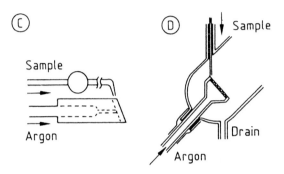

Fig. 40. Pneumatic nebulizers for plasma spectrometry. A: concentric glass nebulizer, B: cross-flow nebulizer, C: Babington nebulizer, D: fritted-disk nebulizer. (Reprinted with permission from Ref. [95].)

spectrometry, the use of concentric glass nebulizers (Meinhard Associates) is typical [108]. They use a fairly low gas flow (1–2 L/min) as compared with the nebulizers used in atomic absorption (up to 5 L/min) and are often self-aspirating. In both cases, however, the sample uptake rates are 1–2 mL/min and the aerosol efficiencies 2–5%. The nebulizers work at pressures of 2–7 bar, which are specified as are their sample uptake rates, being functions of their orifice dimensions (15–35 μm). Meinhard nebulizers can also be made of quartz, in which case even solutions containing dilute HF can be nebulized, without attacking the nebulizer.

In the case of solutions with high salt contents, the danger for clogging and signal drift is high as compared with other nebulizers (Fig. 41) [109]. The maximum admissible salt content is 20–40 g/L but it varies considerably along with the type of salt involved. It is particularly high for easily hydrolyzable compounds such as aluminum salts and phosphates.

Through long-term experiments, concentric glass nebulizers have also been shown to be sensitive to high concentrations of alkali and alkaline earth elements. The changes in nebulization efficiency and sample uptake are connected with ion-exchange phenomena at the glass wall, as mentioned in Ref. [109]. However, the influence of such salts can be reduced considerably by wetting the nebulizer gas flow.

A type of concentric glass nebulizer which is operated at pressures of several tens of bars is known as a high efficiency nebulizer (HEN) [110]. More recently micro-

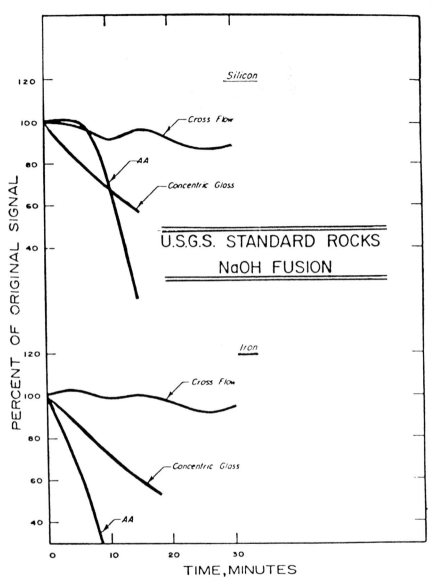

Fig. 41. Operating stability of different types of pneumatic nebulizers having high salt contents. (AA): concentric nebulizer as used in AAS work, solution: ≈16 g dissolved salts (mainly NaCl) per liter. (Reprinted with permission from Ref. [109].)

concentric nebulizers have also been described, for which the internal volume is very small and the liquid channel is connected directly to capillaries. One type that is available, the MCN-100, is very useful for coupling chromatography to plasma spectrometry, as the flow used can be in the sub-µL/min range (Fig. 42) [111].

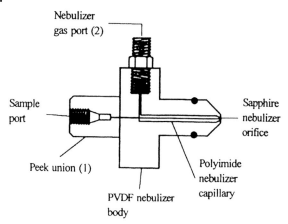

Nebulizer
gas port (2)

Sample
port

Sapphire
nebulizer
orifice

Peek union (1)

Polyimide
nebulizer
capillary

PVDF nebulizer
body

Fig. 42. Microconcentric nebulizer for ICP-OES/MS. (Reprinted with permission from Ref. [111].)

Cross-flow nebulizers

The cross-flow nebulizers are less sensitive to salt deposits at the nozzle. They are used in plasma spectrometry as well as in flame work and require the use of a peristaltic pump for sample feeding. The pump must have a sufficient number of rolls, so as to avoid pulsing, and even then may still introduce a low-frequency noise. However, dual tube pumps with a time delay between both channels, have been introduced, through which pulsing can be further minimized. In the cross-flow nebulizer the sample solution is fed through a vertically mounted capillary and nebulized by a gas flow entering through a horizontal capillary which ends close to the tip of the vertical one. The design of cross-flow nebulizers has been discussed previously by Kranz [112].

A cross-flow nebulizer for dc arc solution and AAS work was described by Valente and Schrenk [113]. For ICP work the nebulizer should have capillaries with diameters < 0.2 mm and a distance of 0.05–0.5 mm between the tips. The capillaries also can be made of metal, glass or plastic and they have similar characteristics to the concentric nebulizer. Types made of Ryton for example enable the aspiration of solutions containing higher concentrations of HF. Then aerosol gas flows are around 1–2 L/min at 2–5 bar. This has been confirmed by optimization measurements described by Fujishiro et al. [114]. They found that by varying the inside and the outside diameters of the capillary from 0.15 to 0.9 mm and from 0.5 to 1.5 mm, respectively, the pressure drop across the sample tube decreases from 150 to 50 mbar. By increasing the gas flow from 0.75 to 1.75 L/min, the pressure drop across the sample tube was found to increase from 150 to 350 mbar. As shown in Fig. 43 [114], the variations in the drops in pressure considerably influence the droplet size.

Babington nebulizer

Here the sample solution is pumped through a wide tube into a groove, where the film running off is nebulized by a gas flow entering perpendicular to the film of

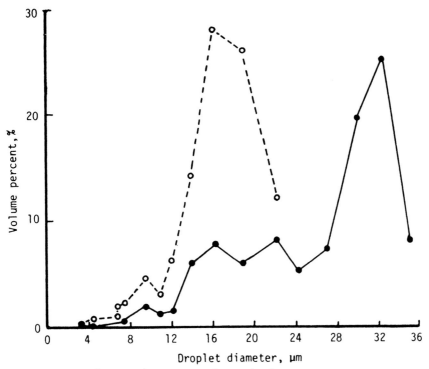

Fig. 43. Influence of pressure decrease across the sample tube of a cross-flow nebulizer on droplet size distribution. Pressure decrease: (●): 95 mm H_2O, (○): 250 mm H_2O. (Reprinted with permission from Ref. [114].)

liquid. As the sample solution does not need to pass through a narrow capillary, solutions containing high salt concentrations and also slurries can be analyzed without the risk of clogging. This type of nebulizer, first described by Babington in 1973 [115], was later used extensively used by Ebdon (see e.g. [116]) especially for slurry nebulization.

Work with slurries requires that the slurries are first nebulized and behave just as solutions with respect to the sample introduction into the aerosol. From electron probe micrographs of aerosol particles sampled on Nuclepore filters under iso-kinetic conditions, it was found that at nebulizer gas flows of 3 L/min, being typical of plasma spectrometry but far below those for flame atomic absorption, particles with a diameter of up to 15 μm can be found in the aerosol (Fig. 44) [117]. This would imply that powders with a grain size of up to about 15 μm could still be nebulized as could a solution. This, however, is not true as the mass distribution in the case of powders may be quite different in the slurry and in the aerosol, as shown for the case of SiC (Fig. 45) [118]. The nebulization limitations for the case of slurry nebulization thus must be investigated from case to case and leads to certain types of restrictions.

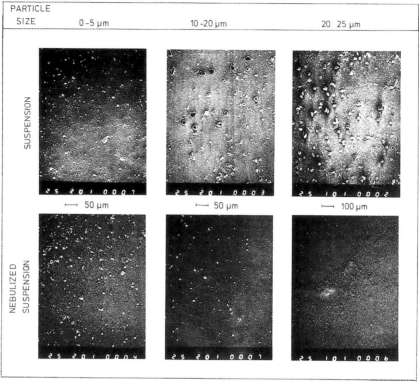

Fig. 44. Scanning electron microscopy images of Al₂O₃ particles sampled from a 1% suspension itself and from the aerosol produced from the suspension. Al₂O₃ (VAW 143–85, Vereinigte Aluminium Werke, Bonn) sieved fractions. GMK nebulizer with an impactor 3 mm from the orifice. Aerosol gas flow rate: 1 L/min, pumping rate: 2 mL/min. (Reprinted with permission from Ref. [117].)

Fritted-disk nebulizers

Nebulizers such as these have high nebulization efficiencies and a low sample consumption (down to 0.1–0.5 mL/min). In particular, they have been shown to be useful for plasma spectrometry in the case of organic solutions or for coupling with liquid chromatography [119]. However, in comparison with the other types of nebulizers they suffer from memory effects. Indeed, the rinsing times are much higher than the 10–20 s as required for the other pneumatic nebulizers. In the case of the so-called thimble nebulizer, as described by Liu et al. [120] (Fig. 46), these disadvantages have been eliminated, mean droplet sizes were reported to be of the 2 μm level, which is lower by a factor of 5 than with the HEN mentioned previously.

Direct injection nebulizers

Direct injection nebulizers, as they were first described by Kimberley et al. [121], are used without a nebulization chamber. In this way sample losses are low as

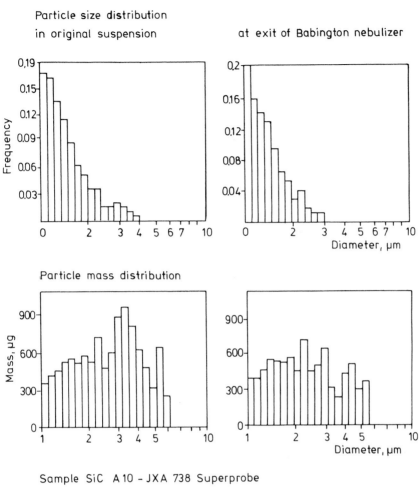

Particle size distribution
in original suspension   at exit of Babington nebulizer

Particle mass distribution

Sample SiC A 10 - JXA 738 Superprobe
G. M. K. nebulizer - 3 bar

Fig. 45. Particle size and mass distribution of a suspended SiC powder and the aerosol generated with a GMK nebulizer [118].

are the dead times. This can only be achieved when the aerosol droplets are fine enough to avoid sample deposition or incomplete evaporation in the source. High nebulization efficiencies can then also be obtained, as shown by the comparative measurements reported by McLean et al. [122]. These features can only be realized when the drop in pressure at the top of the nebulizer is high or when the nebulizer is operated at high pressure.

Oscillating nebulizer

A special type of pneumatic nebulizer is the oscillating nebulizer, described by Wang et al. [123] (see Fig. 47) [124]. Here the droplet distribution is narrow, the

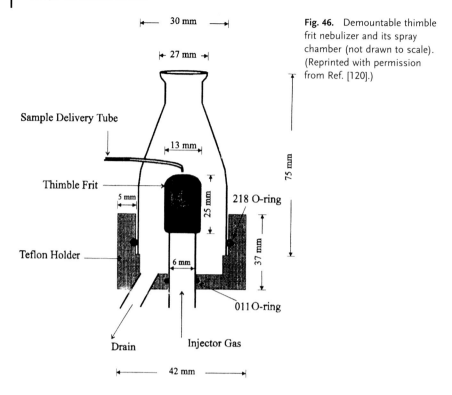

Fig. 46. Demountable thimble frit nebulizer and its spray chamber (not drawn to scale). (Reprinted with permission from Ref. [120].)

droplet size is low, e.g. as compared with the concentric nebulizer and the efficiency is also high.

With pneumatic nebulizers the samples can be taken up continuously. Then after a few seconds of aspiration a continuous signal is obtained and it can be integrated for a certain time. However, small sample aliquots can also be injected directly into the nebulizer [125–128] or into a continuous flow of liquid. The latter

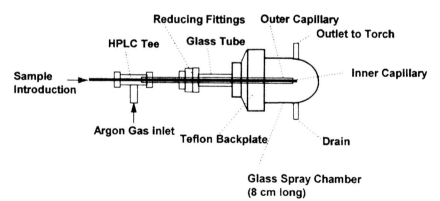

Fig. 47. Oscillating capillary nebulizer with single pass spray chamber. (Reprinted with permission from Ref. [124].)

is known as flow injection analysis (FIA) and has been treated extensively (for the principles, see Ref. [129]). The signals in both cases are transient. They rise sharply and there is some tailing, resulting from generation of the analyte aerosol cloud at the front end and dilution of the aerosol at the signal end, respectively. The signals can last a few seconds up to eventually some 10 s and with sample volumes of 10–20 µL similar signal intensities to continuous nebulization can often be obtained. These techniques enable analysis of small samples (down to around 10 µL), lower the salt deposits on nebulizers and maximize the absolute power of detection.

Also useful in work with the introduction of small-volume samples, denoted as discrete sampling, the use of small cyclone chambers as well as of direct injection nebulizers has to be mentioned [121]. FIA has gained further importance as automation is easy and on-line sample preparation becomes possible. This is specifically due to the use of on-line digestion of slurries at high temperatures, as described e.g. by Haiber and Berndt [130], as well as by on-line preconcentration and matrix removal by coupling with column chromatography. In the case of dissolved Al_2O_3 ceramic powders, matrix removal is easily possible by fixation of the dithiocarbamate complexes of the impurity metals on a solid phase while aluminum is not retained. The traces can then be eluted from the column and e.g. directly determined by ICP mass spectrometry (Fig. 48) [131]. The other solution lies in retaining the matrix on the column, as e.g. is possible in the case of zirconium by the formation of thenoyltrifluoroacetonates [132], which also could be fixed on an RP

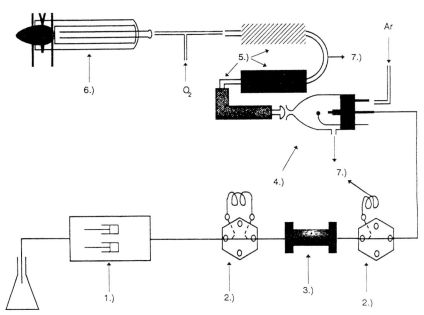

Fig. 48. ICP-MS coupled on-line to liquid chromatography using hydraulic high-pressure nebulization. (1): HPLC pump, (2): sample loop, (3): chromatographic column, (4): HHPN nebulization chamber, (5): desolvation system with heating and two-stage cooling, (6): ICP torch, (7): drain. (Reprinted with permission from Ref. [131].)

18 solid phase, in amounts of up to a few mg [133]. Further examples of on-line matrix removal are given in the literature (see e.g. Refs. [134–136]).

The optimization of pneumatic nebulizers is aimed in particular at selecting the working conditions that give the optimum droplet size and efficieny. The so-called Sauter diameter d_0, i.e. the diameter for which the volume to surface ratio equals that of the complete aerosol, is given by the Nukuyama–Tanasawa equation (see Ref. [137]):

$$d_0 = C/v_G(\sigma/\rho)^{0.5} + C'\{\eta/[(\sigma \cdot \rho)^{0.5}]\}^{C''} \cdot [1000 \cdot (Q_L/Q_G)]^{C'''} \tag{218}$$

v_G is the gas velocity, Q_G the gas flow, Q_L the liquid flow, η the viscosity, ρ the density, σ the surface tension of the liquid and C, C' and C'' are constants.

When the nebulizer gas flow increases d_0 becomes smaller (through v_G and Q_G) and the efficiency of the sample introduction will increase and as a result the signals also do. However, as more gas is passed through the source, it will be cooled and, moreover, the residence time of the droplets will decrease and thus the atomization, excitation and ionization will also decrease. These facts will counteract any increase in the signals as a result of the improved sampling. Accordingly, the highest signal intensity and thus the greatest power of detection will be achieved at a compromise gas flow.

The physical properties of the liquid sample (σ, η and ρ) also influence d_0, the efficiency and the analytical signals. Differences in the physical properties of the liquid samples to be analyzed thus lead to so-called nebulization effects. η has a large influence. It influences d_0 through the second term in Eq. (218). However, in the case of free sample aspiration it also influences d_0 through Q_L, which is given by the Poiseuille law as:

$$Q_L = [(\pi \cdot R_L^4)/(8 \cdot \eta \cdot l)] \cdot \Delta P \tag{219}$$

R_L is the diameter and l the length of the capillary. ΔP is the pressure difference and is given by:

$$\Delta P = \Delta P_U - g \cdot h \cdot \rho - f(\sigma) \tag{220}$$

ΔP_U is the difference between the pressure inside and outside the nebulizer, h is the height and $f(\sigma)$ a correction factor. In order to minimize the nebulization effects one can utilize forced feeding with a peristaltic pump, a result of which is that Q_L is no longer a function of η. One also can operate the nebulizer at a high gas flow, by which Eq. (218) reduces to:

$$d_0 = C/v_G(\sigma/\rho)^{0.5} \tag{221}$$

when Q_G/Q_L becomes >5000. However, the gas flow may then cool the source down through which incomplete atomization or signal depressions or enhancements may lead to even stronger interferences.

Droplet size distributions of aerosols can be determined by stray-light measure-

Fig. 49. Particle size distributions of pneumatic nebulizers for ICP. MAK, high-pressure cross-flow nebulizer; CGN, concentric glass nebulizer; JAB, Jarrell–Ash Babington nebulizer; JAC, Jarrell–.Ash fixed cross-flow nebulizer (Reprinted with permission from Ref. [139].)

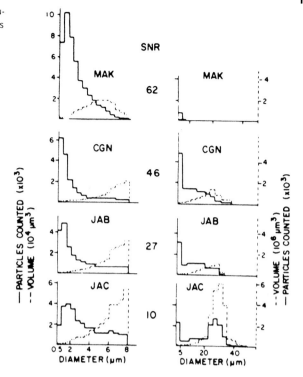

ments (see e.g. Refs. in [122, 137]). Data obtained from sampling with cascade impactors could be influenced by evaporation of the droplets on their way through the device, especially as isokinetic sampling may require the application of reduced pressure [138]. Various nebulizers have been used to provide data, e.g. Olsen and Strasheim (see Ref. [139]). When comparing the different nebulizers used in early work on ICP-OES (Fig. 49), they found that the high pressure cross-flow nebulizer has the highest SNR values, the lowest particle size and the narrowest droplet size distribution. The SNR values found in ICP-OES, for example, seemed to correlate with the fraction of the aerosol present as droplets with diameters below 8 µm. Particle size distributions depend on the nebulizer and its working conditions, on the liquid nebulized and also on the nebulization chamber. These relationships have been studied in depth, also for organic solutions (see e.g. Refs. [140, 141]). In these cases, an aerosol with a lower mean particle size is often produced with a fairly low gas flow pneumatic nebulizer as compared with aqueous solutions and thus resulting in a higher efficiency and power of detection.

Jet-impact nebulization

Apart from pneumatic nebulization through gas entrainment other types of pneumatic nebulization have been realized. In so-called jet-impact nebulization, use is made of the impact of a high pressure jet of liquid on a surface [142].

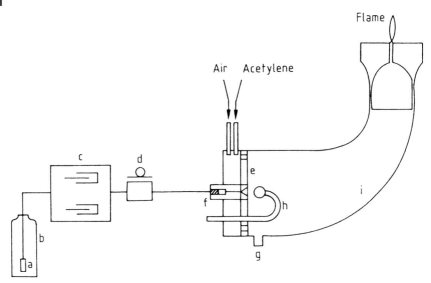

Fig. 50. Hydraulic high pressure nebulization system. (a): Solvent filter (20 μm), (b): solution reservoir (water, organic solvents), (c): high-pressure pump, (d): solvent feed valve/sample loop, (e): plate with concentric holes; (f): high-pressure nebulization nozzle (10–30 μm) with integrated protection filter, (g): drain, (h): impact bead and (i): gas mixing chamber. (Courtesy of H. Berndt)

High-pressure nebulization

This has been described by Berndt [143], and makes use of the expansion of a liquid jet under high pressure at a nozzle (Fig. 50). Here much higher nebulization efficiencies can be obtained even for solutions with high salt solutions and for oils. This is due to the small droplet size, which is in the 1–10 μm size [144]. The nebulization efficiency can be improved further by heating the capillary, so that a mixture of vapor and aerosol is released at the nozzle [145]. In ICP-OES the detection limits can be decreased by an order of magnitude as compared with pneumatic nebulization, provided a heated spray chamber is used. Such systems are ideally suited for speciation work as they allow easy coupling of HPLC and atomic spectrometry [146]. A further advantage lies in the fact that the aerosol properties are fairly independent of the physical properties of the liquid. High-pressure nebulization is very promising for the coupling of on-line pre-enrichement or separations by high performance liquid chromatography (HPLC) with atomic spectrometry.

In the case of plasma spectrometry, however, it is usually necessary to desolvate the aerosol produced, as high water loadings cool the plasma too much and decrease the analytical signals and, further, may also lead to increased matrix effects. Desolvation can be achieved by heating the aerosol leading to solvent evaporation, as previously applied in atomic absorption work by Syti [147] (Fig. 51). Other new ways of cooling in aerosol desolvation are possible by the use of Peltier elements. Also, IR heating may be used to heat up the aerosol. Another new method of

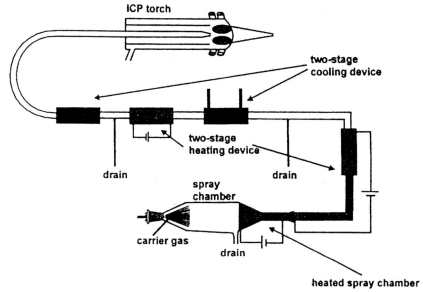

Fig. 51. Desolvation unit for ICP atomic spectrometry.

desolvation lies in the use of membranes, such as those made of Nafion. Here water molecules can diffuse through, while salt aeosol particles as well as droplets are retained (Fig. 52). The principle and features as demonstrated by Yang et al. [148] may be particularly useful for various types of sample introduction used in plasma spectrometry.

3.2
Ultrasonic nebulization

By the interaction of sufficiently energetic acoustic waves of a suitable frequency with the surface of a liquid, geysers are formed by which an aerosol is produced.

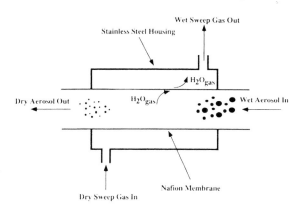

Fig. 52. Aerosol drying by using a single tube Nafion[R] membrane desolvation stage. (Reprinted with permission from Ref. [148].)

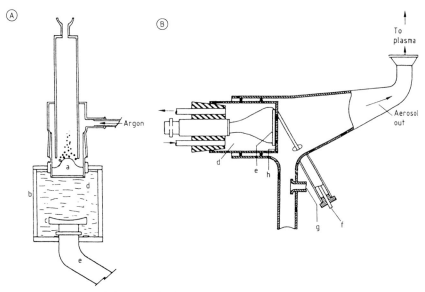

Fig. 53. Ultrasonic nebulization. (A): Discontinuous system (reprinted with permission from Ref. [149]), (B): system with liquid flowing over the transducer (reprinted with permission from Ref. [150]).

The diameter of the aerosol droplets produced depends on the frequency and on the physical properties of the liquid sample. In the case of water for example, aerosol droplets formed at a frequency of 1 MHz have a diameter of about 4 μm. The energy can be focussed with the aid of a liquid lens onto the surface of the sample solution or the liquid sample can be pumped continuously over the transducer, which must then be cooled efficiently (Fig. 53A and B). Ultrasonic nebulization has two considerable advantages over pneumatic nebulization. At first the aerosol particles have a lower diameter and a fairly narrow particle size distribution as compared with pneumatic nebulization (<5 versus 10–25 μm). Therefore, the aerosol production efficiency may be as high as 30% and a high analyte introduction efficiency is achieved. In addition, no gas flow is required for aerosol production and accordingly, the transport gas flow can be freely selected. However, when applying ultrasonic nebulization to plasma spectrometry, it is also necessary to desolvate the aerosol so as to prevent too intensive a cooling of the plasma. After taking these measures ultrasonic nebulization leads to an increase in power of detection.

It should be mentioned that with ultrasonic nebulization memory-effects are generally higher than in the case of pneumatic nebulization. Furthermore, the nebulization of solutions with high salt concentrations may lead to salt depositions. Therefore, specific attention should be paid to rinsing, and the precision eventually achieved is generally lower than in pneumatic nebulization [151].

3.3
Hydride and other volatile species generation

For elements with volatile hydrides or other volatile species, the sampling efficiency
can be increased by volatilization of these species from the samples.

Hydride generation

This can be applied for the elements such as As, Se, Sb, Te, Bi. Sn as well as some
others. Indeed, by in situ generation of the hydrides of these elements (AsH$_3$, etc.)
from the sample solutions the sampling efficiency can be increased from a few
percent in the case of pneumatic nebulization to virtually 100%.

Hydride generation can be performed efficiently by reduction with nascent
hydrogen. This can be produced chemically by the reaction of zinc or NaBH$_4$ with
dilute acids. In the latter case use can be made of a solid pellet of NaBH$_4$ and
placing the sample on it, which might be useful as a microtechnique (see e.g. Ref.
[152]). However, a flow of a solution of NaBH$_4$ stabilized with NaOH can be joined
by one of an acidified sample. This can be done in a reaction vessel, as was first
used in atomic absorption work (see e.g. Ref. [153]) or in a flow-through cell, as
first used in plasma atomic emission spectrometry (Fig. 54) [154, 155]. In this case,
the hydrides produced are separated from the liquid and are subsequently led into
the source by a carrier-gas flow. The hydrides, however, are accompanied by an
excess of hydrogen. In the case of weak sources such as microwave discharges, the
hydrogen may disturb the discharge stability. To avoid this the hydrides can be
separated off, e.g. by freezing them out in a liquid nitrogen cooling trap (b.p.:
$\approx -30\,°C$), during which the hydrogen (b.p.: $< -200\,°C$) escapes, and sweeping the
collected hydrides into the source during a subsequent heating step [156]. The use

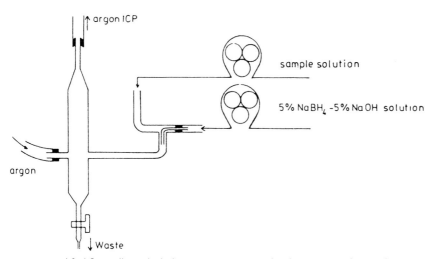

Fig. 54. Modified flow-cell type hydride generator. (Reprinted with permission from Ref. [155].)

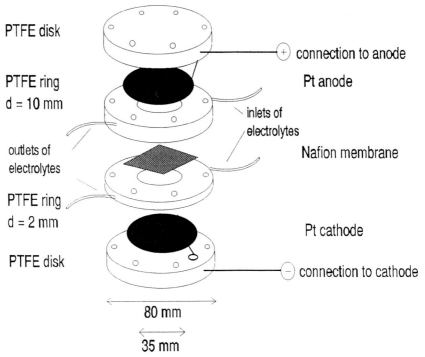

PTFE disk

PTFE ring
d = 10 mm

outlets of
electrolytes

PTFE ring
d = 2 mm

PTFE disk

connection to anode

Pt anode

inlets of
electrolytes

Nafion membrane

Pt cathode

connection to cathode

80 mm

35 mm

Fig. 55. Flow-cell for electrochemical hydride generation. (Reprinted with permission from Ref. [158].)

of membranes may also be useful, through which the hydrogen is selectively removed by diffusion whereas the elemental hydrides are retained [157]. In addition, it has been shown to be effective to lead the reaction gases over concentrated H_2SO_4, so as to dry them before they enter weak sources such as microwave plasma discharges.

Electrolysis can also be used for the generation of hydrogen, which has the advantage that the use of the $NaBH_4$ reagent becomes superfluous. This is advantageous from the point of view of reagent consumption and the related costs, including the efforts of preparing a new solution daily, and because the $NaBH_4$ reagent may also contribute to the blank.

In electrolytic hydride generation, a cell as shown schematically in Fig. 55 can be used [158]. It is made of PTFE rings and disks and can be easily disassembled for cleaning. The electrodes are platinum sheets with a surface of 10 cm² each. The compartments of the anode (10 cm²) and the cathode (2 cm²) both have a solution inlet and outlet and are separated by a Nafion membrane. The cell is held together with six screws. Solutions of the sample acidified with HCl are used as the catholyte and dilute H_2SO_4 as the anolyte. At a cell current of a few amperes, the solutions are continuously fed through the cell compartments, and on the cathodic side hydrogen and the hydrides are generated and led into the gas–liquid separator. At a current of 3.5 A, HCl concentration of 0.5 M and an H_2SO_4 concentration of 2 M as well as equal flow rates for anolyte and catholyte of 2 mL/min, a hydrogen production of about 1.6 L/h is obtained, with which more or less the same effects can be obtained as in chemical hydride generation.

It was found that with electrolytic hydride generation in the case of low-power microwave discharges the detection limits were somewhat lower than in chemical hydride generation using the same plasma emission system. This was shown to be partly due to the smoother generation of hydrogen, but also might be related to the lower blanks stemming from the reagents. Furthermore, such cells can be miniaturized, at the same time keeping the efficiency, and can be coupled on-line with electrolytic preconcentration of the hydride forming elements, constituting a line of further research.

Hydride generation readily allows the power of detection of atomic spectrometric methods to be increased for the determination of elements having volatile hydrides. Moreover, it allows matrix-free determinations of these elements. It should, however, be emphasized that the technique, irrespective of the type of hydride generation used, is prone to a number of systematic errors. Firstly, the hydride-forming elements must be present as inorganic compounds in a well-defined valence state. This may require a sample decomposition step prior to analysis. In the case of water analysis a treatment with $H_2SO_4-H_2O_2$ may be effective [159]. In addition, traces of heavy metals such as Cu^{2+} may have a catalytic influence on the formation and dissociation of the hydrides, as investigated by Welz and Melcher (see e.g. Ref. [160]) in atomic absorption using quartz cuvettes. This effect may also be due to a reaction of the interferent with the $NaBH_4$. The latter can be partly avoided by using fairly high concentrations of acid in the chemical hydride generation, as this was found to be effective for removing interferences of Fe [155]. These interferents can be masked further by complexation with tartaric acid or coprecipitation with $La(OH)_3$. It is advised that calibration should be carried out by standard additions.

In order to increase the power of detection of hydride generation, trapping of the hydrides can be applied, which also has the advantage that hydrogen, which might make low power excitation or ionization sources and atom reservoirs unstable, is removed. This can be done by freezing out the hydrides. However, as the volatilization temperatures of the elements forming voltatile hydrides are much higher than the decomposition temperatures of the hydrides, hot-trapping can also be applied. For instance, as is known from the Marsch method, which is an old method for the isolation and preconcentration of arsenic, AsH_3 decomposes at 600 °C leaving behind a layer of elemental arsenic, which volatilizes at around 1200 °C. Accordingly, for hot trapping the reaction gases containing the hydrogen and the volatile hydrides are led through a quartz capillary into a heated graphite furnace and directed through the sampling hole onto the hot wall of the graphite tube. Thus, the hydride thermally dissociates and the respective elements are deposited, by which a preconcentration by orders of magnitude can take place [161].

In a number of cases the efficiency of the trapping can be increased still further by a pretreatment of the graphite tubes. Indeed Zhang et al. [162, 163] and Sturgeon et al. [164] have shown that Pd can be used for the efficient trapping of hydrides and they explained the mechanism of preconcentration on the basis of the catalytic reactivity of Pd, which promotes the decomposition of hydrides at relatively low temperatures (200–300 °C). Normally $Pd(NO_3)_2$ is used for this purpose. After trapping the elements they can subsequently be released by heating up the furnace.

Mercury cold vapor technique

Similar to hydride generation, the mercury cold vapor technique can be applied. Here Hg compounds are reduced to metallic mercury with nascent hydrogen. The latter is normally formed by using an $SnCl_2$ solution as the reducing agent. However, $NaBH_4$ can also be used as the reagent (see e.g. Ref. [165]). This, however, has the drawback that hydrogen is formed, which dilutes the reaction gases. The elementary Hg is released from the reaction mixture by the aid of a carrier gas flow. The resulting analyte flow can be dried e.g. by passing it over $KClO_4$ and then can be led into an absorption cell or into a plasma source. It can also be trapped on gold e.g. in the form of a gold–platinum gauze. Here the mercury is preconcentrated by amalgamation and subsequently thermally released by resistance heating. Finally, it is transferred to an absorption cell made of quartz or to a plasma source. This approach allows effective preconcenration and in the case of optical and mass spectrometry leads to very sensitive methods for the determination of Hg.

Volatile species formation

This can be applied to many elements by appropriate choice of reactions.

In the case of iodine, iodide present in the sample can be oxidized, e.g. in a flow cell with the aid of $K_2Cr_2O_7$ in acidic solutions and in this way be released into a radiation source for atomic emission spectrometry [166].

Sulfur can be reduced to H_2S by the action of nascent hydrogen [167]. For the determination of chloride, chlorine can be generated in a flow cell by a reaction with $KMnO_4$ and concentrated sulfuric acid [168].

For elements such as bromine, phosphorus, germanium, lead and others, reactions for the generation of hydrides or of similar volatile species can also be found.

For many metals and semi-metals and even for an element such as cadmium, it has recently been described that volatilization can be obtained by vesicle mediation. Indeed, surfactants are able to organize reactants at a molecular level, by which chemical generation of volatile species is enhanced. It was shown, by Sanz-Medel et al. [169], that by adding micelles or vesicles to cadmium solutions it is possible to generate volatile CdH_2 with a high efficiency. This volatile compound can even be transported to a measurement cell where a "cold vapor" of cadmium can be measured.

3.4
Electrothermal vaporization

Thermal evaporation of the analyte elements from the sample has long been used in atomic spectrometry. For instance, it had been applied by Preuß in 1940 [170], who evaporated volatile elements from a geological sample in a tube furnace and transported the released vapors into an arc source. In addition, it was used in so-called double arc systems, where selective volatilization was also used in direct solids analysis. Electrothermal vaporization became particularly important with the work of L'vov et al. [171] and Massmann in Dortmund [172], who introduced electrothermally heated sytems for the determination of trace elements in dry solution residues by atomic absorption spectrometry of the vapor cloud. Since then, the idea has regularly been taken up for several reasons.

- Firstly, an analyte can be released from a solution residue and thus be brought into an atom reservoir, a radiation or an ion source free of solvent. This is particularly useful for the case of sources operated at low power and gas consumption, which are cheap but generally do not tolerate the sudden introduction of moisture. On the other hand, independence of the physical and chemical properties of the sample solvent can be gained, which may introduce physical (nebulization effects), chemical (volatilization effects) or spectral interferences (e.g. those stemming from band spectra of the solvent molecules or their dissociation products).
- Secondly, the thermal evaporation process can be performed with a conversion efficiency of 100%, by which the analyte introduction efficiency into the source may be increased from a few percent in pneumatic nebulization, through around 10–20% in ultrasonic nebulization to nearly 100%.
- Thirdly, it has to be considered that it is often possible, e.g. through selection of the appropriate gas flows, to realize a long residence time in the plasma. This favors volatilization and dissociation of the analyte. Both of these effects enable the high absolute power of detection that can usually be achieved with electrothermal vaporization to be reached, as compared with other sample introduction techniques.

3.4.1
The volatilization process

Electrothermal evaporation can be performed with dry solution residues, resulting from solvent evaporation, as well as with solids. In both cases the analyte evaporates and the vapor is kept inside the atomizer for a long time, from which it diffuses away. The high concentration of analyte in the atomizer results from a formation and a decay function. The formation function is related to the production of the vapor cloud. After matrix decomposition the elements are present in the furnace as salts (nitrates, sulfates, etc.). They dissociate into oxides as a result of the temperature increase. In the case of a device made of carbon or graphite, the oxides are reduced by the carbon in the furnace as:

$$MO(s/l) + C(s) \rightarrow M(g) + CO(g) \tag{222}$$

However, a number of metals tend to form carbides, as they are very stable (a thermodynamically controlled reaction) or because they are refractory (a kinetically controlled reaction). In this case no analyte is released into the vapor phase. The decay of the vapor cloud is influenced by several processes [174], namely:

- diffusion of the liquid sample into the graphite, which often can be prevented by the use of tubes, which are pyrolytically coated with carbon;
- diffusion of the sample vapor in the gaseous phase;
- expansion of the hot gases during the temperature increase (often at a rate of more than 1000 K/s);
- recombination processes, these being minimum when the sample is brought into the electrothermal device on a carrier platform with a small heat capacity, such as was introduced by L'vov et al. [173];
- action of purging gases.

Therefore, in electrothermal devices, transient signals are obtained. They increase sharply and have a more or less exponential decay lasting 1–2 s. Their form has been studied from the point of view of the volatilization processes. The real signal form (see Fig. 56) is also influenced by adsorption and then subsequent desorption of the analyte inside the electrothermal device at the cooler parts.

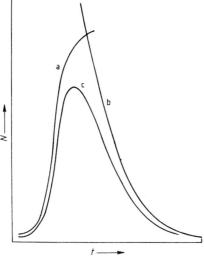

Fig. 56. Signal form in graphite furnace atomic absorption spectrometry.

3.4.2
Types of electrothermal devices

In most cases the furnace is made of graphite, which has good thermal and corrosion resistance. As a result of its porosity, graphite can take up the sample without formation of appreciable salt deposits at the surface. However, apart from graphite, atomizers made of refractory metals such as tungsten have also been used (Fig. 57) (see e.g. Refs. [175, 175a]).

In the case of graphite, tubes with internal diameters of around 4 mm, a wall thickness of 1 mm and a length of up to 30–40 mm are usually used. However, filaments enabling the analysis of very small sample volumes and mini-cups of

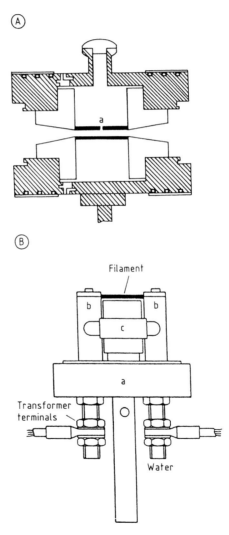

Fig. 57. Graphite atomizers used in atomic spectrometry. (A): Original graphite tube furnace according to Maßmann (a): graphite tube with sampling hide (reprinted with permission from Ref. [172]), (B): carbon-rod atomizer system according to West (a): support; (b): clamps (cooled); (c): graphite rod or cup (reprinted with permission from Ref. [175]).

graphite, which are held between two graphite blocks, have also been described in the literature.

When using graphite tube furnaces for optical spectrometry, the optical path coincides with the axis of the graphite tubes, whereas in the case of filaments and cups, the radiation passes above the atomizer. With furnace currents of up to 100 A and when voltages of up to a few V are used, temperatures of up to and above 3000 K can be reached. The tube is shielded by an argon flow, ensuring stable working conditions for up to several hundred firings. With filaments a lower power is used and the system can be placed in a quartz enclosure so as to keep it away from air. Cups and tubes nowadays are often made of pyrolytic graphite, which prevents the analyte solution from being entrained in the graphite. Thus, chromatographic effects leading to selective volatilization are avoided. Furthermore, the formation of refractory carbides, which hamper the volatilization of elements such as Ti and V, decreases.

Electrothermal furnaces made of refractory metals (tungsten in particular) have been described by Sychra et al. (see Ref. [175a]) for use in atomic absorption work. The heat capacity is generally smaller than in the case of graphite tubes, which results in a steeper rate of heating and cooling. This may be extreme in the case of tungsten probes and cups [176], which are mechanically more stable than probes made of graphite. The signals are then extremely short and the analyte is released over a very short time, which leads to high signal-to-background ratios and extremely low detection limits, as has been shown in wire loop atomization in AAS [177] or wire loops used for microsample volatilization in plasma spectrometry [178]. Therefore, it is difficult to cope with high analyte loadings when using metal devices for electrothermal evaporation. Small amounts of sample are easily lost during heating, as the sample is located only on the surface and thermal effects cause tension inside the salt crystals that are formed during the drying phase. In the case of graphite the sample partly diffuses into the graphite, by which this effect is suppressed.

Graphite has further advantages in that for a large number of elements a reduction is obtained, which leads to free element formation, as is required for atomic absorption spectrometry. Tungsten atomizers can be used to temperatures of beyond 3000 K. However, here oxygen must be excluded from the evaporation device, so as to prevent the volatilization of the device as the oxide. This can be achieved by adding a few percent of hydrogen to the internal and external gas streams by which the number of firings per atomizer becomes higher than 100. Also the presence of chlorine has to be avoided when working with tungsten, as then the more volatile chloride shortens the lifetime of the atomizer. This may be particularly problematic when analyzing biological materials. A decomposition step then needs to be included to remove the chlorine by treatment with higher-boiling oxidizing acids. In general, graphite may enable a still higher number of firings per device to be made. However, the price of high-purity graphite is also high. The use of metal atomizers made of tungsten has another drawback in that the optical spectra of tungsten are very line rich as compared with those obtained with graphite in an inert atmosphere. This will be a particular drawback when using electrothermal vaporization in optical emission spectrometry.

In order to bring the sample rapidly into a hot environment, use is often made of the platform technique, as was first introduced in atomic absorption spectrometry by L'vov [179]. Here the very rapid heating may enable the formation of double peaks to be avoided, which are a result of various subsequent thermochemical reactions, all of which have their own kinetics. Also the high temperature avoids the presence of any remaining molecular species, which are especially troublesome in the case of atomic absorption spectrometry. Thin platforms can be made of graphite, which have a very low heat capacity, or from refractory metals. In the latter case wire loops, on which a drop can easily be previously dried, are often used.

For optical measurements directly in the graphite atomizer, this can be open both during the solvent vapor release and during the analyte volatilization. When the analyte vapor has to be transported into a further source for signal generation, it may be useful to close the cell during the analyte volatilization to guarantee a high transportation efficiency. To this aim pneumatically-driven stubs closing the sampling hole and both gas pressure as well as gas flow control systems are provided in commercial equipment. However, open systems can also be used in electrothermal vaporization with transport of the vapors into a second source. Here, for example, use is made of the fact that the carrier gas pushes the atom vapor cloud into the transport line instead of allowing it to escape through the sampling hole. This even has the advantage that no pressure jumps occur, which makes this approach feasable for coupling with low-power sources, as shown by the case of the MIP in Ref. [180] (Fig. 58).

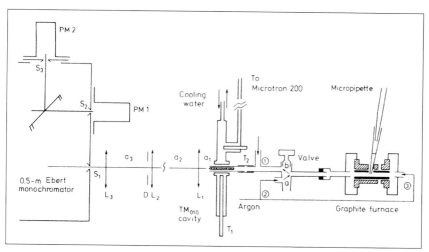

Fig. 58. Graphite furnace evaporation coupled to MIP-OES. Argon gas flows: (1): plasma gas flow, (2): for exhaust of solvent vapor, (3): carrier gas flow. (T$_1$): Coupling rod, (T$_2$): fine tuning rod. Spherical quartz lenses (f: focal length, d: diameter): (L$_1$): f 35 mm, d 19 mm, a_1 41 mm; (L$_2$): f 66 mm, d 19 mm, a_2 249 mm; (L$_3$): f **82 mm**, d 19 mm, a_3 90 mm. (S$_1$): entrance slit, (S$_2$): fixed exit slit, (S$_3$): moveable exit slit (wavelength coverage: $\lambda_a \pm 2$ nm); (PM$_1$, PM$_2$): photomultipliers; (D): diaphragm (6 mm × 6 mm). (Reprinted with permission from Ref. [180])

3.4.3
Temperature programming

When using electrothermal evaporation for sample introduction, the development of a suitable temperature program for the elements to be determined in a well-defined type of sample is of prime importance. In the case of liquid samples a small sample aliquot (10–50 µL) is brought into the electrothermal device with a syringe or with the aid of an automated sampler and several steps are performed.

- Drying stage. The sample solvent is evaporated and the vapors are allowed to escape, e.g. through the sampling hole in the case of a graphite furnace. This step can last from around 10 s to a few minutes and takes place at a temperature near to the boiling point, e.g. at 105 °C in the case of aqueous solutions. This procedure should often consist of several steps, so as to avoid splashing, as is advisable in the case of serum samples.
- Matrix destruction. During this step the matrix is decomposed and removed by volatilization. Often chemical reactions are used to facilitate the volatilization of matrix constituents or their compounds. The temperature must be chosen so that the matrix but not the analyte is removed. This is often achieved by applying several temperatures or even by gradually increasing the temperature by one or several ramping rates. Temperatures during this step are usually between 100 and 1000 °C, depending on the matrix to be removed and on the analyte elements. Thermochemical reagents are often used, which chemically assist the destruction of the matrix and help to achieve complete matrix removal at a lower temperature. Quaternary ammonium salts are often used e.g. for the destruction of biological matrices at relatively low temperatures. Matrix decomposition is very important so as to avoid the presence of analyte in different chemical compounds, which could lead to transient peaks with several maxima. At this level it is, however, very important to avoid analyte losses, which can be done through the formation of volatile matrix compounds as well as by stabilizing the analytes, both of which can be achieved through the use of matrix modifiers (see e.g. Ref. [164]).
- Evaporation stage. The temperature chosen for the analyte evaporation strongly depends on the analyte elements and can range from 1000 K for relatively volatile elements (e.g. Cd or Zn) to 3000 K for fairly refractory elements (Fe, etc.). This step normally lasts around 10 s but not longer, so as to keep the number of firings that can be performed with a single tube as high as possible.
- Heating stage. The temperature is brought to the maximum (e.g. 3300 K over around 5 s with graphite tubes) so as to remove any sample residue from the evaporation device and to minimize memory and cross-over effects.

Direct solids sampling with electrothermal evaporation can be performed by dispensing an aliquot of a slurry prepared from the sample into the furnace. The analytical procedure is then completely analoguous with the one with solutions (see e.g. Ref. [181]). However, powders can also be sampled with special dispensers, as for example the one described by Grobenski et al. [182]. They managed to sample a few mg of powder reproducibly. Analyses with such small amounts put high requirements on the sample homogeneity, so as to prevent errors due to non-representative sampling. In the case of direct powder sampling, the temperature program may often start with matrix decomposition and then proceed as for the case of dry solution residues. In modern atomic spectrometric equipment the sampling into the electrothermal device, the temperature programming, the selection and change of the appropriate gas flows as well as the visualization of the complete temperature programming and the direct signals are usually controlled by computer.

When using matrix modification the aim is to make efficient use of the thermochemical properties of the elements so as to be able to remove the matrix more effectively or to immobilize the analytes. Both should bring the goal of a matrix-free determination nearer, along with its advantages with respect to ease of calibration and minimization of systematic errors. Matrix modification has developed into a specific line of research in atomic spectrometry.

For carbide modification of graphite tubes (for a recent review see Refs. [183, 184], use is made of physical and chemical vapor deposition with metals such as Ta, W, Zr, etc., which leads to the formation of MC-coated graphite tubes or platforms. A solid layer of tantalum or niobium carbide can also be obtained as a result of treatment of the graphite furnace with large quantities of pure salts or a suspension of the element or its oxide in water; however, this can lead to tubes with shorter lifetimes. Alternatively, the surface may be treated with aqueous [e.g. of Na_2WO_4, $(NH_4)_2Cr_2O_7$ or $ZrOCl_2$] or alcoholic solutions of the salts of the elements mentioned.

As further matrix modifiers $Mg(NO_3)_2$ and often $Pd(NO_3)_2$ are used. The can be used separately but are often also used as a mixture. The mechanisms of their stabilizing action, although having been investigated intensively, are not completely known, but seem to relate to the formation of intermetallic compounds with the analytes. In the case of Pd such as is used for the determination of Sn, the selectivity for the stabilization of tin, for example in determinations in organic media, is based on the formation of Pd_3Sn_2, which can be shown by x-ray diffraction.

Salts such as $(NH_4)_2HPO_4$, $Ni(NO_3)_2$ and even organic compounds (e.g. ascorbic acid in the case of Sn) have also frequently been proposed as matrix modifiers.

The development of the temperature program is a most important step in establishing a working procedure for any spectrometric method using electrothermal evaporation. It should be fully documented in all analytical procedures for the determination of a given series of elements in a well-defined type of sample.

3.4.4
Analytical performance

Electrothermal atomization, because of its high analyte vapor generation efficiency (in theory 100%), allows it to obtain extremely high absolute as well as concentration power of detection with any type of atomic spectrometry. In the case of two-stage procedures, where the analyte vapor has to be transported into the signal generation source, diffusional losses of analyte vapor may occur. This has been described in detail, for example, for the case of Cd [185], but it is a general problem. Answers to the problem have been found for a number of cases where use is made of the addition of salts to the analyte solutions, by which nucleation in the vapor flow is promoted.

Owing to the transient nature of the analytical signals, the analytical precision is generally lower, as with the nebulization of liquids. Relative standard deviations range from around 3–5% in the case of manual injection of microaliquots to 1–3% in the case of automated sample dispensing, whereas in pneumatic nebulization they are below 1%. A gain in precision is often possible by measuring peak area instead of peak height. However, peak height measurements enable the best signal-to-background intensity ratios or limits of detection to be reached.

Interferences in electrothermal evaporation may stem from differences in the physical properties between the liquid samples. Indeed, these properties influence the wetting capacities of the graphite or the metal of the electrothermal device. When the device has a temperature profile, this leads to differences in volatilization. In general the use of surface tension modifiers, such as TRITON X, can help to decrease differences from one sample solution to another. Further differences between the anions present may severely influence the evaporation, which is known as the chemical matrix effect. Indeed, the boiling points of the compounds formed dictate the volatilization. It should be considered, here, that in many cases, especially when a graphite atomizer is used, the elemental species themselves evaporate. However, it may be that the reaction with the carbon is too slow and the compounds too volatile, so that often oxides or even halogenides are the volatilizing species. In the case of such kinetically controlled reactions, the boiling points of the elements and their species must be considerd, a brief summary of which is given below.

Elements:
Hg: 357; As: 616; Se: 685; Cd: 765; Zn: 906; Sb: 1380; Te: 1390; Tl: 1457; Pb: 1750; Mn: 2095; Ag: 2212; Sn: 2430; Al: 2447; Cu: 2595; Co: 2880; Ni: 2800; Fe: 3070; Ti: 3280 °C

Chlorides:
$AsCl_3$: 130; $CdCl_2$: 475 °C

Oxides:
PbO: 890; Sb_4O_6: 1559; SnO_2: ≈1850 °C

When using electrothermal vaporization the presence of different species of the same elements, as evoked by an incomplete reaction of a species, can thus lead to the appearance of double peaks. In the case of Cd, this has been documented for a sample that is rich in chloride (Cd evaporates as the chloride at 480 °C) and for a sample rich in nitrate (Cd volatilizes first at 900 °C as Cd or at 1200 °C as CdO) (Fig. 59) [186].

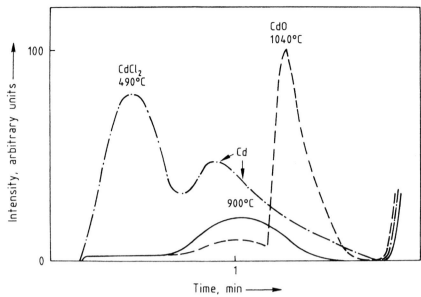

Fig. 59. Thermochemical behavior of Cd and its compounds in graphite furnace evaporation. 3 kW argon–nitrogen ICP; samples: 10 µg/mL Cd(NO₃)₂ in aqueous solution (–), added to 0.1 g/mL of SRM 1571 (– –) and SRM 1577 (–.–.); line Cd II 226.5 nm; sampling volume: 50 µL; temperature program: drying: 95 °C, gradual temperature increase from 350 to 2400 °C (600 °C per min). (Reprinted with permission from Ref. [186])

3.5
Direct solids sampling

These techniques often make use of two-stage procedures. Here a separate source is used for the generation of a sufficiently concentrated sample vapor cloud, the composition of which is a good reflection of that of the sample, and this cloud is lead into a second source. However, several approaches can be used according to the sample properties.

3.5.1
Thermal methods

For powders, electrothermal evaporation may be used as most substances volatilize far below the volatilization temperature of graphite (>3600 °C). This approach is often hampered by the sample inhomogeneity, as it is generally a micromethod, as well as by anion effects causing chemical interferences [187]. Furthermore, transport losses can occur.

In order to solve the problems related to microsamples, suitable solutions may lie in the use of larger furnaces with higher power dissipation so that much larger samples than the usually 2–5 mg admitted to a conventional graphite furnace for atomic absorption can be surpassed by far. Graphite furnaces can even be used where a small crucible is introduced from the side, which then contains the sample (for a description of such systems, see e.g. Ref. [188]). Also when using thermal volatilization from one of the electrodes of a dc arc, the latter normally can contain a few mg of sample, which then must be mixed with graphite powder so as to make them electrically conductive. A rotating arc can thus be used, which can move with a constant velocity over a larger electrode, both improving the precision and the sample size that can be admitted. Rotating arcs can be easily produced by applying magnetic fields.

Transport phenomena occur particularly when transporting the vapors themselves. They disappear completely when the sample is inserted directly into the signal generation source, where it is evaporated thermally. This approach is known from work with graphite or metal probes in atomic absorption, where for example W wire cups and loops are used. The technique is also used in plasma spectrometry with the inductively coupled plasma (ICP), both in atomic emission [189–191] and in mass spectrometry [192]. Its absolute power of detection is extremely high and the technique can be used both for the analysis of dry solution residues as well as for the volatilization of microamounts of solids.

The in situ thermal destruction of the matrix or selective volatilization can also be applied. The latter has been shown, for example, to be useful in the case of geological samples [193]. In addition, selective volatilization of volatile elements from refractory matrices is useful. In this way spectral interferences from matrices with line-rich spectra can be avoided. The approach can also be used for the volatilization of refractory oxide or carbide forming elements that have volatile halogenides (e.g. Ti). Here freons can be added to the carrier gas, through which the elements are converted into volatile fluorides before they can form carbides. The approach appears to be useful, as can be demonstrated from the volatilization of various impurities from Al_2O_3 powders, which seems to increase significantly with the amount of sample used (Fig. 60). In other cases such as ICP-OES, the addition of freons to the working gas has not been found to be very effective, e.g. in the volatilization of particles brought into the plasma. This may be due to the short residence times of the particles in the ICP, through which both thermochemical diffusion of the freons to the elemental species and the thermochemical reactions cannot be completed.

Solid substances such as AgCl and also PTFE powder can be used as thermochemical reagents (see e.g. Ref. [191] for an ICP application). In the case of Ti in the presence of C and PTFE the following reactions have to be considered:

$$TiO_2 + 2C \rightarrow Ti + 2CO$$

$$Ti + C \rightarrow TiC$$

$$Ti + 4F \rightarrow TiF_4$$

At high temperature the last reaction is favored thermodynamically as well as kinetically, the latter because TiF_4 is volatile.

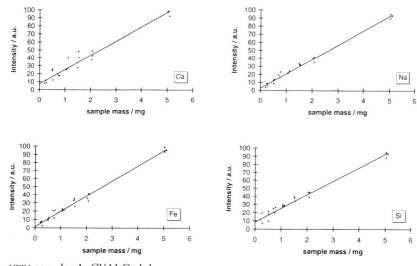

UFIN-sample; AgCl/Al$_2$O$_3$ 1:1

Fig. 60. Increase in signals with sampled amounts in direct solids sampling ETV-ICP-OES with the addition of AgCl [34].

The use of solid thermochemical reagents is well known from early dc arc work and has been continuously refined in spectrochemical analysis. More recently they have also been shown to be very effective for analyte volatilization from refractory ceramic powders introduced as slurries into the graphite furnace in the work of Krivan et al. (see e.g. Ref. [194]). Thermochemical modifiers have also been shown to be efficient when using ETV for sample volatilization only and introduction of the vapor into an ICP (see e.g. [195]).

To find the optimal solid modifier for the determination of analytes with high volatilization temperatures by ETV-ICP-OES or ETV-ICP-MS, thermodynamic calculations are very helpful. For example, from theoretical and experimental data for SiC, which provide information on the physico-chemical properties of SiC and the modifiers, two effects of one (hypothetical) modifier may be discerned: the ability to transform a matrix into a reactive form; and the ability to "attract" silicon contained in that reactive form by another, stronger chemical bonding reagent. It is to be expected that the SiC lattice will be destroyed mainly during the final process and the impurity elements released. Then they can be vaporized independently of either location or chemical form. It is only at this stage that an effective halogenating reaction can be expected. Accordingly, it can be assumed that the extent of matrix destruction achieved is a measure of the effectiveness of a modifier. The results of thermochemical calculations allow the following sequence of chemical modifiers that have been investigated to be arranged, according to increasing effectiveness on the SiC matrix:

$$(C_2F)_n < KF < AgCl < Na_2B_4O_7 < BaCO_3 < Pb(NO_2)_2 < Ba(NO_3)_2$$
$$< Ba(NO_3)_2 + CoF_2 < BaO + CoF_2$$

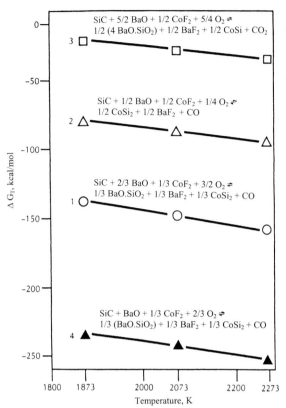

Fig. 61. Thermochemical calculations for the reactions between SiC and the complex modifier BaO + CoF$_2$. (Reprinted with permission from Ref. [196].)

The calculated results show that for the quantitative determination of impurities in SiC powder, a mixture of BaO + CoF$_2$ (1:1) will be the optimal modifier, as shown from the temperature dependence resulting from the thermodynamic calculations for the reaction of SiC with the modifier BaO + CoF$_2$ (Fig. 61) [196]. Here the gain in free chemical energy is definitely maximum. This complex modifier will almost totally destroy the SiC thus enabling the evaporation of the impurities, which is reflected by x-ray diffraction studies that reveal the existence of BaSiO$_3$, CoSi (CoSi$_2$) and BaF$_2$ as reaction products.

3.5.2
Slurry atomization

With powders, stable slurries can often be formed by using suitable dispersing fluids. These fluids should be able to wet the powder and form stable suspensions, both of which depend on their physical properties (viscosity, density and surface tension). For example, the surface tension, which is largely also a function of the pH, is known to influence the surface charge of suspended particles and therewith the stability of the slurry (for ZrO$_2$ powders see e.g. Ref. [117]). The stability of suspensions further depends significantly on the particle size of the powder and its tendency to form particle agglomerates. This tendency depends on the preparation method applied, as is known from synthesized ceramic powders (see e.g. Ref. [197]).

With a Babington nebulizer, as mentioned earlier, slurries of fine powders can be nebulized, as has already been shown by Ebdon and Cave [116] in the early 1970s. With the gas flows (1–2 L/min) and the sample pumping rates (1–2 mL/min) normally used for a Babington nebulizer [198], it was found that slurries prepared with for example 1 g of a ceramic powder (Al_2O_3, SiC or ZrO_2) could be aspirated without deposition of particles in the aspiration sleeves. Therefore, when the slurry is formed by adding water to the powder, it is necessary to treat it in an ultrasonic bath for 10 min. Furthermore, the slurry should be continuously homogenized with a magnetic stirrer during sample aspiration. For many types of powders analyzed (coals, ground biological samples, ceramic powders, fly ashes, etc.), stabilizing the slurries by addition of wetting agents was found to be unnecessary. From experiments with an Al_2O_3 powder with grain size fractions of below 5 μm, between 5 and 15 μm and larger than 15 μm, from measurements with an electron microprobe it was found that no particles larger than 15 μm occur in the aerosol, which marks the first limitation for slurry nebulization of powders with respect to their grain size. This was shown by electron probe micrographs made for dry solution residues of slurries prepared from these powders and for the aerosols collected, and also by laser diffraction measurements under isokinetically controlled conditions, on a Nuclepore filter.

After solvent removal the aerosols produced from slurries deliver solid particles, the diameters of which are those of the powder particles. In slurry nebulization used for flame work or plasma spectrometry, they are injected with a velocity that is less than or equal to the nebulizer gas atom velocities, as viscosity drag forces are responsible for their entrainment into the ICP. The velocity of the gas atoms (v_G) can be calculated from the gas temperature at the location considered (T_G), the injection velocity (v_i) and the temperature at the point of injection (T_i), as $v_G = v_i \times T_G/T_i$ and the acceleration of particles (d^2z/dt^2) as a result of the viscosity drag forces is:

$$d^2z/dt^2 = 3\pi\eta D(v_G - v)/m - g \tag{223}$$

where η is the viscosity of the hot gas, D is the diameter of the solid particles, m their mass and v their velocity, and g is the gravitational constant.

The temperature increase of a particle resulting from the heat uptake from the surrounding gas can be calculated as described by Raeymaekers et al. [117] and in other papers (Refs. [199, 200]). A program, as well as examples, for these calculations is given in Ref. [201], published in *Spectrochimica Acta Electronika*. The gas temperature at a given height z in the plasma $T_G(z)$ can be modelled according to decay functions. Accordingly, the temperature increase (dT_p) of a particle at a point with a certain temperature within a time defined by Eq. (223) can be calculated. By adding the respective amounts of heat taken up, the total amount of heat (q) can be obtained. For a particle with mass m and known latent heats, the mass fraction F

which is evaporated at a certain height in the ICP is then given by $F = q/Q$ with:

$$Q = m \int_{293}^{T_m} c_s \times dT + c_m + \int_{T_m}^{T_d} c_l \times dT + c_d \qquad (224)$$

where T_m is the melting point, c_m the melting heat, T_d the decomposition point, c_d the decomposition heat, c_s and c_l are the latent heats in the solid and the molten phases, respectively. Accordingly, it can be calculated that, in the case of an analytical ICP with a temperature of 6000 K at the point of injection, a 50% evaporation in the analytical zones could be obtained for Al_2O_3 particles with a diameter of 20 μm. For ZrO_2, however, the maximum admittable particle size for a 50% evaporation was found to be 8 μm, which shows that for very refractory powders, even in a high-temperature source such as the ICP, the evaporation and not the nebulization will be the limiting factor for the use of slurry nebulization.

The particle size distribution of powders in the range 0.2–0.5 μm can be determined by automated electron probe microanalysis, as developed for particle characterization work at the University of Antwerp (see e.g. Ref. [202]). Here the exciting electron beam of a microprobe scans a deposit of the aerosol particles collected on a Nuclepore filter under computer control, and from the detection of element specific x-ray fluorescence signals, the diameters of a large number of particles are determined automatically. As shown by results for Al_2O_3, the particle size distributions determined by automated electron probe microanalysis agree to a first approximation with those of stray laser radiation (Fig. 62) [203]. Deviations, however,

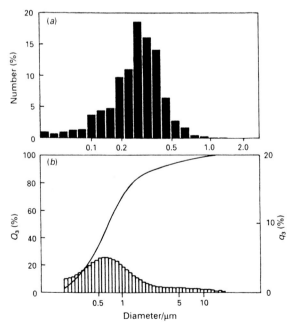

Fig. 62. Particle size distribution for the Al_2O_3 powder AKP-30 (Sumitomo, Japan) by automated electron microprobe analysis (mean diameter = 0.35 μm) (a), and by laser light scattering (mean diameter = 0.59 μm) (b). (Reprinted with permission from Ref. [203].)

occur and are difficult to eliminate, as for example particle aggregates may always form. The powder particle size in work with slurries is particularly critical when the aerosol is produced by pneumatic or ultrasonic nebulization and the particles are to be vaporized, during their passing through a high-temperature source.

For other sample introduction techniques, the particle size is not so critical. This applies when slurries are analyzed by graphite furnace atomic absorption spectrometry. A large number of elements can here be evaporated from a number of types of samples. In a variety of cases, where there is a large difference in volatility between the analyte and the matrix, trace-matrix separations can even be performed in the furnace itself, resulting in reductions in interference. In general, calibration can often be made by standard additions with solutions dried onto the slurry residue in the furnace itself. Whereas Miller-Ihli showed the possibilities of the approach in the analysis of food and biological samples [181] both in the case of furnace AAS but also recently in the coupling of furnace ETV and ICP-MS, much work has been done on industrial matrices by Krivan s group, using graphite furnace AAS [204], including work with diode lasers [205] as well as with ETV from a metal atomizer coupled with ICP-AES [206]. Also inter-method comparison studies on high-purity molybdenum and tantalum as well as on high-purity quartz powders have been reported.

In any case it is often necessary to apply ultrasonic stirring to destroy agglomerates and to disperse powders optimally prior to and even during the slurry analyses. This is e.g. done during slurry sampling by automated syringes into the graphite furnace. The addition of surface active substances such as glycols [207] has been proposed, however, this might introduce contamination when trace determinations are required.

Slurry analysis can also be applied for the analysis of compact samples which are difficult to bring into solution subsequent to their pulverization. Then, however, special attention must be paid to possible contamination resulting from abrasion of the mills. Also in the case of very hard materials, such as ceramics, the abrasion of very resistant mill materials, such as WC, may amount to up to several % of the sample weight. Furthermore, the grinding efficiency depends considerably on the particle size of the starting materials. With a combined knocking–grinding machine with a pestle as well as with a mortar made of high-purity SiC (Elektroschmelzwerk, Kempten, Germany) (Fig. 63), it was possible to grind an SiC granulate material. The mortar was held in a steel enclosure. With this device it was found that SiC granules with grains of dimensions between 1 and 20 mm can be pulverized. Below this grain size the grinding action of the machine was not effective, as it seems that a certain size is required for the pressure to work on the granules. For materials that are not so hard, such as plant tissues, grinding often can be performed for example in agate based grinding equipment consisting of a mortar and balls or disks of suitable dimensions, which are commercially available (see e.g. Spex Industries, Ref. [208]). Grinding instructions for different materials, together with suitable sieving procedures are described in the literature. It must be mentioned that the range of particle sizes for slurry sampling with ETV, as is done in AAS, is generally not as stringent as in the case of slurry nebulization.

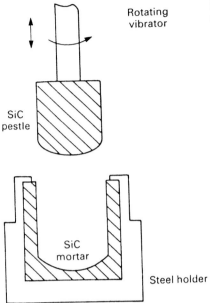

Rotating vibrator

SiC pestle

SiC mortar

Steel holder

Fig. 63. Grinding mill for compact SiC. (Reprinted with permission from Ref. [203].)

3.5.3
Arc and spark ablation

By using an electrical discharge between an electrically conductive sample and a counter electrode, sample material can be ablated. In this case it is advisable to use the sample as the cathode of a dc discharge, as then the anode is hardly ablated because it is only subjected to electron bombardment. A high-melting metal such as tungsten can also be taken as the anode. Consequently, the anode species are not found in the atomic spectra nor are ion signals produced. Ablation can generally occur as a result of thermal processes. This is the case when the heat dissipation in the source is very high, as it is taking place with arcs at atmospheric pressure having high burning currents and fairly low burning voltages. The material thus volatilizes from a molten phase in the burning crater, with the components volatilizing according to their boiling points. When the burning voltages are high, as with discharges under reduced pressures or sparks, the particles impacting on the cathode have high energies and mechanically remove material from the sample, which depends much less on the boiling points of the individual components.

Arc ablation

This has long been proposed for producing an aerosol at the surface of electrically-conducting samples and has been used in combination with various sources. In the version described by Johnes et al. [209], the metal sample acts as the cathode of a dc arc discharge. When using an open circuit voltage of 600 V and currents of 2–8 A, a broad pulse spectrum (mean frequency up to 1 MHz) is observed and a rapid movement of the discharge across the cathode produces a uniform sampling over a well-defined area. The burning voltage of the source is of the order of 60 V and the discharge is ignited by an ignition pulse of ca. 10 kV. The arc burns between the sample, which acts as the cathode, and a hollow anode, the distance between them being around 1–2 mm. The area which is subjected to the discharge is usually restricted to some 6–8 mm with the aid of a BN plate, placed not too close to the sample, which is water-cooled. A flowing gas stream can then transport the aerosol particles, which have a diameter of a few µm only, through the hollow anode into the transport line and further on into the signal generation source.

Remote sampling of up to 20 m makes the system attractive for the analysis of large items. On the other hand, the system is also useful for precision analyses both by atomic absorption and plasma spectrometric methods. The analysis of electrically non-conductive powders is also possible, where a mixture of the powder to be analyzed and copper powder in a ratio of about 1:5 can be made and briquetted into a pellet. This can be realized by using pressures of up to 80 Torr/cm^2, as has been described for glow discharge work by El Alfy et al. [210]. The mass ratio of analyte to copper may, however, differ considerably from one sample base to another.

The volatilization in the case of briquetted powder mixtures can certainly be influenced by sifters, i.e. substances that deliver large amounts of gaseous breakdown products, such as ammonium salts. Here thermally less volatile substances are blown into the arc plasma, where they can be fragmented as a result of the high plasma temperatures. As the arc plasma is almost in thermodynamic equilibrium, high gas temperatures might definitely be expected. However, the introduction of a cold argon carrier gas flow into the arcing regions certainly decreases the gas temperatures, which makes the use of sifters less effective. In the case of NH$_4$Cl, not only the sifter effect but also the effect of halogenation may be positive with respect to the volatilization of refractory substances through plasma reactions. Furthermore, the use of plugs such as CsCl for work with powders briquetted to pellets may be useful. They can help to stabilize the burning voltage of the arc when there are variations in the sample composition. This may then lead to a uniformity in thermal volatilization.

The use of internal standardization may also be very helpful, both with respect to the analytical precision as well as for obtaining low matrix effects. In many cases the sample matrix element can be taken as the reference element. However, when an internal reference is added, such as is possible when mixing powders and briquetting pellets, a reference element with thermochemical behaviour (i.e. boiling points of the elements but also of the volatile compounds eventually formed in the arc plasma) similar to the analytes should be selected.

The ablation rates are considerably enhanced by using the jet electrode, where the argon being used as carrier is blown through the electrode, which has a narrow bore gas channel. This results not only in a very efficient transport of the ablated material away from the arc channel, but its particle size might even be favorably

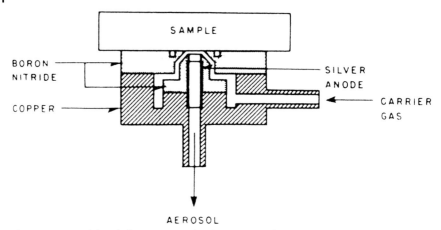

Fig. 64. Direct solids nebulizer. (Reprinted with permission from Ref. [209].)

decreased, as a result of an immediate dilution of the atomic vapor produced, which makes analyte condensation into large droplets less likely.

In the case of an ac discharge, the thermal nature of the sample volatilization may decrease in favor of sputtering, as the thermal energy released is lower. Therefore, ablation of the anode begins to occur, which may be kept low by taking very hard and high-temperature resistant anode materials such as tungsten. This approach may be useful for the ablation of very weak and easily volatile metals such as lead alloys.

With an optimized device, as described by Jones (Fig. 64) analysis times of about 30 s, including 2 s for preflushing, 10–20 s for preburn and 10 s for measuring, were possible. The signal versus time curves are the same as those found for atmospheric pressure discharges in argon. With a capillary arc as the analytical atomic emission radiation source (dc arc with 5 A discharge current and 50 V burning voltage in argon), the calibration curves are linear over a large concentration range. Self-reversal of atomic spectral lines has not been found and the matrix effects are low. The precision is also high, as relative standard deviations of 1% are obtained and the sampling area, which must be available as a flat surface is not larger than 10 mm in diameter.

In the case of arc ablation wandering effects may make the sampling irreproducible. To avoid these difficulties, however, the use of magnetic fields, as known from classical dc arc spectrometry, may be very helpful. Indeed, when applying a magnetic field perpendicular to the discharge column of the arc, the arc can be rotated with a frequency depending on the magnetic field strength. Thus the sample ablation can be made to be more reproducible.

Sparks

These were proposed in the 1970s for sample ablation use only [211]. Both the high voltage sparks, with mechanical or electronic interruption and a burning voltage of

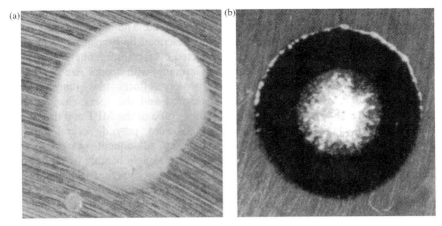

Fig. 65. Burning spots resulting from a diffuse (a) and a concentrated spark discharge (b) for the case of low-alloyed steel samples. (Reprinted with permission from Ref. [212].)

up to 10 kV, as well as medium voltage sparks, which are now in standard use for the atomic emission spectrometric analysis of metals, can be made use of. In the latter case the burning voltages are between 500 and 1000 V and it is often necessary to provide a high-frequency spark across the spark gap for re-ignition of the discharge after a spark train. Therefore with medium voltage sparks repetition rates of up to 1 kHz can be used. Through suitable provisions in the discharge circuit, unidirectional sparks can be obtained, by which material ablation at the counter electrode is suppressed, especially when very hard metals (such as tungsten) are used and the electrode is water cooled.

For the favorable use of sparks as ablation devices, it must be guaranteed that condensed sparks and not diffuse spark discharges are obtained. The latter can be recognized immediately from the burning spot (Fig. 65), and they often occur particularly in the case of aluminum samples. They are mostly due to the presence of molecular gases, set free by desorption or by the presence of oxide layers, but may also be caused by leaks in the sparking chamber. Here the provision of a Viton-0-ring in the petri dish bearing the sample is very useful, as is the use of gas-tight electrode mountings and small but easily purgeable sparking chambers. Work is usually carried out with a geometry that is point to plane with a sample surface, which is flat as a result of turning off and/or polishing. In the spark chamber, it is again useful to direct the carrier-gas flow onto the spark burning spot so as to remove the ablated material easily and to dilute it before large particulates can be formed. Furthermore, the form of the spark chamber should be optimized from the flow dynamics point of view so as to minimize the risk of deposition of larger ablated particles. A miniaturized spark chamber is schematically shown in Fig. 66 [213].

It was found that an increase in voltage as well as in spark repetition rate leads to a considerable increase in the ablation rates. Indeed, in the case of aluminum

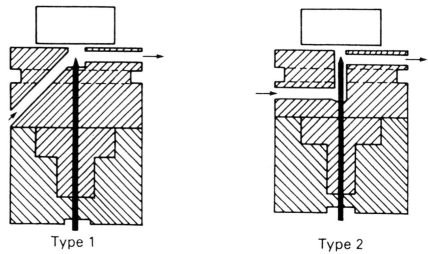

Type 1   Type 2

Fig. 66. Miniature spark chambers for spark ablation of metal samples. (Reprinted with permission from Ref. [213].)

samples it can be shown by trapping of the ablated material in concentrated HNO_3 that the ablation rates increased with the burning voltage of 0.5–1.5 kV from 1 to 5 µg/min in the case of a 25 Hz spark [215].

The particle diameter in the case of a medium voltage spark is of the order of a few µm, through which the particles can easily be volatilized in many sources, ranging from flames, through microwave discharges to inductively coupled plasmas. When a high-voltage spark is used the particle diameter may be considerably increased, which is due to an increase in the ablation of larger particles as a result of the highly energetic impacting gas and metal species. This hampers the volatilization of the material in a high-temperature source such as the ICP and may lead to flicker noise [215]. With the formation of larger particles, selective effects can also become important. Hence low volatility elements may be present in the larger particles and highly volatile elements present more so in the small-size particles, which may stem from vapor condensation. This may also be the case when the elements are distributed irreproducibly in the sample. The latter e.g. is the case in supereutectic alloys.

After aerosol sampling on Nuclepore filters, it was shown, by x-ray fluorescence spectrometry, that the composition of the aerosol produced for aluminum samples was in good agreement with the composition of the compact samples [216]. The particle size in the case of a 400 Hz spark was in the 1–2 µm range and the particles are mostly spherical, which is a good argument for their formation by condensation outside the spark channel. In the case of supereutectic Al–Si alloys ($c_{Si} > 11\%$ w:w), the smaller particles (especially Si) were indeed found to be enriched in some elements. Therefore, it is better to use a medium voltage spark at a high sparking frequency. In this way, small-sized particles are obtained for a large

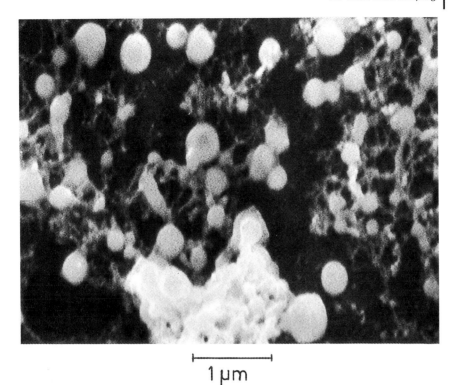

1 µm

Fig. 67. Scanning electron micrograph of aerosol particles collected on a Nucleopore filter (pore size: 1 µm). Spark at 1 kV and 25 Hz; Al-alloy 442, tube length: 0.9 m. (Reprinted with permission from Ref. [216].)

variety of matrices, as can be seen from electron probe micrographs of aerosols sampled isokinetically on a Nucleopore filter [216] (Fig. 67).

When sparks are used for the ablation of electrically conducting solids, less changes with variations in the matrix composition than in the case of arc ablation occur. This is due to the fact that thermal volatilization plays less of a role. However, in the case of brass, it is seen from x-ray analyses of the ablated material on a Nucleopore filter, for samples of the crater wall and the burning crater, that zinc volatilizes more than copper (Table 5), which makes the method difficult to apply to these samples.

In spark ablation, a spark at constant density is obtained in a matter of seconds, and thus, particularly in the case of small spark chambers, preburn times are accordingly low. In plasma emission as well as in plasma mass spectrometry a linear dynamic range of more than 4 orders of magnitude can be obtained and RSDs are a few percent in the case of absolute measurements. However, as shown by the results in Table 6, they can easily fall to below 1%, when using an internal standard element (Fe in the case of steel samples). The matrix effects from the sampling

Tab. 5. Selective volatilization effects in laser and spark ablation as measured by x-ray fluorescence electron probe microanalysis (according to Refs. [212, 217].)

Laser ablation			Spark ablation		

a) Brass (S 40/2)

	Concentration (%)			Concentration (%)	
	Cu	Zn		Cu	Zn
bulk	58 ± 1	42 ± 1	bulk	30 ± 3	40 ± 3
crater wall	66 ± 1	34 ± 1	burning spot	71 ± 3	29 ± 2
droplet	65 ± 1	35 ± 1	particles	51 ± 6	49 ± 7

b) Steel (E-106)

	Concentration (%)				Concentration (%)		
	Fe	Ni	Cr		Fe	Ni	Cr
bulk	51 ± 2	20 ± 1	28 ± 1	bulk	69 ± 2	18 ± 1	13.2 ± 0.6
crater wall	50 ± 2	20 ± 1	30 ± 1	burning spot	70 ± 2	17 ± 1	13.4 ± 0.7

source are low, as will be shown in combination with ICP-OES (see Refs. [209, 216]). They are lower than in arc ablation, as here differences stemming from the thermal volatility of the elements and their compounds play a lesser role. The cross-contamination in the source is also low. Spark chambers with special features for small samples, such as wires and chips, have also been developed. Here the sample is cooled less efficiently and thermal volatilization has to be limited by using low-frequency sparks and an appropriate choice of the reference element.

The gliding spark, which is formed through an hf discharge superimposing sparks along the surface of electrically non-conducting samples, can also ablate

Tab. 6. Analytical precision of spark ablation ICP-OES for a BAS 410/1 steel sample, $c_{Cu} = 3.6$ mg/g. Line pair $I_{Cu\ 324.7\ nm}/I_{Fe\ 238.2\ nm}$. 400 Hz medium voltage spark, 1.5 kW argon ICP, transport gas flow: 1.2 L/min, 0.5 m Paschen–Runge spectrometer, measurement time: 10 s [213].

101.89
102.18
101.11
103.72
103.54
102.91
100.87
101.84
101.69
$\sigma_{relative} = 0.009$

electrically non-conducting materials such as plastics and even ceramics. This has been proposed for use in the classification of plastics, as the halogens can also be excited in atomic emission sources coupled with gliding sparks as the sampling devices [218]. However, the ablation rates are much lower than in the case of sparks used for metal analysis and thus only poor detection limits have been obtained so far. This certainly applies when turning to more robust sample types, such as ceramics.

Spark ablation becomes very abrasive, when it is performed under a liquid. Indeed, Barnes and Malmstadt [219] have already used this effect to increase the material ablation and thus to reduce errors stemming from sample inhomogeneity in classical spark emission spectrometry. In addition, the approach is also useful for the dissolution of refractory alloys, which are highly resistive to acid dissolution [220]. Indeed, when sparking under a liquid, ablation rates of up to around 3 mg/min can easily be achieved. With measurements on high alloy Ni–Cr steels, from electron probe micrographs it can be concluded that selective volatilization might here become problematic. Under a standing liquid, it was found that very stable colloids can be formed when a complexing agent such as EDTA is present in the liquid.

3.5.4
Laser ablation

By the interaction of the radiation from high-power laser sources with solid matter, the latter can be volatilized. This occurs partly as a result of thermal evaporation through the local heating of the sample as a result of the absorption of the laser radiation. However, material volatilization also occurs partly as a result of the highly energetic atoms from the laser vapor cloud or the laser plasma impacting on the sample surface. Laser ablation is independent of the fact of whether the sample is electrically conductive or not and thus has become increasingly important for solids analysis. This is enhanced by the fact that stable laser sources with high energy density have become more widely available in recent years. The feasibility of the approach has been treated in several classical textbooks, but the analytical figures of merit of the technique have recently drastically improved as a result of the novel laser sources that are now available [221, 222].

Laser sources

Laser sources make use of population inversion. When radiation enters a medium both absorption or stimulated emission of radiation can occur as a result of the interaction with matter and the change in flux at the exit is given by:

$$dF = \sigma \cdot F(N_2 - N_1) \, dz \qquad (225)$$

where dz is the length of the volume element in the z direction, σ is the cross section for stimulated emission or absorption and F is the flux. The sign of

$(N_2 - N_1)$ is normally negative, as the population of the excited state (N_2) is given by:

$$N_2 = N_1 \cdot \exp[-(E_2 - E_1)/kT] \tag{226}$$

and is smaller than the population of the ground state (N_1). In the case of population inversion $N_2/N_1 > 1$, and the medium acts as an amplifier. When it is brought between two reflecting mirrors of which one is semi-reflectant, the energy can leave the system. When R_1 and R_2 are the reflectances of the mirrors, the minimum population inversion required for amplification is given by:

$$(N_2 - N_1)_{th} = 1/(2\sigma l) \ln[1/(R_1 R_2)] \tag{227}$$

as then the losses by reflection over a double pathway $2l$ are compensated for. Population inversion requires the presence of a three- or four-term system in the energy level diagram of the medium. The excited level is populated during the pumping process. This may make use of the absorption of radiation, of an electrical process, of adiabatic expansions or of chemical reactions. The excited level can be depopulated to a level that is just below it, for example as a result of non-radiative processes such as thermal decay. The laser transition can go back to the ground state or to a slightly higher state (Fig. 68). Laser radiation is monochromatic and coherent. The beam has a radiance and a divergency α_D ($\approx \lambda/D$, where D is the beam diameter). The medium is in a resonator the length (d) of which determines the resonance frequency, $v = nc/2d$, where n is the mode.

Solid state lasers are of particular interest for laser ablation. Here the laser medium is a crystal or a glass, which is doped with a transition metal. The medium is optically pumped by flashlamps (discharges of 100–1000 J over a few a few ms) or continuously with a tungsten halogenide lamp. When using a flashlamp, both the laser rod and the flashlamp must be cooled so as to provide a frequency of a few pulses per second. Both the laser rod and the flashlamps are placed in a resonator, which can be the space between two flat mirrors, or an ellipsoid in which the laser rod is in one focus and the flashlamp in the other. Also a cylindrical biellipsoidal mirror can be used where one laser rod is in the common focus and a flashlamp is in each of the two other focus points. Here the pumping efficiency is lower but the thermal damage to the laser rod is also lower. A ruby laser uses a rod of Al_2O_3 doped with 0.05% (w:w) Cr, the wavelength of which is in the visible region (694.3

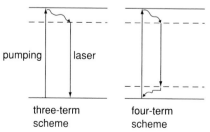

three-term scheme   four-term scheme   **Fig. 68.** Three- and four-level scheme of a laser medium.

nm). It is thermally very robust but requires a high pumping energy and its energy conversion efficiency is low. The Nd:YAG laser, which uses an yttrium aluminum garnet doped with Nd is very widely used. Its wavelength is 1.06 µm and it has a much higher energy conversion. Gas lasers with CO_2 or Ar can be pumped electrically with a high power output as well, whereas so-called dye lasers and semiconductor lasers are mostly used for selective excitation (see later).

The lasers can be operated in the continuous mode or in the pulsed mode, depending on the type of pumping applied. With flashlamps only pulsed operation is possible. However, when pulsing with W-X lamps or electrically, continuous operation is also possible. In addition, a pulsed laser can be operated free-running. However, a lot of irregularly spaced spikes of <1 µs then appear, which start about 100 µs after the pumping pulse. Lasers can also be operated in the Q-switched mode. Here they are forced to deliver their energy spikes very reproducibly, as a result of a periodic interruption of the radiation path. In order to realize this, opto-acoustic or electro-acoustic switches are often used. The former make use of a change of the refractive index of gases such as SF_6 with pressure variations, which may be produced periodically at a suitable frequency (Boissel switch), whereas the latter are based on periodic changes of the transmittance of crystals when they are brought into ac electric fields. To this aim crystals of ADP ($NH_4H_2PO_4$), KDP (KH_2PO_4) and KD^+P (KD_2PO_4) are used.

As the interaction of laser radiation with solids very much depends on the wavelength of the radiation, frequency doubling resulting from non-linear effects e.g. in $LiNbO_3$ or quartz is often used. By doubling the frequency the degree of reflection can often be drastically reduced in favor of the degree of absorption. For this reason in the case of Nd:YAG lasers even a quadrupling of the frequency is used. As the intensity of non-linear effects is low, the original laser radiation must have very high radiant densities. Apart from frequency doubling and quadrupling, Raman shifting can also be used to shift the radiation further towards lower wavelengths in the UV range.

Interaction of laser radiation with solids

When a laser beam with a small divergence impinges on a solid surface, part of the energy is absorbed (10–90%) and material evaporates locally [223]. The energy required therefore varies between about 10^4 W/cm^2 (for biological samples) and 10^9 W/cm^2 (for glasses). Hence as a result, a crater is formed, the smallest diameter of which is determined by the diffraction of the laser radiation, and can be approximated by:

$$d = 1.2f \cdot \alpha_D \cdot v \tag{228}$$

where f is the focal length of the lens used for focussing the laser radiation onto the sample (between 5 and 50 mm, the minimum of which is dictated by the risk of material deposits) and α_D is the divergency (2–4 mrad). Typical crater diameters are of the order of about 10 µm. They also depend on the energy and the Q-switch used. Also the depth of the crater relates to the laser characteristics (wavelength, radiant density, etc.) as well as to the properties of the sample (heat conductance, latent heat, evaporation heat, reflectance, etc.). The ablated material is ejected away from the surface with a high velocity (up to 10^4 cm/s); then it condenses and is again volatilized by the absorption of radiation. In this way a laser vapor cloud is formed, the temperature and the optical density of which are high and for which the expansion velocity, the composition and the temperature again very much depend on the laser parameters and the gas atmosphere and pressure. It is possible to perform different types of optical and mass spectrometry directly on the laser vapor cloud and also to conduct the ablated material into another source.

Analytical performance of laser ablation

Laser ablation is a microsampling technique and thus enables microdistributional analyses to be made. Laterally resolved measurements can be taken with a resolution of around 10 μm [224]. However, the sampling depth can also be varied from 1 to around 10 μm by suitably adjusting the laser. The amounts of material sampled can be of the order of about 0.1–10 μg. As a result of the high signal generation efficiencies possible, e.g. with mass spectrometry or in laser spectroscopy, the absolute power of detection will be very high. Furthermore, the laser sources now available allow high precision to be obtained (relative standard deviations in the region of 1%), being limited only by variations in the reflectivity along the sample surface and the sample homogeneity. Different ways have been investigated to improve the precision. One method makes use of the acoustic signal of the laser, which can be converted into an electrical signal and used as a reference signal. Matrix atomic emission lines or ion signals have also been successfully used as reference signals, as shown by the measurements made when laser ablation is combined with ICP-OES [217].

With advanced Nd:YAG lasers at atmospheric pressure, as utilized when coupling with ICP-OES or MS, selective volatilization is moderate. However, in the case of brass, it is as high as in spark ablation and causes problems in calibration [225]. In recent work favorable working conditions in laser ablation studies were also shown to apply at reduced pressures of around 10–100 mbar [224, 226, 227]. For a number of cases analytes in very differing matrices were giving signals which fitted astonishly well with the same calibration curves from OES, and a nearly matrix independent calibration could be possible. This would be very welcome in the analysis of compact ceramics, for which no other direct analysis methods exist. By careful optimization of the laser working conditions, it now is possible to obtain very reproducible sample material from plastics, as shown by Hemmerlin and Mermet [228].

An interesting development is the system where the laser beam is moved over the sample surface by swinging the focussing lens (Fig. 69). In this way material sampling from a larger part of the sample is possible. Bulk information from the sample is then obtained and the method could be an alternative to conventional spark emission spectrometry. The high precision of the approach, which again can

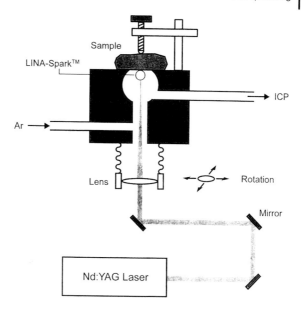

be significantly improved when applying internal standardization, and the low detection limits, these being in the μg/g range, when coupled to ICP-OES [209] testify to the prospects of the method.

The heat conductance through the sample and in the plasma is responsible for the fact that with the Nd:YAG lasers available today, the crater diameters are still much wider than the values determined by the diffraction limitations. When using conventional lasers with pulses in the ns and ps range the plasma shields the radiation, whereas with the femtosecond lasers that are now available a free expanding plasma is obtained, where the heating of the plasma appears to be less supplemented by the laser radiation. This leads to less fractionated volatilization of the solid sample and differences in crater shape, which need to be investigated further [229].

3.6
Cathodic sputtering

In discharges under reduced pressure the atoms and ions undergo little collision in the gas phase. Therefore in an electric field they can gain the high energies required to remove material from the grating positions in solids by impact and momentum transfer. As positive ions in particular can take up the required energies, the phenomenon takes place when the sample is used as a cathode and it is known as cathodic sputtering. Its nature can be understood from the properties of low-pressure discharges, both with dc and also with rf discharges. The models developed in physical studies allow the analytical features of cathodic sputtering to be understood and will be discussed briefly (for an extensive treatment, see Ref. [230]).

Discharges under reduced pressure

When providing a dc voltage across two electrodes positioned in a gas atmosphere under reduced pressure, the ionization in the vicinity of the electrodes produces ions and some free electrons. The latter in particular may easily gain energy in the field and cause secondary ionization of the gas by collision. In addition, secondary electron emission occurs when they impact on the electrodes. Field emission may take place near the electrodes (at field strengths above 10^7 V/cm) and when the cathode is hot, glow emission can also take place. In such a discharge energy an exchange takes place as a result of elastic collisions. When a particle with a mass m_1 collides with one having a mass m_2, the fraction of the kinetic energy of particle 1 being transferred equals:

$$\Delta E/(E_1 - E_m) = 2m_1 \cdot m_2/(m_1 + m_2)^2 \tag{229}$$

where ΔE is the transferred amount of kinetic energy, E_i the kinetic energy of particle 1 and E_m the mean kinetic energy of particles 1. When an electron collides with an atom ($m_e \ll m_{atom}$):

$$\Delta E/(E_1 - E_m) \approx 2m_1/m_2 \quad \text{and} \quad 10^{-5} < 2 \times m_1/m_2 < 10^{-3} \tag{230}$$

Whereas when the masses of both collision partners are equal, the efficiency of energy transfer is optimal:

$$\Delta E/(E_1 - E_m) \approx \tfrac{1}{2} \tag{231}$$

In the case of charge transfer leading to ionization, only electrical charge and no energy is transferred. Also recombination may occur, the probability of which increases quadratically with the gas density. At increasing dc voltage across the electrodes, a dc current is built up accordingly. The characteristic (Fig. 70) includes the region of the corona discharge (a) and the normal region (b), where the discharge starts to cover the whole electrodes (c). Here the current can increase at practically constant voltage, whereas once the discharge covers the whole of the electrodes (restricted discharge) the current can only increase when the voltage is increased drastically (d) (abnormal part of the characteristic). The burning voltage, the positive space charge and the field gradient in front of the cathode are very high. Once the cathode has been heated to a sufficiently high temperature, thermal emission may start and the arc discharge region is entered, where the burning voltage as well as the space charge decrease rapidly (e). The arc discharge has a burning voltage of ca. 50 V, thus reaching the order of magnitude of the ionization energy of the filler gas, the current and the heat devlopment become high (several A) and the characteristic is normal.

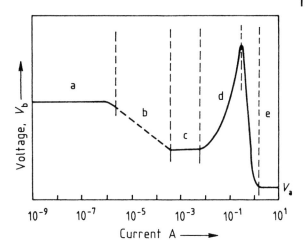

Fig. 70. Current–voltage characteristic of a self-sustaining dc discharge. V_b: breakdown voltage, V_n: normal cathode fall of potential, and V_a: arc voltage. (Reprinted with permission from Ref. [231].)

Analytically relevant discharges under reduced pressure (glow discharges) [232, 233] are operated in the region of 0.01–1 mbar (for mass spectrometry) and 1–10 mbar (in optical atomic spectrometry). The burning voltage is then between 400 and a few thousand volts and the currents are between 0.05 and 2 A. When the whole electrode, mostly consisting of the sample, is exposed to the discharge the characteristic is abnormal. A normal characteristic could be due to an increase in the electrode surface exposed to the discharge, however, also to increased thermionic emission.

In a glow discharge we recognize the cathode dark space in the immediate vicinity of the cathode. Here the energies are too high for there to be efficient collisions. Other regions are the cathode layer where intensive emission takes place, a further dark space and the negative glow where the negative space charge is high and thus excitation as well as ionization through electron impact occurs (Fig. 71).

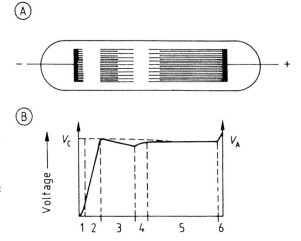

Fig. 71. Geometry (A) and potential distribution (B) of a dc electrical discharge under reduced pressure. (1): Aston dark space, (2): Hittorf dark space, (3): negative glow, (4): Faraday dark space, (5): positive column, (6): anode region.

As the potential outside the cathode fall region hardly changes, the length of a discharge tube will not really have any influence on the electrical characteristic. Apart from these collisions of the first kind, also collisions of the second kind between excited gas species and atoms released from the cathode may occur. This process in the case of an argon discharge is particularly important when argon metastables, with energies at 11.7 eV, are involved. They can cause ionization and excitation in one step (Penning ionization) and this process is of specific importance when it has resonance character. As the elements volatilizing from the cathode and their excitation or ionization is analytically most important, the negative glow of the plasma will be the most analytically relevant region.

A glow discharge plasma is not in local thermodynamic equilibrium (LTE), as the number of collisions is too low to thermally stabilize the plasma. Thus the electron temperatures are high (5000 K for the electrons involved in recombination and >10 000 K for the high-energy electrons responsible for excitation through electron impact) but the gas temperatures (below 1000 K) are low.

Rf discharges are now widely used for sputtering [234], but the principle of these discharges goes back to the work of Wehner et al. [235]. They proposed the use of high (radio) frequency potentials to power a low pressure plasma. The placement of a high voltage on the surface of a non-conductor, e.g. by an electrode at the rear side of the non-conductor, induces a capacitor-like response. The surface acquires the applied potential only to be neutralized by charge compensation by (depending on the polarity) ions or electrons. The result is no net current flow and an unsustained discharge. Rapid polarity reversals of voltage pulses allow for rapid charge compensation and reapplication of the desired high voltage, overcoming the inherent decay time constant. To achieve a "continuous" discharge, pulse frequencies of the order of 1 MHz are required.

A necessary by-product of the capacitor-like response is the self (dc) biasing of the electrode, such that it gets an average negative bias potential sufficient to maintain the discharge processes, establishing it as a cathode. Analytical radio-frequency glow discharges are operated at the 10–40 W level at an operating pressure of up to 10 mbar. The average kinetic energies of atoms leaving the surface at the 0.1 mbar level was found to be 10–14 eV, as measured with Langmuir probes [236] and floating plasma potentials of around 40 V were obtained as well. Accordingly, the sputtering in these sources certainly qualitatively can be treated as in the case of dc glow discharges. Rf glow discharges became very important for direct solids analysis, as here electrically non-conducting samples are also analyzed, such as compact ceramics and glasses, with respect to their bulk as well as to their in-depth composition.

Material ablation by cathodic sputtering

In abnormal glow discharges the working gas ions, after having passed through the cathodic fall, have very high energies and even after neutralization they can knock atoms out of the cathode when impacting, which is denoted as cathodic sputtering.

The models developed for cathodic sputtering start from ideal solids, these being monocrystals without defects, where in fact real samples in atomic spectrochemical analyses are polycrystals which are actually chemically heterogeneous. Furthermore, the models available are only valid for monoenergetic beams of neutrals impacting on the sample under a well defined angle, whereas in fact both ions and atoms with widely different energies impact at different angles.

The ablation is characterized by a sputtering rate q (in µg/s) and a penetration rate w (in µm/s). The latter is the thickness of the layer removed per unit of time and relates to the sputtering rate as:

$$w = (10^{-2} \cdot q)/(\rho \cdot s) \tag{232}$$

s is the surface of the target (in cm^3) and ρ its specific weight (in g/cm^3). The sputtering yield (S) indicates the number of sputtered atoms per incident particle and is given by:

$$S = (10^{-6} \cdot q \cdot N \cdot e)/(M \cdot i^+) \tag{233}$$

N is the Avogadro number, e the charge of the electron (in coulombs), M the atomic mass and i^+ the ion current (in A).

When a monocrystalline sample without defects is subjected to atom or ion bombardment, an ion or atom can be released from the grating, when the energy of the impacting particle at least equals the bond energy of the analyte species in the solid. This displacement energy (E_d) is given by:

$$E_d = \Sigma(E_{\text{Van der Waals}} + E_{\text{coulomb}} + E_{\text{covalent}} + \cdots) - E_{V(T)} \tag{234}$$

The sum of the binding energies thus has to be lowered by the vibration energy. As the energy of the impacting ions is high, any particles can be displaced and released from the solid sample into the glow discharge plasma.

From classical sputtering experiments with a monoenergetic ion beam under a high vacuum, it was found that the sputtering yield increases with the mass of the incident ions and that it first increases with the pressure but then decreases. Furthermore, in the case of polycrystals the sputtering yield was found to be maximum at an incident angle of 30°. For monocrystals it was maximum in the direction perpendicular to a densely packed plane (often the 111 plane). The results of these experiments can only be explained by the impulse theory. According to this theory a particle can be removed from a grating position, when the displacement energy is delivered by momentum transferred from the incident particles. Provided collisions are still of the "hard-sphere" type, where there is no interaction between the positive nuclei, the smallest distance between two atoms is:

$$R = (1/\mu \times V^2) \times 2Z_1 \times Z_2 \times e^2 \tag{235}$$

where $\mu = m \times M/(m + M)$ with m and M the masses of the incident and the displaced particles, Z_1 and Z_2 the number of elementary charges e per atom and V the velocity of the impacting particles. The cross section σ_d for displacement is:

$$\sigma_d = \pi R^2 (1 - E_d/E_{max}) \tag{236}$$

The maximum fraction E_{max} of the energy transferrable from an incident particle is given by:

$$E_{max}/E = 4m \cdot M/[(m + M)^2] \tag{237}$$

The ablation rate thus will be proportional to the number of particles which deliver an energy equal to the displacement energy. It should, however, be taken into account that a number of incident particles are reflected (f_r) or adsorbed at the surface. In addition, particles with a small mass can penetrate into the grating and be captured (f_p). Other incident particles enter the grating and cause a number of collisions until their energy is below the displacement energy. The overall sputtering yield accounting for all processes mentioned can finally be written as:

$$S = [(\alpha \cdot E)/E_d]^{1/2}(f_p + A \cdot f_r)\phi \tag{238}$$

$\alpha = 2m \cdot M/(m + M)^2$ and $\phi = f(m, M)$. Accordingly, the cathodic sputtering increases with the energy of the incident particles and is inversely proportional to the displacement energy. It will be maximum when $m = M$. This explains why sputtering by removed analyte particles which diffuse back to the target is very efficient (self-sputtering).

Also the dependence of the sputtering yield on the orientation of the target with respect to the beam of incident particles can be easily explained. In a monocrystal, there is a focussing of momentum along an atom row in a direction of dense packing. Indeed, when D_{hkl} is the grating constant and d the smallest distance between atoms (or ions) during a collision, the angle θ_0 under which particles are displaced from their grating places relates to the angle θ_1 between the direction of the atom row and the connection from the center of the displaced atom to that of the atom approaching to the closest possible distance in the next row, and is given by:

$$\theta_1 = \theta_0 \cdot (D_{hkl}/d - 1) \tag{239}$$

This focussing of momentum, denoted as Silsbee focussing, takes place when:

$$f = \theta_{n+1}/\theta_n < 1 \quad \text{or} \quad D_{hkl}/d < 2 \tag{240}$$

Also the energy distribution of the ablated atoms and ions can be calculated (see Ref. [230]).

Analytical performance

The model described above delivers the theoretical background for understanding the features of cathodic sputtering as a technique of sample volatilization and is very helpful for optimizing sources using cathodic sputtering with respect to their analytical performance (see e.g. Ref. [232]). When using sputtering for sample volatilization, however, it should be noted that some unique features only can be realized when working under sputtering equilibrium conditions. Indeed, when initiating a discharge the burning voltage normally is so high as to be able to break through the isolating layer of oxides and gases adsorbed at the electrode surface. When these species are sputtered off and the breakdown products are pumped away, the burning spot can start to penetrate with a constant velocity into the sample and the composition of the ablated material can become constant. The time required for obtaining sputtering equilibrium (burn-in time) depends on the nature and on the pretreatment of the sample as well as on the filler gas used and its pressure. All measures which increase the ablation rate will shorten the burn-in times. The ablated material is deposited at the edge of the crater wall and partly taken into the interspace between the anode and cathode when a restricted glow discharge lamp with two vacuum connections is used. These ablations finally limit the total burning time without sample change and interspace rinsing to about 10 min depending on the discharge parameters and types of sample. As for different types of discharge lamps with flat cathodes, the current–voltage characteristics may, for example, differ considerably (Fig. 72) [237], the preburn times may depend on the type of glow discharge lamp used. A feasable way of shortening the preburn time is the use of high-energy preburns, with which preburn times down to around 10 s can be reached. This can be achieved either by increasing the current density or by decreasing the gas pressure through which the burning voltage increases.

The burning crater itself has a topography which depends on the solid state structure of the sample. It reflects the graininess, the chemical homogeneity and the degree of crystallinity. Inclusions and defects in the crystal structure can disturb the sputtering locally. These effects can be observed on micrographs, comparing the craters obtained with a glow discharge and a spark, respectively [233]. The roughness of the burning crater can be measured with sensing probes. It constitutes the ultimate limitation of the in-depth resolution that can be obtained when applying sputtering to study how the sample composition varies with the distance to the sample surface.

The burning craters are curved. They are normally slightly convex as the field density in the middle of the sample might be lower than at the edges, where the sputtered sample is removed more efficiently. This is usually pronounced in a discharge with a flat cathode according to Grimm, where there is a higher vacuum in the interspace. It is less pronounced in a discharge lamp with a floating anode tube, which accordingly should perform better for depth-profiling work.

In order to increase the sputtering, gas jets have been shown to be very effective. They are particularly useful when glow discharges are used as atom reservoirs.

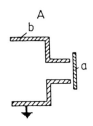

Conventional source with restrictor tube as anode (b) and sample as cathode (a)

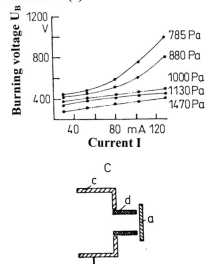

Conventional source with isolated restrictor tube (c) and anode (e)

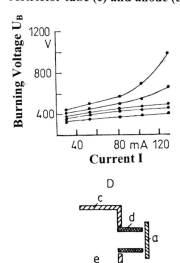

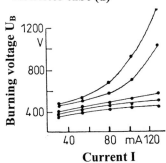

Electrically non-conducting restrictor tube (d)

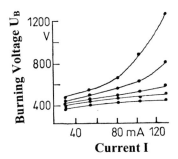

Electrically non-conducting restrictor tube (d) with holder (c) and anode (e)

Fig. 72. Restrictor tube configurations and current–voltage characteristics of flat cathode dc glow discharges. (A): According to Grimm; (B): floating restrictor; (C): restrictor made of isulating material; (D): restrictor made of isolating material and isolated anode. (Reprinted with permission from Ref. [237].)

(a)

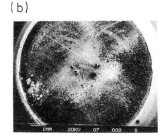 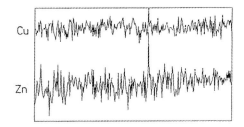

(b)

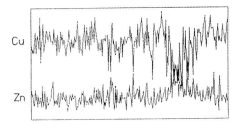

Fig. 73. Electron microprobe (EPMA) line scans (x-ray intensity in arbitrary units, versus beam location) of burning spots of jet-assisted Grimm-type glow discharge. (a) 0.5 mm jets and (b) 0.2 mm jets; sample: brass; sputtering time: 5 min; gas flow: 210 mL/min at 530 Pa argon pressure, burning voltage: 1 kV; current: 58 mA. The scanned line is highlighted in white in the left-hand photograph. (Reprinted with permission from Ref. [238].)

Then of course through the jet action places of increased sputtering occur (Fig. 73) and the analyte removal and redeposition may then become element specific under particular conditions , as can be shown for brass samples (Fig. 73) [238]. Bogaerts et al. [239] set up models enabling a calculation of the species densities in an analytical glow discharge as used as an ion source for mass spectrometry. They then calculated trajectories and also burning crater profiles, which were in particularly good agreement with those obtained experimentally. In their work they used three-dimensional models based on fluid dynamics and Monte Carlo simulations, respectively. In the fluid model the energy gained for the species from the electrical field is balanced by the different energy loss mechanisms for different species. The Monte Carlo simulations cope with the non-equilibrium situation of the plasma species for a statistically significant number of particles of different energies.

The achievable ablation rates depend on the sample composition, the discharge gas and its pressure. As a filler gas a noble gas is normally used. Indeed, in the case of nitrogen or oxygen, chemical reactions at the sample surface would occur and disturb the sputtering, as electrically non-conductive oxide or nitride layers would be formed. Furthermore, reactions with the ablated material would produce molecular species, which emit molecular band spectra in optical atomic spectrometry or produce cluster ion signals in mass spectrometry. In both cases severe spectral interferences could occur and hamper the measurement of the analytical signals. The relationship between the ablation rates and the sample composition can be understood from the impulse theory. In most cases argon ($m = 39$) is used as the sputtering gas, and then the sequence of the ablation rates would agree with the mass sequence:

C < Al < Fe < steel < copper < brass < zinc.

Within the series of the noble gases, helium is not suitable, as due to its small mass its sputtering efficiency is negligible. The sputtering rates further increase in the sequence neon < argon < krypton < xenon. The last two gases are rarely used because of their price. The use of neon may be attractive because of its high ionization potential and for the case of argon, spectral interferences occur. In addition, the gas pressure is a very important parameter which considerably influences the electrical characteristic. Indeed, at low gas pressure the burning voltage is high as is the energy of the incident particles. At high pressure the number density of potential charge carriers is higher and the voltage decreases. The number of collisions will increase, by which the energy of the impinging particles decreases. The resulting decrease in the sputtering rate with the gas pressure for the case of a glow discharge with a planar cathode and abnormal characteristic [240], accordingly, can be described well by:

$$q = c/\sqrt{p} \qquad\qquad\qquad (241)$$

where c is a constant and p is the pressure in mbar.

Many studies have been done on material volatilization by cathodic sputtering in analytical glow discharges used as sources for atomic emission, atomic absorption and for mass spectrometry (for a treatement see Refs. in [241]). Also studies on the trajectories of the ablated material have been performed, as described e.g. for the case of a pin glow discharge [242]. The species number densities obtained for defined working conditions can already be calculated with high reliability for the case of a dc glow discharge [243]. Accordingly, it could be expected that, similar to burning crater profiles, ablation rates can also be calculated, both for dc and rf glow discharges.

Ablation rates have also been measured for rf glow discharges and compared with those of dc glow discharges. With 5–9 mbar of argon and a power of 30 W, penetration rates in the case of a dc glow discharge are of the order of 2 μm/min. This is, for the case of an average density metal (5 g/cm³), a few mg/min and is of the same order of magnitude or even higher than in a 50 W dc discharge. Also in the case of rf discharges, the burning craters have curvatures that limit the depth resolution. They show the same profiles more or less as dc discharges (Fig. 74) [244]. Both the ablation rates and the form of the burning craters were found to be influenced by the addition of helium to an argon discharge.

In addition, the influence of magnetic fields on dc and rf discharges has been investigated, also with respect to the sputtering properties of the plasmas. In the

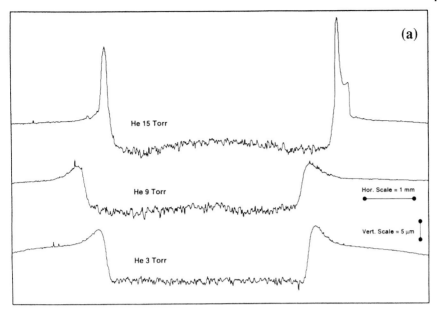

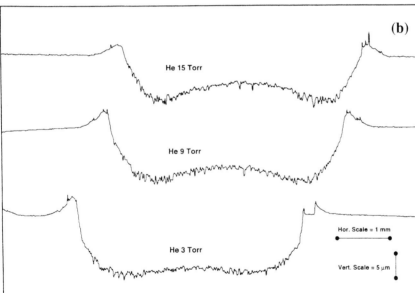

Fig. 74. Crater profiles in an rf glow discharge for a steel sample when adding He to a 5 Torr (a) and a 9 Torr (b) argon plasma. (Reprinted with permission from Ref. [244].)

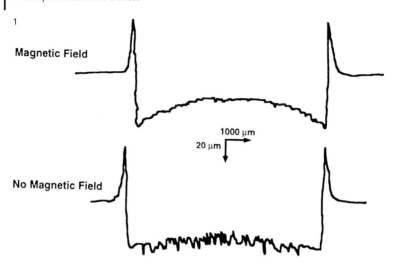

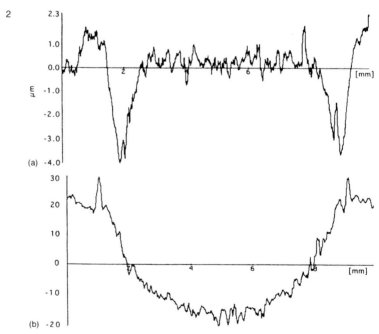

Fig. 75. Influence of a magnetic field on the crater profiles. (1): Dc discharge, sample: electrolytic copper plate, discharge voltage: 1 kV, argon pressure: 4 Torr, cathode-restrictor distance: 0.29 mm, sputtering time: 15 min, 500 Gauss. (reprinted with permission from Ref. [245]) (2): Rf discharge, sample: quartz, argon pressure: 2 Torr (a) and sample: electrolytic copper plate, argon pressure: 2 Torr (b) (reprinted with permission from Ref. [246]).

case of dc discharges, the application of a magnetic field, even through thin plate samples (1–2 mm of copper plates) was considerable. The ablation rates were found to increase by a factor of 2 as compared with the abscence of a magnetic field. Also the curvature of the burning crater increases (Fig. 75) [245]. This can easily be understood as the plasma, in the case of a flat Grimm-type glow discharge, takes on a ring structure, causing a more intensive ablation towards the edge of the burning crater. In the case of an rf argon glow discharge, operated at 2 mbar and 90 W a fairly ring-shaped plasma is obtained, resulting in the appropriate burning profile in the case of a quartz sample [246]. When applying a magnetic field through the positioning of a magnet behind the sample, the plasma contracts, resulting accordingly in a change of the burning crater profile. This effect could be ideally adapted to provide flat profiles as required in depth-profiling, which is now also very important for glass and plastic coatings on surfaces.

4
Atomic Absorption Spectrometry

4.1
Principles

As a method for elemental determinations atomic absorption spectrometry (AAS) goes back to the work of Walsh in the mid-1950s [247]. In AAS the absorption of resonant radiation by ground state atoms of the analyte is used as the analytical signal. Accordingly, a source delivering the resonant radiation of the analyte is required as well as an atom reservoir into which the analyte is introduced and atomized. The absorption of resonance radiation is highly selective as well as very sensitive and thus, AAS became a powerful method of analysis, which is now used for trace elemental determinations in most analytical laboratories for a wide variety of applications. Its methodological aspects and applications have been treated in several textbooks (see e.g. Refs. [248–250]).

A primary source is used which emits the element-specific radiation. Originally continuous sources were used and the primary radiation required was isolated with a high-resolution spectrometer. However, owing to the low radiant densities of these sources, detector noise limitations were encounterd or the spectral bandwidth was too large to obtain a sufficiently high sensitivity. Indeed, as the width of atomic spectral lines at atmospheric pressure is of the order of 2 pm, one would need for a spectral line with $\lambda = 400$ nm a practical resolving power of 200 000 in order to obtain primary radiation that was as narrow as the absorption profile. This is absolutely necessary to realize the full sensitivity and power of detection of AAS. Therefore, it is generally more attractive to use a source which emits possibly only a few and usually narrow atomic spectral lines. Then low-cost monochromators can be used to isolate the radiation.

Accordingly, it was very soon found that using sources for which the physical widths of the emitted analyte lines are low is more attractive. This is necessary so as to obtain high absorbances, as can be understood from Fig. 76. Indeed, when the bandwidth of the primary radiation is low with respect to the absorption profile of the line, a higher absorption results from a specific amount of analyte as compared with that for a broad primary signal. Primary radiation where narrow atomic lines are emitted is obtained with low-pressure discharges as realized in hollow cathode lamps or low-pressure rf discharges. Recently, however, the availability of narrow-band and tunable laser sources, such as the diode lasers, has opened up new per-

Fig. 76. Importance of physical line widths in atomic absorption spectrometry. (a): absorption signal for elemental line; (b): spectral bandpass of monochromator; (c): emission of hollow cathode lamp.

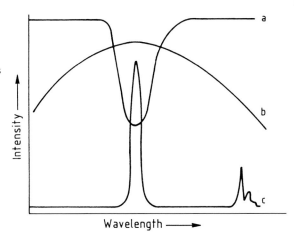

spectives, as shown in recent work [251, 252]. Here only the analytical line is present and the use of a monochromator is superfluous. When tuning is applied the absorption profile can be scanned, which allows the dynamic range to be increased, as measurements are taken in the side-wings of the lines as well as at their maxima. Also, correction for non-element specific background absorption then becomes very easy, and switching from one element to another becomes feasible, certainly when several diode lasers are provided. The restriction, however, still lies in the laser sources due to the limited spectral range that can be covered at this moment and whuch only goes down to the green. However, a recent breakthrough reaching down to the 400 nm range has been achieved, which should increase the prospects of this approach.

In the present commercially available instrumentation, virtually always only line sources are used as primary sources, which generally leads to a high analytical sensitivity and power of detection as well as to a high dynamic range within which the Lamber–Beer law, as expressed in Eq. (47) applies. Commercially available diode AAS systems already exist. A linear relationship between the absorption and the concentration can only apply when all radiation passing to the detector is absorbed to the same extent by the analyte atoms (the ideal case). In a real case, the calibration curve displaying the relationship between the concentration and the absorption bends off towards the concentration axis, as a result of the presence of non-absorbed radiation. For a primary source emitting narrow lines (line widths being below 1/5 of those of the absorption lines) non-absorbed radiation mainly consists of contributions from non-absorbed lines of the cathode material or of the filler-gas that fall within the spectral bandwidth of the monochromator. Furthermore, at high concentrations a decrease in dissociation gives rise to lower absorbances and a suppression in ionization leads to higher absorbances. Thus the calibration curve starts to bend towards the concentration axis and towards the absorption axis, respectively.

For AAS the analyte must be present in the atomic vapor state. Therefore, the use of an atomizer is required. Both flames and furnaces are used and the appropriate methodologies are known as flame AAS and graphite furnace AAS, respectively. The shape of the atomizer with respect to obtaining the highest possible atomic vapor cloud density is still an important field of research. To this aim, attempts can be made to increase the efficiency of the sample introduction device as well as to increase the residence time of the atomic vapor in the atomizer or to prevent a diffusion of the analyte out of the absorption volume. In order to achieve these aims, special methods of atomization have been developed, which are based on volatile compounds formation, as is done with the hydride technique. AAS is generally used for the analysis of liquids and thus solids must first be brought into solution. Therefore, wet chemical decomposition methods are of use, but involve all the care that is normally required in trace analysis to prevent element losses or contamination and the according systematic errors. For direct solids analysis a few approaches also exist.

4.2
Atomic absorption spectrometers

Classical atomic absorption spectrometers contain a primary source, an atom reservoir with the necessary sample introduction system and a monochromator with a suitable detection and data acquisition system, as schematically shown in Fig. 72.

4.2.1
Spectrometer

The radiation of the primary source (a) is lead through the absorption volume (f) and subsequently into the monochromator (h). As a rule the radiation densities are measured with a photomultiplier (i) and the measured values are processed electronically. Usually a Czerny–Turner or an Ebert monochromator with a low focal length (0.3–0.4 m) and a moderate spectral bandpass (normally not below 0.1 nm) is used.

In most instruments, the radiant flux is modulated periodically. This can be achieved by modulating the current of the primary source or with the aid of a rotating sector (g) in the radiation beam. Accordingly, it is easy to differentiate between the radiant density emitted by the primary source and that emitted by the flame. Both single beam and dual-beam instruments (see also Fig. 77) are used. In the latter the first part of the radiation of the primary source is led directly into the monochromator, whereas the second part initially passes through the flame. In this way fluctuations and drift can be compensated for insofar as they originate from the primary radiation source or the measurement electronics. Furthermore, the spectrometer can be provided with equipment for a quasi-simultaneous measurement of the line and background absorption [253].

Therefore, a second source which emits continuous radiation is used. The radiation of this secondary source (b) is passed with rapid alternation relative to the radiation of the other primary source, through the absorption volume, eventually

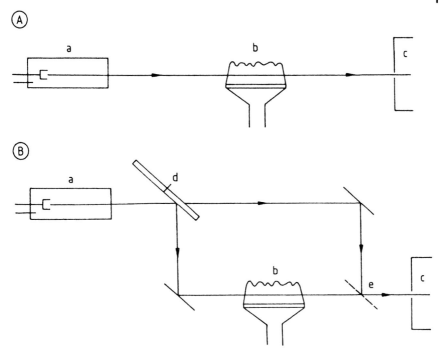

Fig. 77. Flame atomic absorption spectrometer. A.: Single beam and B: dual-beam system. (a): hollow cathode lamp; (b): flame; (c): monochromator; (d): rotating mirror; (e): semi-transparant mirror.

by reflection at a semi-transparent mirror. The radiant flux of the continuum source will not be significantly decreased by atomic absorption, due to the low spectral resolution of the monochromator, however, it will be weakened by broad-band absorption from molecules or stray radiation. With the aid of this principle non-element specific background absorption can be compensated for. A detailed discussion of this and other approaches for background correction will be discussed in Section 4.6.

Apart from single-channel instruments, with which only measurements at a single wavelength in one channel can be performed, dual-channel instruments are also used. They contain two independent monochromators which enable measurements to be taken simultaneously at two different wavelengths. They are of use for the simultaneous determination of two elements, where e.g. one element can be the analyte and the second a reference element. Two lines with widely different sensitivities can also be used so as to determine one element over a wide concentration range.

Multichannel spectrometers which would have a large number of measurement channels and allow the simultaneous determination of a large number of elements, as is done in atomic emission spectrometry, have as yet not found a way into AAS. However, work over a number of years with high-intensity continuous sources and high-resolution Echelle spectrometers for the case of multielement AAS determinations deserves some mention (see e.g. Ref. [254]). This was fostered by the fact that Fourier transform spectrometry and multichannel detection with photodiode arrays opens up new prospects for the simultaneous detection of a larger number of spectral lines and by the considerable improvements in high-intensity sources.

4.2.2
Primary radiation sources

The primary radiation sources used in AAS have to fullfil several conditions:

- they should emit the line spectrum of the analyte element, or several of them, with line widths that are smaller than those of the absorption lines in the respective atom reservoirs;
- they should possibly have a high spectral radiance density (radiant power per surface, space angle and wavelength unit) in the center of the spectral line;
- the optical conductance (radiant surface multiplied by the usable space angle) must be high;
- the radiances of the analytical lines must be constant over a long period of time.

These conditions are fullfilled by discharges under reduced pressure, such as hollow cathode discharges and for some elements by high-frequency discharges.

In a commercially available hollow cathode lamp (Fig. 78), the cathode has the form of a hollow cylinder and is closed at one side. The lamp is sealed and contains a noble gas at a pressure of a few mbars. At discharge currents of up to 10 mA (at about 500 V), a glow discharge is sustained between this cathode and an anode at a removed distance away. The atomic vapor is produced by cathodic sputtering and excited in the negative glow contained in the cathode cavity. Lines of the discharge gas are also emitted, which may lead to interferences in AAS. In most cases high-purity argon or neon is used. Because of mechanical reasons, it may also be necessary to manufacture the hollow cathode mantle from a material other than that of the internal part of the hollow cathode, which as a rule is made of the analyte, and thus a further atomic spectrum could be emitted.

For a number of elements, lamps where the hollow cathode consists of several elements may also be used. The number of elements contained in one lamp is limited because of the risk of spectral interferences.

Electrodeless discharge lamps are preferred over hollow cathode lamps for a small number of elements. This applies to volatile elements such as As, Se and Te.

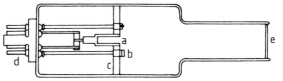

Fig. 78. Hollow cathode source for atomic absorption spectrometry. a: Hollow cathode; b: anode; c: mica isolation; d: current supply; e: window (usually quartz).

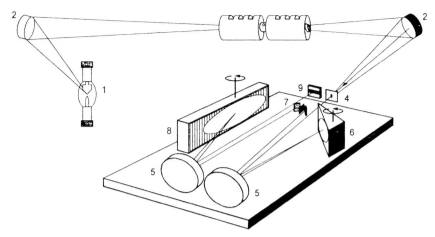

Fig. 79. Set-up for continuous source-AAS with a double Echelle spectrometer (DEMON): (1): xenon arc lamp, (2): off-axis ellipsoidal mirrors, (3): longitudinal Zeeman graphite furnace, (4): variable entrance slit, (5): off-axis parabolic mirrors, (6): prism in Littrow configuration, (7): deflection mirrors and variable intermediate slit, (8): Echelle grating and (9): linear array CCD detector. (Reprinted with permission from Ref. [255].)

In the hollow cathode lamps of these relatively volatile elements self-absorption at low discharge currents may also be considerable and even self-reversal may take place. This is not the case with electrodeless discharge lamps. They consist of a quartz balloon in which the halogenide of the element is present. The analyte spectra are excited with the aid of a high-frequency (MHz range) or a microwave field (GHz range), supplied e.g. through an external antenna.

The development of high-intensity continuous sources is very straightforward. Indeed, Heitmann et al. [255] have reported on the use of a xenon arc lamp (Hanovia 95 9C1980/500 W) with a special design of electrode and working at a pressure of about 17 atm in cold conditions. This lamp operates in a hot spot mode, which leads to an increased radiation intensity especially in the UV region. In addition, an ellipsoidal mirror is used for focussing the radiation into the graphite furnace. Subsequently, the exiting radiation is focussed onto the variable slit width of a double Echelle spectrometer (Fig. 79). With this instrument an extremely high practical resolving power, $\lambda/\Delta\lambda$, of up to $\approx 110\,000$ can be obtained, combined with a stable and compact design. In the apparatus an off-axis parabolic mirror (5) (focal length 302 nm) is used to form a parallel light beam and to reflect the incoming radiation beam onto a prism (6) (apex angle 25°). The prism is mounted in a Littrow mounting and is used as a predisperser. Once again, the returning beam is reflected by the off-axis parabolic mirror, and only the preselected radiation passes through the intermediate slit (7).

The main part of the dual Echelle monochromator, which the authors refer to as DEMON, is arranged symmetrically to the premonochromator. It consists of an Echelle grating (8) (75 grooves/mm, length 270 mm, blaze angle 76°) operating in high orders. For wavelength selection both components (prism and grating) are rotated by means of stepping motors. A spectral range from 190 nm (136th order) to 900 nm (29th order) is covered and the instrumental bandwidth (with a slit width of 23 μm) was found by the authors to be 1.9 pm at 200 nm and 8.6 nm at 900 nm. Finally, the spectral radiation is recorded by a UV sensitive linear array CCD detector (9) (L 172, 512 pixel, size 43×480 μm^2, WF Berlin). The system is controlled by an 80 486/33 MHz personal computer running a data acquisition program developed in-house. With this instrument it is easily possible to measure the Zeeman effect of absorption lines using a continuous radiation detector. This could be shown both for the normal Zeeman effect of the $^1S_0 \rightarrow {}^1P_1$ transition of cadmium at 228.8 nm when using a longitudinal magnetic field and a field strength of 610 mT, as well as for the anomalous Zeeman effect of the $^3P_2 \rightarrow {}^3S_1$ transition of selenium at 196.0 nm. Both profiles agree well with the calculated line profiles, in the latter case after summing up the individual components. Further time-resolved absorption spectra can be readily measured, this being possible simultaneously at a large number of individual wavelengths. Such time-resolved spectra e.g. of a Pb 247.638 nm line (1 μg) and a Pd 247.642 nm line (2 ng) with and without a magnetic field show that lines, which through the use of different atomization temperatures can be slightly separated, can be clearly separated when using a magnetic field [256]. This demonstrates that continuous source AAS combined with a high-luminosity and high-resolution spectrometer shows prospects for even atomic absorption spectrometry becoming a multielement method.

Further development of atomic absorption spectrometry in the other direction lies in the use of photodiode lasers as primary sources, as discussed in a review article by Niemax et al. [257].

Diode lasers are based on semiconductor materials doped with elements from Group III–V and the technology has now been developed for low power systems and also certainly partly for high-power devices such as CO_2 lasers used for welding and cutting. Low-power semiconductor lasers, often called etalon-type diode lasers, are very small (300 μm $\times$ 300 μm $\times$ 150 μm) and are able to convert electrical power into optical radiation with high efficiency (typically 10–30%). Their active region is a few micrometers wide, less than 1 μm thick and as long as the laser strip. As the refractive index of the semiconductor material is high, the ends of the laser substrate act as resonator mirrors. Originally AlGaAs devices, working in the near-IR (770–810 nm) were used, which provided a power of about 3 mW and here the laser range came down to about 670 nm. Later on InGaAsP devices with a wavelength down to 630 nm came on the market at lower prices. These diode lasers are not available for each wavelength of the spectral range 630–1600 nm but leave some wavelength gaps, despite the fact that their wavelength of operation can be changed. However, now new II-VI type diodes (ZnSe) working in the range 470–515 nm and more recently GaN lasers going down to the blue region are being tested in research laboratories, which makes the outlook for analytical spectroscopy with monochromatic sources very attractive.

Diode lasers have interesting spectroscopic properties. They produce radiation of single spatial and longitudinal modes at a maximum injection current and side modes at low diode current, down to <1% of the radiation power at maximum current. If the temperature and the diode injection current are kept constant, the spectral line width is narrow (typically 40 fm), resulting in a high spectral resolving power ($\approx 10^7$), this being much higher than is necessary for a complete resolution of atomic or molecular spectral lines in the gas phase. As this spectral band width is 10–100 times lower than the Doppler widths of atomic spectral lines, they can be used for Doppler-free isotopically selective analysis.

A diode laser spectrometer in a basic setup consists of a laser diode powered by a low-power supply and it is attached to a Peltier element, the temperature of which must be precisely controlled with the aid of an appropriate temperature sensor and electronic circuit. Typical operating data for a 50 mW AlGaAs laser diode are 2 V and 130 mA at room temperature. As the diode laser radiation is divergent and the beam has an elliptical shape, the beam must be collimated by a large-aperture lens system with a numerical aperture of ≈ 0.6. The wavelength of the diode lasers can be tuned by either temperature or current. Temperature tuning is slow because of the inertia of heat transfer from a heat sink to the semiconductor chip, whereas current tuning can be very rapid. Smooth tuning ranges stemming from changes of the laser cavity length (typically ≈ 0.06 nm/K) are limited by sudden hops to other wavelengths due to the fact that the temperature shift of the gain curve is larger (typically ≈ 0.25 nm/K) than the change in the cavity length. The wavelength distance of the hop depends on the laser and its cavity length and is usually of the order of 0.25 nm. The overall temperature tuning range of a diode laser is given by the shift in the gain curve and is ≈ 20 nm if the diode temperature varies from -30 K to $+50$ K and the free range without hops is ≈ 0.25–0.4 nm. Although a few types of diode laser on the market have overlapping wavelength areas which can be continuously tuned by temperature, most do not. The wavelength gap can be narrowed by variation of the current, but some wavelengths simply cannot be reached with a particular diode laser.

Because diode lasers are compact and inexpensive, several units can be integrated and run simultaneously in a single spectrometer. Such parallel operation is necessary for simultaneous determinations of different species, because even if a single diode can be tuned to the analytical lines of various species, temperature tuning is too slow for many analytical applications. In addition, the duty cycle of a continuous-wave measurement is seriously reduced if several species must be measured sequentially with a single diode laser.

Given the limited tuning range of a single diode laser, many different types of diode lasers must be used in a spectrometer in order to cover the whole analytical range that is now accessible, from about 400 to 1600 nm. The simultaneous use of many diode lasers in a spectrometer is only limited by the necessity of dispensing with a number of independently operating power supplies. With several laser diodes, each tuned independently by temperature and locked to the wavelengths of various analytes of interest, the analytical signals can be discriminated by modulating the wavelengths of the laser diodes at different frequencies. The detector signal must then be processed by using narrow bandwidth lock-in amplifiers or a Fourier analyzer and a monochromator is no longer required, as shown for a six-diode set-up in Ref. [258].

The analytically usable range of laser diodes can be extended by applying frequency doubling, i.e. by using the second harmonic generated by laser radiations in intracavity arrangements with nonlinear crystals. Thus, from 100 mW of fundamental power around 0.1 mW in the range 315–500 nm can be generated by focussing $\approx$ 50 mW of 500 to 1000 nm fundamental radiation into an $LiIO_3$ crystal.

Whereas with 0.1 mW only absorption measurements are appropriate, in fluorescence or laser induced ionization, saturation of the excited level can often only be attained with high spectral densities. Given the present limited wavelength range of commercial diode laser, only the alkali elements Li, K, Rb, and Cs can be efficiently excited by fundamental radiation, as their resonance lines are between 670 and 895 nm. Accordingly, simultaneous laser atomic fluorescence spectrometric determinations of Li and Rb using a furnace as the atom reservoir were possible down to the pg/mL level [259]. Once lasers down to the blue wavelength range become available, there will thus be new possibilities for many elements.

Because of the wide analytical range already accessible with second harmonic generation, many elements routinely determined by conventional AAS in analytical flames or furnaces can also be determined by AAS with diode lasers. The availablility of laser diodes with lower wavelengths will only make the approach cheaper, as then second harmonic generation will become superfluous. The elements now accessible with $\lambda > 630$ nm with resonance lines are already manifold: Li, Na, Al, K, Ca, Sc, Ti, V, Cr, Mn, Fe, Co, Ni, Cu, Ga, Rb, Sr, Y, Zr, Nb, Mo, Ru, Rh, Pd, Ag, Cd, In, Sn, Cr, Ba, La, Hf, Ta, W, Re, Ir, Pt, Tl, Pb, Nd, Sm, Eu, Gd, Ho, Tm, Yb and Lu. Also U and some of the actinides can be determined. Important elements such as Be, Mg, As and Hg with diodes emitting in the blue region will eventually become accessible.

The fact that the wavelength of diode lasers can be current modulated up to the MHz range allows reduction of $1/f$ noise and measurements can be taken near the shot noise limit, i.e. down to absorbances of 10^{-7}–10^{-8} in gas or vapor cells [260], and with second harmonic generation resulting in a low power down to 10^{-4} e.g. for a furnace [261]. So for the Cr 425.44 nm line, an absorption signal with the second harmonic for the case of a 0.5 µg/mL solution in flame AAS with a 5 kHz wavelength modulation is still very easily detected (Fig. 80). The modulation also corrects for background shifts resulting from current increases as a result of wavelength scanning. As in conventional AAS, absorbances down to 10^{-3} can be measured, and the improvement in the power of detection of laser AAS will be considerable. This turns out to be true, as for Cr the detection limit with laser AAS in the case of a flame, is as low as 5 ng/mL when using ordinary pneumatic nebulization. Thus for furnace AAS and wavelength modulation the detection limits can compete with those of ICP-MS [262].

Non-metals such as H, O, S, noble metals and the halogens cannot be determined by flame or furnace AAS, but have long-life excited states from which strong absorption transitions can be induced by the red and near-IR radiation of diode lasers. Furthermore, also elements such as Se and Hg can be determined by diode laser radiation absorption from metastable states. Metastable states can be produced in low-pressure plasmas such as dc, microwave and high-frequency discharges,

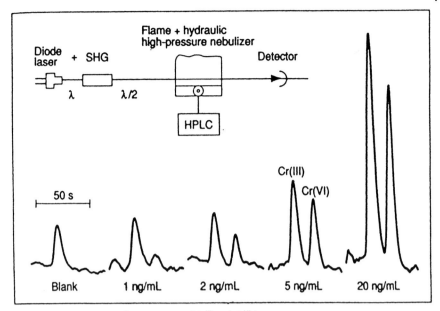

Fig. 80. Instrumentation for speciation of Cr^{III} and Cr^{VI} by wavelength modulation laser AAS in an analytical flame and the corresponding chromatograms obtained for deionized water. (Reprinted with permission from Ref. [257].)

where under optimum conditions the population of metastable states can amount to up to 0.1–1% of the total number densities. As there are very strong transitions from these states to higher levels in the red and near-IR, sufficiently high power for good S/N ratios in wavelength modulation laser AAS can be delivered at the fundamental wavelength by commercially available diode lasers for e.g. F, Cl, Br and O. This use of diode laser AAS with a low-pressure microwave induced plasma in a resonator as described by Beenakker is shown in Fig. 81 [257]. When modulating both the plasma and the wavelength in particular, the power of detection is high, as a result of correction for background absorption and reflections in the optical pathway. For instance 60 ppt of $C_2Cl_2F_4$ can be detected by absorption of metastable chlorine with the aid of a helium plasma.

For element-specific detection in gas chromatography, diode laser AAS is very powerful. When using several diode lasers simultaneously, signals for several elements can be determined at the same time and the composition of molecular species determined, e.g. Cl at the 837.60 nm line and Br at the 827.24 nm line and this with detection limits of down to 3 ng/mL or with respect to the injection volume 0.1 pg/s or 1 pg absolute.

As the line widths of diode lasers are considerably narrower than those of atomic spectral lines excited in a thermal atomizer, spectra can be recorded at very high resolution. When performing the atomization at reduced pressure (e.g. at 100–500 Pa), pressure broadening is low as compared with the Doppler broadening. As the

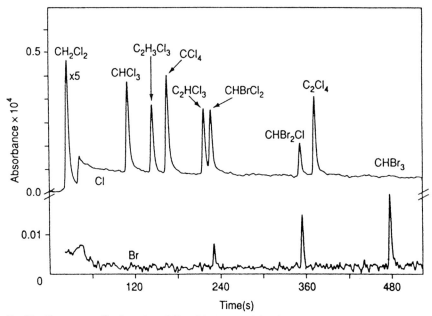

Fig. 81. Element-specific detection of Cl and Br in a modulated low-pressure helium plasma coupled to a gas chromatograph. (Reprinted with permission from Ref. [257].)

isotope shifts are often larger than the Doppler broadening, signals of different isotopes such as ^{235}U and ^{238}U can often be differentiated making isotopic analyses of enriched uranium possible [263]. Doppler-free spectroscopy with two laser beams in co- or counter-propagating directions can also be applied, which becomes even more isotopically selective, as isotope patterns hidden in the Doppler profile can then be visualized.

Instrumentation for diode laser based AAS is now commercially available and the method certainly will expand as diode lasers penetrating further into the UV range become available, especially because of their analytical figures of merit that have been discussed and also because of their price. In diode laser AAS the use of monochromators for spectral isolation of the analyte lines becomes completely superfluous and correction for non-element specific absorption no longer requires techniques such as Zeeman-effect background correction atomic absorption or the use of broad band sources such as deuterium lamps.

4.3
Flame atomic absorption

In flame AAS the sample solutions are usually nebulized pneumatically into a spray chamber and the aerosol produced is lead together with a mixture of a burning gas and an oxidant into a suitable burner.

4.3.1
Flames and burners

Flames make use of propane, acetylene and hydrogen as the burning gases together with air or nitrous oxide as the oxidant gas. With these mixtures, temperatures between 2000 and 3000 K can be obtained as listed below:

- air–propane: 1930 K
- air–acetylene: 2300 K
- air–hydrogen: 2045 K
- nitrous oxide–acetylene: 2750 K
- oxygen–acetylene: 3100 K.

The air–acetylene flame is most often used. Its temperature is high enough to obtain sufficient atomization for most elements that can be determined by AAS, and it is not so high that ionization interferences become significant. Only in the case of the alkaline and alkaline earth elements do the latter occur. The analytical conditions can be optimized by changing the composition of the burning gas mixture. The sensitivity for the noble metals platinum, iridium, gold, etc. can be improved by increasing the oxidative properties of the flame (excess of air), which was also found to lower interferences. Therefore, the use of a reducing flame was found to be more advantageous for the determination of the alkali elements.

For more than 30 elements, however, the temperature of this flame is too low. This applies to elements that have very stable oxides such as V, Ti, Zr, Ta and the lanthanides. These elements require the use of a nitrous oxide–acetylene flame, as was introduced into flame AAS by Amos and Willis in 1966 [264], which is more suitable. However, this flame has a higher background emission, due in particular to the radiation of the CN bands.

The hydrogen–air flame is of little use. For the determination of Sn, however, its sensitivity is higher than the acetylene–air flame. Because of its higher transmission at short UV wavelengths the hydrogen–air flame is also used for the determination of As and Se with the 193.7 nm and the 196.0 nm lines, respectively.

The propane–air flame has the lowest temperature and finds use for the determination of easily atomized elements such as Cd, Cu, Pb, Ag, Zn and specifically the alkali elements.

In flame work the burners have one or several parallel slits of length 5–10 cm. Thus an appropriate absorption path for the primary radiation is provided. In special laminar burners the density of the analyte atoms at relatively high observation positions can be kept high, which is advantageous as measurements are then possible in these higher observation zones, where more complete atomization is achieved. Multislit burners may be useful so as to prevent deposits in the case of solutions with high salt contents. Owing to the high gas flows used and the narrow slits of the burner, aimed at preventing any analyte flow as a result of diffusion, the residence times can become low. Thus in order to avoid that as a result of this, the dissociation of molecules in flames would be hampered, special precautions such as the use of impact surfaces above the flame can be helpful. Additionally, the high-temperature flames require the use of special burners. Detailed information on flames is available in classical textbooks (see e.g. Ref. [265–267]).

The dissociation of the molecules in which the analytes are present is of crucial importance for atomic absorption. When introducing the sample solutions as an aerosol into the flames it is assumed that drying and evaporation of the solid particles is completed in the flame itself [248]. In the gas phase mostly diatomic molecules will be present. Triatomic species will be confined to the monohydroxides of some alkaline and alkaline earth metals, monocyanides as well as some oxides such as Cu_2O. Most other polyatomic compounds dissociate very quickly at temperatures far below the temperatures in analytical flames. The dissociation of molecular compounds at high temperatures is treated in Ref. [19]. The dissociation energies of diatomic molecules are in the range of 3–7 eV. In analytical flames compounds with dissociation energies of <3.5 eV will usually dissociate completely, whereas when dissociation energies are >6 eV the compounds are difficult to dissociate. The degree of dissociation may change considerably with the flame temperature [see Eqn. (90)].

It should be kept in mind that the solvent used may have a considerable influence on the flame temperature. The influence of the sample matrix on the flame temperature is low and therefore its consequences for the degree of dissociation are negligible. The equilibrium itself is only a function of the temperature but the speed with which the equilibrium is reached depends on the relevant reaction mechanisms and the reaction velocities.

Dissociation reactions, in which the flame gases are involved, play an important role. Typical examples are the dissociation of oxides, hydroxides, cyanides and hydrides. Indeed, the concentrations of O, OH, CN and H in a flame depend on the reactions between the components of the flame. For instance in the primary combustion zone of a nitrous oxide–acetylene flame, the concentration of atomic oxygen is about by 3 orders of magnitude higher than the thermodynamically expected concentrations [268]. Therefore, interferences cannot generally be explained by a consideration of the oxygen available in the flame. Whereas for the dissociation of oxides, the mechanism will hardly have an influence on the equilibrium itself, this is different in the case of the halogenides. Indeed the anion here does not stem from the flame gases but from the sample solutions. Because the amount of halogenide bound to the analyte is negligible, as long as the acid concentration in the solution exceeds 1 mol/L, the degree of dissociation can be obtained as:

$$\alpha_D = 1/(1 + n_X/K_D) = 1/(1 + p_X/K_D) = 1/(1 + 10 \cdot c_X/K_D) \tag{242}$$

when n_X/K_D is expressed in cm^{-3}, p_X/K_D in Pa and c_X/K_D in mol/L.

Alkali metal halogenides are usually evaporated completely and Eq. (242) gives the slope of the analytical curve as a function of the halogenide concentration in the measurement solution. Halogenides with concentrations of >1 mol/L therefore will have a depressing influence on the analytical signals for elements where the dissociation energies of the monohalogenides have a dissociation constant $K_D < 10^3$ Pa.

Fig. 82. Burner–nebulizer assembly for flame atomic absorption spectrometry. a: Burner head with mixing chamber; b: nebulizer; c: impactor bead; d: impact surfaces; e: nebulizer socket. (Courtesy of Bodenseewerk PerkinElmer, Überlingen.)

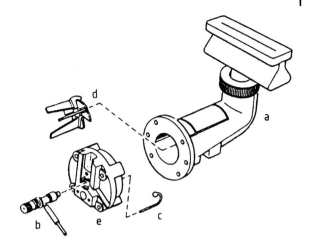

4.3.2
Nebulizers

Pneumatic nebulizers are usually used in flame AAS. The nebulizer is mounted in the spray chamber–burner assembly, as shown for the nitrous oxide or air–acetylene burner system (Fig. 82). Here air or nitrous oxide are fed into the mixing chamber through the nebulizer. The burning gas is lead directly to the burner. A self-aspirating concentric nebulizer is often used, but other types that can be used also include the Babington and cross-flow types (see Section 3.1). This nebulizer may be made of glass or of corrosion-resistant metals such as Pt-Ir but also of plastic such as Ryton. The sample solution consumption usually amounts to around 5 mL/min in the case of gas flows of 1 to a few L/min. In flame AAS the nebulization system must be able to generate small droplets (<10 μm), as only such droplets can be transported and completely vaporized in the flame. This is a prime condition for the realization of high sensitivity and low volatilization interferences. The AAS nebulizers generate not only small droplets but also larger ones (up to more than 25 μm). For the fragmentation of the latter the nebulizer should be positioned in a mixing chamber provided with impact surfaces. In pneumatic nebulization the sampling efficiency remains restricted (some %) as fragmentation is limited and all particles >10 μm pass into the waste. In flame AAS using pneumatic nebulization as a rule the sample solution is aspirated continuously. After 5–7 s a stable signal is obtained and signal fluctuations, depending on the concentration, are of the order of 1%.

Apart from continuous sample aspiration also flow injection and discrete sampling can be applied (see Section 3.1), both of which deliver transient signals. In the latter case 10–50 μl aliquots can be injected manually or with a sample dispenser into the nebulization system, as was first proposed by Ohls et al. [125] and described by Berndt et al. (see Ref. [126]). The approach is especially useful for preventing clogging in the case of sample solutions with high salt contents, for the analysis of microsamples as required in serum analysis or when aiming at the

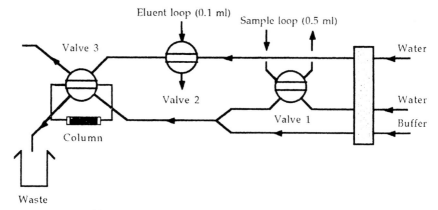

Fig. 83. FIA manifold as used in preconcentration work with AAS, ICP-OES and ICP-MS. (Reprinted with permission from Ref. [269].)

highest absolute power of detection, as is required in combined analytical procedures for trace analysis. In flow injection the analyte solution is injected into a carrier flow with the aid of a valve provided with a loop (Fig. 83) [269].

Flow injection procedures are very useful for performing trace analyses in highly concentrated salt solutions. Fang and Welz [270] showed that the flow rate of the carrier solution can be significantly lower than the aspiration rate of the nebulizer. This allows even higher sensitivities than with normal sample delivery can be obtained. Despite the small volumes of sample solution, the precision and the detection limits are practically identical with the values obtained with continuous sample nebulization. The volume, the form of the loop (single loop, knotted reactor, etc.) and the type and length of the transfer line between the flow injection system and the nebulizer considerably influence the precision and detection limits that are attainable.

When combined with solid-phase extraction, flow injection in flame AAS also enables on-line trace matrix separations to be performed. Here the matrix can be complexed and the complexes kept on the solid phase while trace elements pass on towards the atomizer. For the case of the trace analysis of ZrO_2, after dissolution it was thus possible to keep up to 4 µg of Zr as a TTA (thenoyltrifluoroacetone) complex on the column, while impurities such as Fe were eluted and determined with a high efficiency [133]. This opens up a new line of research on the use of on-line trace-matrix separations for any type of complex samples.

4.3.3
Figures of merit

Power of detection

The lowest elemental concentrations that can be determined by flame AAS found in the literature are often given in terms of the so-called "characteristic concentrations" (in µg/mL). For aqueous solutions of such a concentration, an absorption of 1% is measured which corresponds to an absorbance of 0.0044.

The noise comprises contributions from the nebulizer as well as from the flame. However, detector noise limitations can also occur. The last can be minimized by operating the hollow cathode lamp at a sufficiently high current.

In order to obtain a maximum power of detection, the atomization efficiency should be as high as possible. Therefore, an optimization of the form of the spray chamber and also of the nebulizer gas flow is required. Furthermore, the primary radiation should be well selected by the monochromator and the amount of non-absorbed radiation reaching the detector should be minimized by selection of the appropriate observation zone with the aid of a suitable illumination system.

The detection limits of flame AAS are particularly low for fairly volatile elements, which do not form thermally stable oxides or carbides and have high excitation energies, such as Cd and Zn. Apart from these and some other elements such as Na and Li the detection limits in flame AAS are higher than in ICP-AES (see Table 20 in Section 10).

Analytical precision

When integrating the absorbance signals over 1–3 s, relative standard deviations down to 0.5% can be achieved. Injection of discrete samples into the nebulizer or utilizing flow injection analysis results in slightly higher RSDs being obtained. However, the RSDs soon increase when leaving the linear part of the calibration curves and on applying linearization by software.

Interferences

The interferences in flame AAS consisit of spectral, chemical and physical interferences.

Spectral interferences of analyte lines with other atomic spectral lines are of minor importance as compared with atomic emission work. Indeed, it is unlikely that resonance lines emitted by the hollow cathode lamp coincide with an absorption line of another element present in the atom reservoir. However, it may be that several emission lines of the hollow cathode are within the spectral bandwidth or that flame emission of bands or a continuum occur. Both contribute to the non-absorbed radiation, by which the linear dynamic range decreases. Also, the non-element specific absorption (see Section 4.6) is a spectral interference.

Incomplete atomization of the analyte causes so-called chemical interferences. They are due to the fact that atomic absorption can only occur with free atoms. Thus reactions in the flame which lead to the formation of thermally stable species decrease the signals. This fact is responsible for the depression of calcium signals in serum analysis by the proteins present, as well as for the low sensitivities of metals that form thermally stable oxides or carbides (Al, B, V, etc.) in flame AAS. A further example of a chemical interference is the suppression of the absorbance of earth alkali metals as a result of the presence of oxyanions (X) such as aluminates or phosphates. This well-known "calcium-phosphate" interference is caused by the reaction:

$$M{-}O{-}X \underset{\text{excess}}{\rightleftharpoons} X + M{-}O \overset{T\uparrow}{\rightleftharpoons} M + O \overset{+C}{\rightleftharpoons} M + CO \tag{243}$$

In hot flames such as the carbon rich flame, the equilibrium would lie on the right side. However, in the case of an excess of oxyanions (O–X) the equilibrium is shifted to the left and no free M atoms are formed. This can be corrected for by adding a metal (R) which forms still more stable oxysalts and releases the metal M again. To this aim La- and Sr-compounds can be used according to:

$$M{-}O{-}X + R{-}Y \underset{\text{excess}}{\rightleftharpoons} R{-}O{-}X + M{-}Y \tag{244}$$

When $LaCl_3$ is added to the sample solutions, the phosphate can be bound as $LaPO_4$.

With alkali metal elements the free atom concentrations in the flame can decrease as a result of ionization, which occurs particularly in hot flames. This leads to a decrease of the absorbances for the alkali metal elements. However, it also may lead to false analysis results, as the ionization equilibrium for the analyte element is changed by changes in the concentration of the easily ionized elements. In order to suppress these effects, ionization buffers can be added. The addition of an excess of Cs because of its low ionization potential is most effective for suppressing changes in the ionization of other elements, as it provides for a high electron number density in the flame.

Physical interferences may arise from incomplete volatilization and occur especially in the case of strongly reducing flames. In steel analysis, the depression of the Cr and Mo signals as a result of an excess of Fe is well known. It can be reduced by adding NH_4Cl. Further interferences are related to nebulization effects and arise from the influence of the concentration of acids and salts on the viscosity, the density and the surface tension of the analyte solutions. Changes in physical properties from one sample solution to another influence the aerosol formation efficiencies and the aerosol droplet size distribution, as discussed earlier. However, related changes of the nebulizer gas flows also influence the residence time of the particles in the flame.

4.4
Electrothermal atomic absorption

The use of furnaces as atomizers for quantitative AAS goes back to the work of L'vov and led to the breakthrough of atomic absorption spectrometry towards very low absolute detection limits. In electrothermal AAS graphite or metallic tube or cup furnaces are used, and through resistive heating temperatures are achieved at which samples can be completely atomized. For volatile elements this can be accomplished at temperatures of 1000 K whereas for more refractory elements the temperatures should be up to 3000 K.

The high absolute power of detection of electrothermal AAS is due to the fact that the sample is completely atomized and brought in the vapor phase as well as to the fact that the free atoms are kept in the atom reservoir for a long time. The signals obtained are transient, as discussed earlier.

4.4.1
Atomizers

Apart from graphite tube furnaces, both cups and filaments are used as atomizers in electrothermal AAS [271]. The models originally proposed by L'vov et al. [171] and by Massmann [172] were described in Section 3.4. In the case of the latter, which is most widely used, the optical beam is led centrally through the graphite tube, which is closed at both ends with quartz viewing ports mounted in the cooled tube holders. Sample aliquots are introduced with the aid of a micropipette or a computer controlled dispenser through a sampling hole in the middle of the tube.

Normal graphite furnaces have a temperature profile and thus differences in the spreading of the analyte over the graphite surface may lead to changes in the volatilization behavior from one sample to another. This effect can be avoided by using a transversally-heated furnace, where the temperature is constant over the whole tube length (Fig. 84). The latter furnace, which has been proposed by Frech et al. [272], can accordingly allow a number of volatilization effects to be avoided. In electrothermal AAS specific problems may arise from recombination of the atomized analyte with oxygen and other non-metals, and so then the free atom concentration, which is being measured in AAS, decreases. This can take place particularly when the volatilized analyte enters a fairly cool plasma, as is normally the case in a tube furnace. This can be prevented, as proposed by L'vov [179], by dispensing the sample on a low-mass graphite platform located in the tube furnace. The low heat content then allows very rapid heating and volatilization of the analyte into a furnace plasma of about the same temperature as the sample carrier which lowers the risks of recombination. The latter effect can be very efficiently used when introducing the sample as a dry solution aliquot onto a graphite probe in a heated atomizer.

As discussed earlier, tungsten furnaces, as first proposed by Sychra et al. [175a], are useful for the determination of refractory carbide forming elements, which in the case of a graphite furnace may suffer from poor volatilization, but they are more

Fig. 84. Spatially isothermal graphite furnace for atomic absorption spectrometry using side-heated cuvettes with integrated contacts. (a): Cuvette contact area clamped in terminal blocks, (b): injection port, (c): aperture for fiber optics. (Reprinted with permission from Ref. [272].)

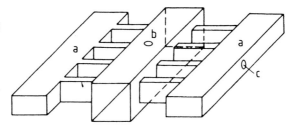

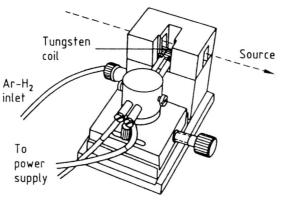

Tungsten
coil

Ar-H₂
inlet

To
power
supply

Source

Fig. 85. Schematic diagram of the tungsten filament atomizer. (Reprinted with permission from Ref. [273].)

difficult to use in the case of real samples. Further, carbothermal reduction of analyte oxides does not take place here. Recently, Berndt et al. [273] have shown that tungsten coils are suitable atomizers for dry solution aliquots, especially in the case of matrix-free solutions as obtained in combined analytical procedures involving a separation of the analyte from the matrix elements. Owing to the relatively large coil surface (Fig. 85) the salt problems in the case of real samples may be lower than with a metal tube furnace.

Despite the progress made in graphite furnace AAS, the basic mechanisms have not been fully established as yet. This applies to the processes responsible for the atomization itself as well as for the transfer of free atoms to the absorption volume and their removal from this volume. The time-dependence of the atom population in the absorption volume can be described by:

$$dN/dt = n_1(t) - n_2(t) \tag{245}$$

where $n_1(t)$ represents the number of atoms entering the atom reservoir and $n_2(t)$ the number of atoms leaving the absorption volume. The first attempts at modelling were made by L'vov et al. [274] and Paveri-Fontana et al. [275]. Accepting a continuous increase in the atomization temperature, which applies mostly at the beginning of the heating, and with $A = dT/dt$:

$$N_t(t) = A \cdot t \tag{246}$$

and:

$$\int_0^{\tau^1} n_1(t)\, dt = N_0 \tag{247}$$

where τ_1 is the time required to transfer the total number of atoms N_0 into the absorption volume. Accordingly,

$$n_1(t) = 2 \cdot N_0 t/(\tau_1)^2 \tag{248}$$

When accepting that at the start of the absorption signal no atoms have been removed from the absorption volume, the second term in Eq. (245) can be dropped and integration of the equation provides:

$$N = N_0/(\tau_1^2) \cdot t \tag{249}$$

Whereas the beginning of the absorption signal can be calculated relatively well, this is much more difficult for the decay. For an open atomizer (such as a graphite rod or probe) diffusion is predominant, which is no longer the case for tube atomizers, where adsorption–desorption processes are also important. In the "stabilized temperature platform furnace" (STPF), where a rapid furnace heating and a platform are used according to Slavin et al. [276], the integrated absorption (A_{int}) becomes independent of the atom formation rate. Also there are no longer any mechanisms other than diffusion involved in the transport, as the analyte first becomes atomized after the gas expansion is virtually finished. In this way losses through the sampling hole are limited considerably and the atom losses can be calculated as e.g. have been done by van den Broek and de Galan [277]. When using the integral:

$$N(t) = \int_0^t S(T') R(t - t') \, \mathrm{d}t' \tag{250}$$

here $S(T)$ is a function of the atom supply, $R(t)$ is a function of the atom removal and t' is the transition variable. The absorbance at any time t is proportional to $N(t)$ and for the integrated absorption

$$A_{\text{int}} = \int_0^\infty C \cdot N(t) = CN_0 t \tag{251}$$

C is a proportionality factor and t_r is the mean residence time of the atoms. When $S(t)$ is practically zero, this equation almost gives $N(t)$ and

$$N(t) = N_{ti} R(t) \tag{252}$$

Here N_{ti} is the number of atoms being present at the time t_i, with $t_i < t$. Assuming that diffusion towards the ends of the tube is the only mechanism for atom losses, this equation reduces to:

$$N(t) = N_{ti} \cdot e^{-k(t - t_i)} \tag{253}$$

where $k = 8(Dl^{-2})$, D is the diffusion coefficient in cm^2/s and l is the length of the graphite tube in cm. For a number of elements good agreement between model and experimentally determined signal forms could be found.

4.4.2
Thermochemistry

The dissociation equilibria between the analyte elements and their compounds are very important, as they determine the fraction of analyte available as free atoms for AAS. They are accordingly important both for the analytical sensitivity achievable and especially so with respect to systematic errors.

The thermochemical behavior of the sample is of prime importance for the height of the absorbance signal as well as for its form. The acids present in the sample solution are normally removed during the drying and matrix decomposition steps. The residues eventually present during the absorption measurement lead to non-element specific absorption. This specifically occurs in the case of acids with high boiling points such as $HClO_4$, H_2SO_4 or H_3PO_4 and stems from ClO, SO, SO_2 or PO molecular bands. These species, however, may also be produced by the dissociation of the respective salts. Further problems may be caused by the oxides of the analytes, which result from the dissociation of the salts. This fraction of non-dissociated oxide in its turn is lost for the AAS determination and at the same time may give rise to non-element specific absorption. When the dissociation of the salts and the reduction of the analyte oxides is not completed before the absorption measurement several peaks may occur in the absorption signal. This can often be avoided by the platform technique, which facilitates a sharp rise in the heating of the furnace and lowers the risks of analyte deposition at the cooler parts of the furnace.

The thermochemistry of the elements is particularly important when a reliable destruction of the matrix is to be achieved and possibly removal of the matrix elements without risking analyte losses. Also the use of thermochemical reagents such as quarternary ammonium salts $(R_4N^+Cl^-)$ should be mentioned. They allow organic samples to mineralized at low temperatures (below 400 °C) and prevent losses of elements which are volatile or form partly volatile organic compounds. This may be helpful in the case of Cd as well as of Zn, which forms volatile organozinc compounds in a number of organic matrices. Furthermore, the removal of NaCl, for example, which is present in most biological samples, may be beneficial so as to prevent matrix interferences. However, this must be done at low temperatures so as to prevent analyte losses and can be realized by the addition of NH_4NO_3, according to:

$$NH_4NO_3 + NaCl \rightleftharpoons NH_4Cl + NaNO_3$$

As the excess of NH_4NO_3 dissociates at 350 °C, NH_4Cl sublimates and $NaNO_3$ decomposes below 400 °C, all NaCl is removed at a temperature below 400 °C. Without the addition of NH_4NO_3 this would only be possible at the volatilization temperature of NaCl (1400 °C), by when analyte losses would be inevitable. Furthermore, in the case of the graphite furnace elements such as Ti and V form thermally stable compounds such as carbides, which lead to negative errors, because in this way fractions of the analytes are bound and do not contribute to the AAS signal.

Here the use of pyrolytic graphite coated graphite tubes is helpful, as the diffusion of the analyte solution into the graphite and thus the risk of the carbide formation are decreased. Alternatively, flushing the furnace with nitrogen can be helpful. Indeed, in the case of Ti a nitride is then formed which in contrast to the carbide can be dissociated easier. Other thermochemical means to decrease interferences, as discussed earlier, are known as matrix modification. The addition of a number of substances, such as Pd-compounds or $Mg(NO_3)_2$, has been shown to be successful for the realization of a matrix-free vapor cloud formation (see e.g. Ref. [278]). The mechanisms involved also relate to surface effects in the furnace (see e.g. Ref. [279]) and are in themselves a specific field of research.

The development of the temperature program is the main task when establishing a working procedure for furnace AAS. The selection of the different temperature steps but also the use of all types of thermochemical effects are most important so as to minimize the matrix influences without causing analyte losses.

The use of radiotracers is very helpful for the understanding as well as for the optimization of the analyte volatilization in furnace AAS, and with this element losses and their causes at all levels of the atomization processes can be quantitatively followed. This has been studied in detail for a number of elements such as As, Pb, Sb and Sb in furnace atomization by Krivan et al. (see e.g. Ref. [280]). The results, however, may differ considerably from those when the furnace is used as an evaporation device only and the vapor produced is transported into a second system for signal generation, as has been studied extensively by Kantor et al. (see e.g. Ref. [281]). Here the transport efficiencies were calculated for the case of transport of the vapors released through the sampling hole, and similar considerations could be made when releasing the vapors end-on.

4.4.3
Figures of merit

Analytical sensitivity and power of detection

In electrothermal AAS these are both higher by orders of magnitude than in flame AAS. This is due to the fact that in the furnace a higher concentration of atomic vapor can be maintained as compared with flames. Furthermore, dilution of the analyte by the solvent is avoided, the solvent being evaporated before the atomization step, as is dilution due to large volumes of burning gases. For most elements the characteristic masses, being the absolute amounts for which an absorbance of 0.0044 or a 1% absorption is obtained are lower by orders of magnitude than in flame AAS [271].

Detection limits in flame AAS: 0.1–1 ng/mL (Mg, Cd, Li, Ag, Zn); 1–10 ng/mL (Ca, Cu, Mn, Cr, Fe, Ni, Co, Au, Ba, Tl); 10–100 ng/mL (Pb, Te, Bi, Al, Sb, Mo, Pt, V, Ti, Si, Se); >100 ng/mL (Sn, As).

Detection limits in furnace AAS (20 µL): 0.005–0.05 ng/mL (Zn, Cd, Mg, Ag); 0.05–0.5 ng/mL (Al, Cr, Mn, Co, Cu, Fe, Mo, Ba, Ca, Pb); 0.5–5 ng/mL (Bi, Au, Ni, Si, Te, Tl, Sb, As, Pt, Sn, V, Li, Se, Ti, Hg).

When sample aliquots of 20–50 µL are used, the detection limits in terms of concentration are often in the sub-ng/mL range. They are particularly low for elements such as Cd, Zn and Pb which have a high volatility and high excitation energies, but also for Mg, Cu, Ag, Na, etc.

Graphite furnace AAS remains of course a relative method of analysis. However, efforts have been made to establish a direct absolute method of analysis. Estimates of detection limits have already been made that are in good agreement with experimental results [282]. This necessitates calculating the exact reciprocal linear dispersion of the spectrometer and correcting for Doppler broadening in the source as well as for superposition of adjacent lines, and calculating the diffusion coefficients for the atomic vapor in the furnace as well as atom losses through ionization. Furthermore, the fundamental parameters must be sufficiently reliable and matrix interferences must be excluded. This last condition is readily fulfilled by the stabilized temperature platform furnace technique. As is shown by an extract of these results in Table 7, the agreement between measured and calculated characteristic masses is clearly remarkable.

Interferences

In furnace AAS interferences are much higher than in flame AAS. This applies both for physical as well as for chemical interferences.

Physical interferences   These may stem from differences in viscosity and surface tension, resulting in different degrees of wetting of the graphite tube surfaces with the sample solution, and also as a result of changes in the diffusion of the sample solution into the graphite. The first may be suppressed by the addition of surfactant substances to the sample solutions, such as Triton X. The second effect can be avoided by using pyrolytic graphite coated graphite tubes, where diffusion of the analyte into the graphite tubes is lower. This measure is also very effective at preventing carbide-forming elements from reacting with the graphite, which are thus lost and no atomic absorption signal is thus generated. Electron-probe microanalysis studies, as shown by many papers from Ortner and coworkers (see e.g. Ref. [283]), is very useful at improving the optimization of ETAAS methods in this respect.

The influence of concomitants on the volatilization can be minimized by the use of platform techniques and of an isothermal furnace, as already discussed. The first such technique goes back to the work of L'vov. Here a thin graphite platform, often made from pyrolytic graphite is slid through the furnace and kept in place through the provision of two grooves in the graphite tube along its axis. As heating occurs first at the wall the gaseous phase is hotter than the platform, promoting dissociation of molecular species, as the analyte is brought directly to high temperature. In the case of the isothermal furnace, going back in its origins to the work of Frech (see e.g. Ref. [272]), temperature profiles across the furnace do not occur. In both cases the formation of double peaks is largely avoided, as thermochemical reactions are speeded up or cooler parts of the furnace where other species volatilize are eliminated. Both measures are particularly helpful in the case of direct solids

Tab. 7. Calculated and experimental values for the characteristic masses (mass for 1% absorption) in graphite furnace atomic absorption spectrometry [282].

Element	Characteristic mass (pg)	
	calculated	*experimental*
Ag	1.02	1.2
Al	6.1	7.3
As	7.9	10.3
Au	6.5	7.3
Be	0.37	0.45
Bi	18.4	23.6
Cd	0.44	0.42
Co	4.11	4.8
Cr	1.57	2.1
Cs	2.37	5.2
Cu	2.24	3.2
Fe	3.18	3.6
Ga	9.5	10.1
Ge	10.5	15.3
Hg	68.8	69
In	8.6	9.3
K	0.59	0.7
Li	0.43	0.8
Mg	0.32	0.31
Mn	1.4	1.7
Na	0.42	0.5
Ni	5.82	7.5
P	2800	3300
Pb	8.5	10.9
Pd	15.0	15.4
Rb1.	2.2	2.4
Sb	12.6	15.0
Se	9.5	11.8
Si	12.4	15.2
Sn	13.5	20.0
Te	9.7	11.7
Tl	9.9	10.0
Tm	2.7	3.1
Yb	0.88	1.3
Zn	0.42	0.4

sampling for powders. For biological samples residuals from the organic matrix may remain after thermal matrix removal and could lead to volatilization difficulties such as double peak formation. Through both of these measures this is often effectively avoided.

In addition, undissociated molecules, which may be oxides arising from the dissociation of ternary salts (MgO, ZnO, etc.), but also radicals and molecular species arising from solvent residues, such as OH, PO, SO_2, etc., may cause non-element specific absorption. In addition, Rayleigh scattering of primary radiation by non-evaporated solid particles may occur, which also leads to radiation losses. Both phenomena necessitate the application of suitable techniques for background correction in most analyses of real samples by furnace AAS (see Section 4.6). Furthermore, emission by the furnace itself may give rise to continuum radiation, which can lead to systematic errors.

Chemical interferences Such interferences may be caused by losses of analyte during the ashing step. They occur particularly in the case of volatile elements such as As, Sb, Bi and Cd, or elements forming volatile compounds (e.g. halogenides). Here thermochemical decomposition reactions, which can occur in view of the anions that are present, are very important. Accordingly, in the course of the ashing step the removal of NaCl, which is present in most biological samples, is often most critical. Further element losses can occur as the result of the formation of thermally stable compounds such as carbides or oxides that cannot be dissociated any further and hamper the analyte element that is to be atomized and which then produces an AAS signal. Because of this, elements such as Ti and V are particularly difficult to determine by furnace AAS. As discussed, the use of pyrolytically coated graphite can lower this effect and some of these sources of systematic errors can be reduced in furnace AAS by the use of thermochemical reactions (see earlier).

Differences in the volatility of the elements and their compounds may also be used for speciation work. Accordingly, it is possible, for example, to determine different organolead compounds such as $Pb(CH_3)_4$ and $Pb(C_2H_5)_4$ directly by furnace AAS (see e.g. Refs. [284, 285]) or after on-line coupling with chromatographic separations. This approach can be used for other species as well but must be developed carefully for each special case.

As a rule calibration in furnace AAS is done by standard additions, as many of the interferences mentioned cause specific changes to the slope of the calibration curves. This makes sense as background changes can be readily measured through different techniques and in all cases with a sufficient time resolution to cope with the transient nature of the analytical signals and of the background absorption as well.

4.5
Special techniques

4.5.1
Hydride and cold-vapor techniques

For the determination of traces and ultratraces of Hg, As, Se, Te, As and Bi the formation of the volatile mercury vapor or of the volatile hydrides of the appropriate elements is often used, respectively. This allows a high sampling efficiency to be achieved and accordingly a high power of detection. The absorption measurement is often performed in a quartz cuvette. Hg, for instance, can be reduced to the metal and then transported with a carrier gas into the cuvette. For the absorption measurement no heating is required. The other elements are reduced to the volatile hydrides, which are then transported with a carrier-gas flow (argon or nitrogen) into a cuvette. In order to dissociate the hydrides thermally, the quartz cuvette must be heated to a temperature of 600–900 °C.

In both cases flow-cells facilitate continuous formation of the volatile compound and a subsequent separation of the reaction liquid from the gaseous compound (see e.g. Ref. [286]). Here the use of drying stages with for example $NaClO_4$ is very effective at lowering the vapor content in the determination of Hg with the cold-vapor technique, which in its turn guarantees a low non-specific absorption, e.g. through stray light by small droplets or adsorption by molecular bands. Flow-cell systems are very useful (see Section 3.3) for routine work and allow operation of an automated system with sample turnables. They have practically replaced earlier systems, where a volume of $NaBH_4$ solution was allowed to react with a given volume of acidified sample as described e.g. in Ref. [287].

Trapping of the analyte vapor released would thus be very useful in order to increase further the absolute power of detection. In the case of Hg, this can be achieved by passing the Hg vapor over a gold sponge and binding it as the amalgamate [288]. Subsequent to the collection step the Hg can be released completely by heating. For the hydrides, cold trapping can be applied. Here the hydrides can be condensed by freezing in a liquid nitrogen cooled trap and hence freed from the excess of hydrogen formed during the reaction [289]. The isolated hydrides are released by warming after a removal of hydrogen. By the trapping, enrichment factors of far above 100 can be realized. However, as described by Sturgeon et al. [290] and others, hot-trapping can also be applied. This is possible because the hydrides have decomposition temperatures which are lower than the boiling points of the respective elements. In the case of AsH_3 the decomposition temperature is below 600 °C, whereas the volatilization temperature for As is above 1200 °C. Accordingly, trapping can be performed in the graphite furnace itself. The furnace only has to be brought to a temperature of about 600–700 °C to decompose the volatile hydrides and the elements then condense in the furnace. Thus they can be pre-concentrated and volatilized at temperatures above 1200 °C during the absorption measurement step.

The cold-vapor technique for Hg allows detection limits of <1 ng to be obtained when using 50 mL of sample and they can be improved still further by trapping. With the hydride technique detection limits below the ng/mL level can be achieved for As, Se, Sb, Bi, Ge, Sn, etc. Accordingly, the levels required for analyses used to control the quality of drinking water can be reached.

Hydride techniques, however, can suffer from many interferences (see Section 3.3). In AAS these interferences can not only occur as a result of influences on the hydride formation reaction but also as a result of influences of concomitants on the thermal dissociation of the hydride. Interferences from other volatile hydride forming elements can also occur [291]. Recently it has been found that still more elements can form volatile hydrides, as demonstrated e.g. by Cd. Here the hydride is normally unstable above liquid nitrogen temperatures. However, through the use of didodecyldimethylammonium bromide (DDAB) the cationic vesicles were found to stabilize the volatile species of cadmium formed after reaction with $NaBH_4$ [292], because the detection limit using ICP-AES thus obtained, as compared with aqueous solutions, could be improved by a factor of 5. It was proposed that the CdH_2 is stabilized in the vesicle, whereas it immediately decomposes outside producing a cold vapor of cadmium, which can be measured.

4.5.2
Direct solids sampling

In both flame and furnace AAS direct solids sampling can be applied. In flames, sampling can be carried out in a boat or a cup, as introduced by Delves [176]. However, new approaches such as the combustion of organic samples and the introduction of the released vapors into a flame is possible, as shown by Berndt [293]. In the case of volatile elements such as Pb, detection limits obtained in this manner are in the μg/g range. Direct powder sampling with furnaces was introduced into atomic spectrometry in the 1940s by Preuss and in furnace AAS it can easily be performed with a powder sampling syringe, as described by Grobenski et al. [182].

The most feasible approach for direct powder analyses with flame and furnace AAS, however, is the work that has been done with slurries. Slurry atomization was introduced for flame AAS by Ebdon and Cave [116] in 1982. Although it has been shown to be useful for flame work, this is only for the determination of relatively volatile elements in samples of which the matrix can easily be thermally decomposed. This is, for example, the case in a number of biological samples, such as powdered plant or tissue materials.

Slurry nebulization can be used specifically to improve powder sampling in furnace AAS. This has been shown to be successful for all types of materials such as plant and animal tissue, powdered food materials and also, as was subsequently developed by Krivan's group, for the analysis of refractory powders. It was shown to be of use for the direct analysis of SiC powders [294]. In a recent study, Hornung and Krivan [295] reported on the analysis of high-purity tungsten trioxide and high-purity tungsten blue oxide and determined Ca, Co, Cr, Cu, Fe, K, Mg, Mn, Na, Ni and Zn directly in powders, of which 0.28–13.5 mg of tungsten trioxide and 0.40–70 mg of tungsten blue oxide were introduced with a micro spatula onto a graphite platform in a transversally heated graphite furnace. The high background absorbance resulting from the oxides could be considerably decreased by reducing the powders during the pyrolysis stage with the aid of a hydrogen flow. Calibration in all cases could be done by pipetting 10 μL amounts onto the platform after the removal of the matrix residues and peak area measurements.

Direct solids sampling, however, be it directly with powders or with slurries, should be used very carefully. Indeed, the amounts of sample used are often of the order of a few mg and accordingly sampling errors may occur as a result of sample inhomogeneity. Therefore, the use of fairly large furnaces in which larger amounts of powder can be handled has been proposed. In direct powder analysis calibration may be a problem because the particle size has an impact on the analyte release from the powder particles.

With compact metallic samples direct analysis by AAS can also be applied. Here the use of cathodic sputtering combined with AAS of the atom vapor cloud produced has been described previously by Gough (e.g. [296]). They used jet-enhanced sputtering so as to produce high ablation rates (several mg/min) and thus the correspondingly high analyte number densities in the glow discharge plasma. This approach has been made commercially available as an accessory to AAS equipment [297]. The feasibility for direct analyses of steels has been shown especially for samples that are difficult to bring into solution and in which refractory oxide forming elements such as Zr have to be determined [298].

For the case of both electrically conducting and electrically non-conducting samples, laser ablation combined with AAS may be useful. In this case AAS measurements can be performed directly at the laser plume. Measurement of the non-element specific absorption will be very important, because of the presence of particles, molecules and radicals as well as due to the emission of continuum radiation. In addition, the absorption measurements should be made in the apprppriate zones. When applying laser ablation for direct solids sampling, the atomic vapor produced can also be led into a flame for AAS work, as has previously been described by Kantor et al. [299] in their early work.

4.5.3
Indirect determinations

For the determination of elements of which the most sensitive lines are in the VUV spectral region, conventional AAS cannot be applied. Indirect methods can be used here, as they can for the determination of chemical compounds. Sulfates, for instance, can be determined by their precipitation as $BaSO_4$, and then the excess of barium used can be determined with AAS. A similar approach can be used for the determination of the halogenides after their precipitation with silver.

Other indirect determinations can be performed by making use of chemical interferences. Based on the calcium phosphate interference in flame AAS for example, phosphate can be determined by measuring the decrease in the absorption signal obtained for calcium. However, for real samples such indirect techniques should be used very carefully in view of possible systematic errors.

4.5.4
Flow injection analysis

Flow injection analysis as introduced by the work of Ruzicka [129] has been used extensively in combination with atomic absorption spectrometry, the aim being to:

• enable automated dilution of the samples to be made so as to extend the linear dynamic range;

- perform on-line matrix removal so as to avoid interferences;
- enable on-line preconcentration to be performed where the power of detection of the method is to be improved.

In a flow injection system the sample flow and a reagent flow are continuously brought together so as to allow a chemical reaction to take place. This reaction produces a gaseous compound, which has to be separated off as in hydride generation, or forms a complex, which can be adsorbed onto a solid phase to be isolated and preconcentrated. In the latter case, elution with a suitable solvent is carried out and the analytes are led on-line into the AAS system.

In the case of flame atomic absorption a continuous recording of the analyte signal can be performed and transient signals are obtained. As a result of the flow pattern in the tubing, the analyte signals will normally display a steep rise and some degree of tailing, which depends on the analytes, their compounds formed and the nature of the column filler material as well as on the dimensions and the material from which the column holder is made. In the determination of elements that form complexes, for example with dithiocarbamates, such as many heavy metals [300], determination in highly concentrated salt solutions is straightforward through fixing the complexes on a suitable solid phase column, while releasing elements such as Na, K, Ca, or ammonium ions.

With graphite furnace atomic absorption only an off-line determination is possible or by a semi-automated coupling, which adapts the analysis cycle with respect to the frequency and time required for a determination to the flow-rates in the system.

4.5.5
Diode laser atomic absorption spectrometry

Through the availability of tunable diode laser sources the set-up used for atomic absorption (see Section 4.2.2) can be simplified considerably. In this way, several additional advantages can be realized. As through wavelength tuning, measurements in the wings of the absorption profile can also be made, and possible improvements to the linear dynamic range in a number of cases could be expected.

Through the use of more than one diode and tuning them to several analyte lines, multielement determinations and the use of an internal standard become simple. In AAS work the latter also enables instrumental drift to be overcome and/or the short-term precision to be improved as well. For the detection, all that is required is pulsing of the primary source and the use of lock-in amplification or a Fourier analyzer.

Furthermore, tuning of the primary source enables off-peak measurements and background correction to be made directly.

As discussed earlier, present limitations of diode laser atomic absorption lie in the fact that the lower wavelengths are not yet accessible and that the whole wavelength range cannot be covered continuously as there are wavelength gaps between the wavelength ranges of the diodes presently available.

An interesting approach to the use of diode laser atomic absorption lies in the use of discharges under reduced pressure as atom reservoirs. In the case of helium low pressure microwave discharges for instance, metastable levels of elements such as halogens, hydrogen, oxygen or sulfur are excited, which can be probed with commercially available diode lasers as a result of the availability of a higher level in the analyte term scheme. Such a principle can be applied successfully for element-specific detection in gas chromatography [301–303] and also for the direct analysis of polymers with respect to halogens [304].

4.6
Background correction procedures

In AAS, systematic errors are often due to non-element specific absorption, which necessitates the use of background correction procedures. The absence of non-element specific absorption can only be expected in the analysis of sample solutions with low matrix concentrations by flame AAS, but in furnace AAS work background correction is required particularly for matrix loaded solutions. The determination of the non-element specific absorption can be performed in several ways.

4.6.1
Correction for background absorption with the deuterium lamp technique

This approach was introduced by Koirthyohann and Pickett in 1965 [253] and is now provided with almost every AAS system. The total absorption resulting from the presence of the element and the background absorption are measured with the hollow cathode lamp radiation, but in addition a continuum source is used with which only the background absorption can be measured. This is possible as the monochromators used in AAS work have a large spectral bandwidth as compared with the physical width of the resonance line emitted by the hollow cathode source and also with respect to the width of the absorption line. Accordingly, all radiation emitted by the hollow cathode lies within the absorption profile of the line and at the same time the elemental absorption does not significantly contribute to the absorption in the case of the continuum source (Fig. 86), which extends homogeneously over the whole spectral bandwith. An electrical discharge in deuterium can be used as the continuum source, as it emits fairly smooth continuum radiation in the UV spectral range up to 400 nm.

Alternative measurements of the absorption of the hollow cathode lamp radiation and of the radiation of the continuum source can also be taken in a rapidly alternating mode, and accordingly a quasi-simultaneous measurement of line and background absorption, as required with transient signals, becomes possible. This can be realized by a rotating mirror and is sufficiently rapid to enable a high time resolution, as shown by Berndt et al. [305].

This type of background correction assumes that the background absorption has

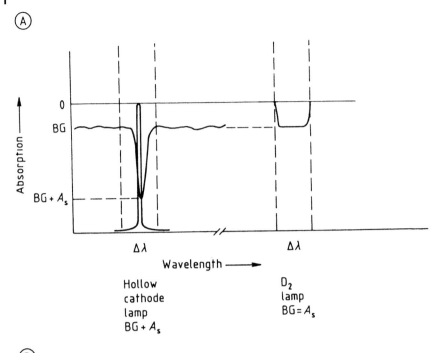

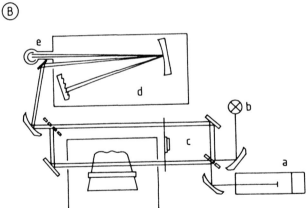

Fig. 86. Principle of background correction with the D_2-lamp technique (A), $\Delta\lambda$: spectral bandwidth of monochromator, BG: non-element-specific "background" absorption, A_S: element-specific absorption signal. (B), Optical diagram of an atomic absorption spectrometer with D_2 lamp for background correction (b) (Model 2380) (Courtesy of PerkinElmer Bodenseewerk GmbH).

a continuum nature within the spectral bandwidth of the monochromator. This is not the case when the background absorption arises from molecular bands, which have a rotation–vibration hyperfine structure. These can arise from radicals produced by a dissociation of the solvent (OH, SO_2, SO_3, N_2^+, CN, etc.) but also from molecular oxides (MO). As such contributions occur particularly with the complex matrices of real samples, where, for example, residues of the chemicals used for dissolution or metal oxides are present, a background correction with a D_2 lamp often leads to systematic errors, which cannot be corrected for by calibration with standard additions.

Furthermore, the radiant density of the D_2 lamp in a large part of the spectrum is fairly low. Hence, the procedure limits the number of analytical lines which can be used and the number of elements that can be determined. As the spectral radiance of the D_2 lamp is generally low as compared with that of a hollow cathode lamp, the latter must be operated at a low radiant output (low current), which means that detector noise limitations and poor detection limits are soon encountered. Finally, as work is carried out with two primary radiation sources, which are difficult to align as they have to pass through the same zone of the atom reservoir, this may lead to further systematic errors.

4.6.2
Background correction with the aid of the Zeeman effect

Zeeman AAS makes use of the splitting of the atomic spectral lines into several components under the influence of a magnetic field. When a magnetic field B (up to 10 kG) is applied, the shift in wavenumber (ΔT_m) of the so-called σ-components with respect to the original wavelength, where the π-components may remain, is given by:

$$\Delta T_m = M \cdot g \cdot L \cdot B \tag{254}$$

M is the magnetic quantum number, L is the Lorentz coefficient $[e/(4\pi mc^2)$ in $cm^{-1} \cdot G^{-1}]$ and g the Landé factor, being a function of the total quantum number J, the orbital momentum quantum number L and the spin quantum number S:

$$g = 1 + [J(J+1) + S(S+1) - L(L+1)]/[2J(J+1)] \tag{255}$$

The intensities of the σ-components (for which $\Delta M = \pm 1$) and the π-components (for which $\Delta M = 0$) are a function of ΔJ (0 or 1) and ΔM (0, ± 1) for the transitions. In the normal Zeeman effect, which occurs in the case of singlet transitions (e.g. with alkali earth metals and metals of the IIb Group such as Cd and Zn) $g = 1$ and there are single components, whereas in all other cases there are groups of components (anomalous Zeeman effect). In the case of a transverse magnetic field (perpendicular to the observation direction), a spectral line splits into three lines. These are one π-component at the original wavelength; for this component $\Delta M = 0$ which is polarized parallel to the field. In addition, there are two σ-components (σ^+ and σ^-) for which $\Delta M = \pm 1$. They are polarized in directions perpendicular to the field. With a longitudinal field (parallel to the direction of observation) there is no π-component ($\Delta M = 0$ is forbidden) and the σ-components ($\Delta M = \pm 1$) are circularly polarized.

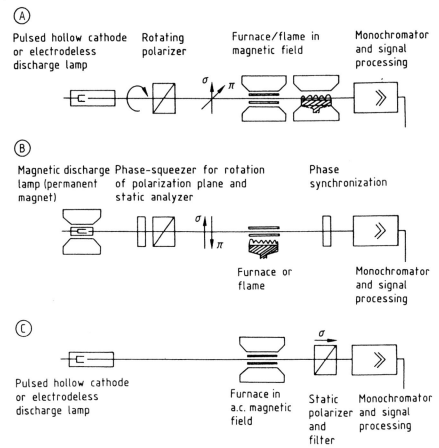

Fig. 87. Approaches for Zeeman atomic absorption. (A): Rotating polarizer and permanent magnet applied to the atomizer; (B): permanent magnet around the primary source; (C): longitudinal field of ac magnet applied to the atomizer (Reprinted with permission from Ref. [307].)

In order to use the Zeeman effect for background correction [306] several approaches can be applied (Fig. 87) [307]. A magnetic field around the primary source or around the atom reservoir can be provided, by which either the atomic emission lines or the absorption lines are subjected to Zeeman splitting. Use can be made of a constant transverse field and the absorption for the π- and the σ-components measured alternately with the aid of a polarizer and a rotating analyzer. However, an ac longitudinal field can be used and with the aid of a static polarizer only the σ-components are measured, once at zero and once at maximum field strength.

When I_1 and I_2 in both instances (be it the σ- or the π-component in the case of a transverse field or the total radiation and the σ-component in the case of the longitudinal field) are the intensities of the total signal and the background signal, respectively, for each of them:

$$I = I_0 \exp(-k_b) \cdot \exp(-k_a) \tag{256}$$

where I_0 in each of the cases is the intensity of the incident radiation and k_a and k_b are the absorption coefficients for the background and the line. The net absorbance, which is proportional to the concentration can be calculated as:

$$\ln(I_2/I_1) = (k_{a_1} - k_{a_2}) + (k_{b_1} - k_{b_2}) + \ln(I_{2_0}/I_{1_0}) \tag{257}$$

Accordingly, by subtracting in both cases the two absorption signals from each other, the background absorbance measured under the line can be eliminated. This assumes that both signals have constant intensities through the whole analytical system and that both have the same absorption coefficients for the background. In order to have a linear calibration curve, high sensitivity and an accurate background correction, $(k_{a_1} - k_{a_2})$ must be large and relate linearly to the concentration, the absorption coefficients for the background must be equal $(k_{b_1} = k_{b_2})$ and the incident beams must be constant through the whole system $(I_{2_0} = I_{1_0})$.

Different set-ups for have been used for Zeeman atomic absorption spectrometry. Indeed, a permanent magnet or an electromagnet with a dc or an ac field can be used around the source or around the atom reservoir. The set-up that has a permanent magnet may be the cheapest. However, the field is constant and must be transverse There must also be an alternating polarization system (e.g. a rotating analyzer), where the principle of which means it has a low transmittance and the ratio of I_{1_0} and I_{2_0} is difficult to keep constant. The set-up with an electromagnet has the advantage that the magnetic field can be changed by changing the current. Accordingly, the splitting can be optimized with respect to the element being determined and to the background structure. The magnetic field can be applied at the atomizer, which is possible both with a permanent magnet as well as with an electromagnet. In principle the magnetic field can also be placed around the primary source, which is possibly best in the case of a permanent magnet. Then both a flame and a furnace can be used as the atomizer and the according exchange is easier. Moreover, a larger furnace can then be used, which is very useful for direct solids sampling. Several types of the set-ups discussed, which have been realized in commercial Zeeman AAS equipment are discussed in Ref. [307].

Analytical advantages

Zeeman AAS has several analytical advantages. First, the accuracy of the background correction in the case of a structured background is better than with the D_2 lamp technique. However, when the background structure arises from molecular bands, it should be borne in mind that molecular bands may also display the Zeeman effect. Systematic errors resulting from this fact may be larger when one line component is measured in a strongly changing field.

The detection limits in Zeeman AAS could be expected to be lower than in the case of the background correction with a D_2 lamp. Indeed, here the system uses only one source. Accordingly, it can be operated at high intensity, through which detector noise limitations are avoided. This advantage will certainly be most pronounced when one component is measured in an alternating field.

Another consequence of the use of one primary source will be the better stability of the system. The analytical sensitivity in Zeeman AAS, however, will be inferior to that of conventional AAS. This disadvantage is lowest for a field which can be varied from case to case.

In Zeeman atomic absorption spectrometry the linear dynamic range will also be lower than in the case of background correction with the D_2 lamp. This is related to the fact that a difference between two absorbances is taken, which in the case of a magnetic field of non-optimal strength may actually lead to a bending away of the calibration curve. These effects again are less pronounced when measuring one component in an ac field. Nowadays Zeeman AAS is widely used, for instance, for trace determinations in biological samples.

4.6.3
Smith–Hieftje technique

This technique for background correction [308] makes use of the fact that resonant atomic spectral lines emitted by a hollow cathode lamp may display self-reversal when the lamp is operated at a high discharge current. During the first part of the measurement cycle the hollow cathode lamp is operated at a low current. The self-reversal then does not occur and the resonant radiation is absorbed both by the analyte atoms and by background-producing species. In a second part of the measurement cycle, the current is briefly pulsed to above 500 mA, through which a very high self-reversal occurs. Then the intensity at the original analytical wavelength becomes low and the intensities in the side wings remain high, which causes most of the background absorption to occur. By subtraction of both absorbances the net atomic absorption signal is obtained.

Similar to Zeeman AAS the Smith–Hieftje technique can be used for lines in the whole spectral range and again uses only one primary radiation source, thus both the alignment and the stability are optimum. Moreover, the technique is simple and hence much cheaper than Zeeman AAS. In addition, radiation losses as a result of the use of polarizers in some Zeeman atomic absorption systems or limitations to the volume of the atom reservoir do not occur here. The linearity of the calibration curves and the accuracy in the case of structured background absorption should also be better than with the Zeeman technique, as the Smith–Hieftje technique is not subjected to limitations due to Zeeman splitting of molecular bands. However, as the self-reversal is not complete the technique can only be used for fairly low background absorbances, the sensitivity is decreased as the self-absorption is at the most 40% and special provisions have to be taken for pulsing the lamps at high currents.

4.6.4
Coherent forward scattering

Intensity of scattered radiation

Scattering of radiation is a one-step process in which two photons are involved, one being absorbed by the atom and one being emitted. The intensity of scattered radiation is particularly high because when monochromatic radiation is used as the primary radiation the wavelength equals that of a resonance line of the scattering atoms. When the latter are brought into a magnetic field the scattered radiation becomes coherent in the direction of the primary beam and the scattering atomic vapor becomes optically active (magneto-optical effect). Depending on whether a transversal or a longitudinal magnetic field is used a Voigt or a Faraday effect is observed, respectively.

In a system for coherent forward scattering, the radiation of a primary source is led through the atom reservoir (a flame or a furnace), across which a magnetic field is applied. When the atom reservoir is placed between crossed polarizers scattered signals for the atomic species occur on a zero-background. When a line source such as a hollow cathode lamp or a laser is used, determinations of the respective elements can be performed. In the case of a continuous source, such as a xenon lamp, and a multichannel spectrometer simultaneous multielement determinations can also be performed. The method is known as coherent forward scattering atomic spectrometry [309, 310]. This approach has become particularly interesting since flexible multichannel diode array spectrometers have became available.

Intensities of the scattering signals

These can be calculated both for the case of the Voigt and the Faraday effect [310].

In the case of the Faraday effect (with the field parallel to the observation direction), there are two waves which are polarized parallel to the magnetic field. When n_+ and n_- are the refractive indices, $n_m = (n^+ + n^-)/2$ and $\Delta n = (n^+ - n^-)/2$ the intensity $I_F(k)$ at the wavenumber k is given by:

$$I_F(k) = I_0(k) \times F[\sin(k \cdot l \cdot \Delta n)] \times \exp(k \cdot l \cdot n_m) \tag{258}$$

$I_0(k)$ is the intensity of a line of the primary radiation with wavenumber k and l is the length of the atom reservoir. The sinusoidal term relates to the rotation of the polarization plane and the exponential term to the atomic absorption. As both n_m and Δn are a function of the density of the scattering atoms, $I_F(k)$ will be proportional to the square of the density of scattering atoms (N), according to:

$$I_F(k) = I_0(k) \cdot N^2 \cdot l^2 \tag{259}$$

For the Voigt effect, the scattered radiation has two components. One is polarized parallel to the magnetic field (normal component) and the other perpendicular to the field (abnormal component). When n_0 and n_e are the respective refractory indices the intensity of the scattered radiation $I_V(k)$ can be calculated as for the Faraday effect.

Multielement method

Coherent forward scattering (CFS) atomic spectrometry is a multielement method. The instrumentation required is simple and consists of the same components as a Zeeman AAS system. As the spectra contain only some resonance lines, a spectrometer with just a low spectral resolution is required. The detection limits depend considerably on the primary source and on the atom reservoir used. When using a xenon lamp as the primary source, multielement determinations can be performed but the power of detection will be low as the spectral radiances are low as compared with those of a hollow cathode lamp. By using high-intensity laser sources the intensities of the signals and accordingly the power of detection can be considerably improved. Indeed, both $I_F(k)$ and $I_V(k)$ are proportional to $I_0(k)$. When furnaces are used as the atomizers typical detection limits in the case of a xenon arc are: Cd 4, Pb 0.9, Tl 1.5, Fe 2.5 and Zn 50 ng [309]. They are considerably higher than in furnace AAS.

The sensitivity of CFS atomic spectrometry is high as the signals are proportional to the square of the atom number densities. The dynamic range is similar to that in atomic emission spectrometry and is of the order of 3 decades. It should be considerably better than in atomic absorption spectrometry, where the linear dynamic range as a consequence of Lambert–Beers' law and of limitations through the line profiles is restricted to 1–2 decades. However, this again depends on the primary source and especially in the case of continuous sources limitations can occur. Information on matrix effects for real samples is still scarce. As scattering by molecules and undissociated species is expected to be low, background contributions may be low as compared with AAS.

4.7
Fields of application

Solutions

The different methods of AAS and also the related CFS are very powerful for the analysis of solutions. The instruments are simple and easy to operate. Accordingly, they are now used in almost all analytical laboratories. In particular, when one or only a few elements have to be determined in a large number of samples, as is e.g. the case in clinical analysis or in food analysis, AAS methods are of great use as compared with other methods of elemental analysis.

To a limited extent atomic absorption spectrometry can also be used for multi-element determinations. Several manufacturers introduced systems with multi-lamp turrets, where different lamps can be held under preheated conditions. Here rapid switching from one lamp to another enables sequential multielement determinations to be made by flame atomic absorption, for a maximum of around 5 elements. Simultaneous determinations are possible with multielement lamps, however, the number of elements that can be brought together and used as a hollow cathode lamp with a sufficiently stable radiation output and lifetime is rather limited. The use of continuous sources facilitates flexible multielement determinations for many elements in principle. It is necessary to use high-resolution spectrometers (e.g. Echelle spectrometers) with multichannel detection. CCDs offer good chances of realizing high and flexible multielement capacity without much loss of power of detection as compared with single-element AAS. This applies to flame AAS, however much less so to furnace AAS, as a result of the individual thermochemical behaviors of elements and compounds. In furnace AAS some of these restrictions can be removed by the use of the stabilized temperature furnace platform concept, as described by Slavin et al. [276].

Solids

When solids have to be analyzed, the samples must be brought into solution and sample decomposition methods have to be used. They range from simple dissolution in aqueous solutions to a treatment with strong and oxidizing acids or eventually fluxes at high temperature and/or pressure.

From a series of acids, the use of HCl and HNO_3 or mixtures of them is in most cases free of problems from the point of view of the AAS determination. Simple acid concentration matching of standard and sample solutions is required, when applying calibration with synthetic standard solutions or solutions of reference samples. Often high-boiling acids such as H_2SO_4, H_3PO_4 and $HClO_4$ are used, with the aim of increasing the temperatures in the reaction mixtures and accordingly also the reaction and dissolution velocities. Complications may arise from the oxide residues after solution drying and the dissociation of the salts. These may lead to the formation of less-dissociated compounds with the analytes as well as to molecular bands causing structured background absorption.

The dissolution speed and efficiencies may also be increased considerably by high-pressure digestion, as is done in PTFE vessels or quartz vessels with a steel enclosure and resistive heating. A digestion of organic materials that is very efficient and leaves no residue is often possible as well as a complete dissolution of refractories such as Al_2O_3, SiC. In the case of PTFE vessels, blank contributions and even memory effects often occur. Also the digestion times may be quite long (up to 20 h or more including the cooling time) and the amounts of materials, which can be decomposed at once, are often very low. The latter may hamper the power of detection and also introduce restrictions from the point of view of sample inhomogeneities.

As an alternative to resistive heating, microwave assisted digestion is finding more and more acceptance [311]. The mechanisms of digestion are under discussion, but it may be that in a number of cases dipole molecules dissociate earlier as a result of the coupling of the dipoles with the microwave field. In any case the heat is coupled directly to the reaction mixture and is not led via conduction to the medium, which increases the transfer efficiency and accordingly shortens the sample digestion times. Therefore, all reactions have to be done in plastic vessels. Through the development of special vessels, now higher pressures (up to 80 bar) can also be tolerated under completely safe conditions. The latter is guaranteed by temperature and pressure monitoring as well as by the presence of overpressure switches and vapor release systems. Microwave-assisted heating for biological samples, such as plants and tissues, as well as for environmental samples, such as sludges and sediments and even soils, now can be performed without leaving any organic or inorganic residues, and thus can be used successfully prior to flame as well as to furnace AAS.

Both in the case of digestion under resistance heating or of microwave-assisted digestion it is possible to perform the digestion on-line with AAS analyses. The samples must be brought into a powder form and stable suspensions must be produced which are pumped through a digestion line after the addition of the digestion reagents. The approach has already been shown to be viable for pulverized biological materials [130].

Refractory samples are often only present as granulates and are hard to grind. Indeed, here the abrasion of the mills (be it even a hard material such as WC) is high and this leads to contamination and consequently to interferences in the AAS determination. In this case melting with fluxes is the only method of digestion. For refractory powders, slags, minerals and rocks melting with alkali ($NaCO_3$, K_2CO_3, $Na_2B_4O_7$, etc.) or with acidic salts ($K_2S_2O_8$, $KHSO_4$, etc.) is very useful. They have to be used in an excess of 5–10 times the amount of sample (ca. 100 mg), which leads to high amounts of salts in the analysis solutions. With graphite furnace AAS this may lead to serious volatilization interferences and with flame AAS to nebulizer blocking. Furthermore, the melting procedure must be performed in high-temperature resistive vessels such as platinum, glassy carbon, iron, nickel or china crucibles. These materials may lead to contamination, which can also be entrained from the solid reagents, these being much more difficult to purify than acids. The latter can indeed be very efficiently purified by isothermic distillation.

In some cases such as in the analysis of ZrO_2, use is made of ammonium salts. This has the advantage that the melting procedure can be performed in quartz crucibles limiting the risk of contamination, that the salts are relatively pure as they can be prepared from pure chemicals (NH_3 gas and $ClHSO_3$, e.g. see Ref. [312]), and that the excess can be volatilized at low temperatures which makes the risk of analyte losses through volatilization low. For example, with NH_4HSO_4, the melting procedure can be performed at 400 °C and the excess of flux volatilized at below 500 °C.

A further possibility to reduce the salt contents is their separation from the analysis solutions by techniques such as ion exchange or adsorption of the analytes, after initial complexation onto columns filled with suitable solid phases.

In all sample dissolution and sample pretreatment work for AAS, attention must be paid to all problems that may arise in trace elemental analytical chemistry. This includes precautions for avoiding contamination from the reagents, the vessels used and from the laboratory atmosphere. Also, all factors that may cause analyte losses as a result of adsorption or volatilization must be studied in detail, as discussed by Tschöpel and Tölg (see e.g. Ref. [313]).

Sample types

Clinical chemistry, the metals industry, the analysis of high-purity chemicals and pure substances, environmental analysis and life sciences are the most important fields of application for AAS methods.

In clinical analysis, flame AAS is very useful for serum analysis. Ca and Mg can be determined directly in serum samples after a 1:50 dilution, even with micro-aliquots of 20–50 μL [314]. In the case of Ca, La^{3+} or Sr^{2+} are added so as to avoid phosphate interferences. Na and K are usually determined in the flame emission mode, which can be realized with almost any flame AAS instrument. The burner head is often turned to shorten the optical path so as to avoid self-reversal. For the direct determination of Fe, Zn and Cu, flame AAS can also be used but with a lower sample dilution. Determination of trace elements such as Al, Cr, Co, Mo and V with flame AAS often requires a pre-concentration stage, but in serum and other body fluids as well as in various other biological matrices some of these elements can be determined directly with furnace AAS. This also applies to toxic elements such as Ni, Cd and Pb, which often must be determined when screening for work place exposure. When aiming towards the direct determination of the latter elements in blood, urine or serum, matrix modification has found wide acceptance in working practices that are now legally accepted for work place surveillance, etc. This applies e.g. for the determination of Pb in whole blood [315] as well as for the determination of Ni in urine (see e.g. Ref. [316]).

For metal samples flame and furnace AAS are both important methods of analysis. They find use in the characterization of raw materials and for product analysis. In combination with matrix removal in particular, they are further indispensible for the characterization of laboratory standards. These are used to calibrate direct methods such as x-ray spectrometry and spark emission spectrometry for production control. However, it should be mentioned that specific elements of interest such as B, the rare earths, Hf and Zr, can be determined much better by plasma emission spectrometry.

When determining trace elements in steel, removal of the Fe by extraction with diethylether from a strong aqueous acid solution is very appropriate. A preconcentration of trace impurities as dithiocarbamates, and eventual adsorption onto activated carbon has been proposed for trace determinations in aluminum subsequent to sample dissolution (see e.g. Ref. [317]). However, multitrace determination methods such as plasma atomic spectrometry are a better alternative to AAS methods in this instance.

Trace determinations in different metal bases (e.g. lead, gallium and indium) can make use of pre-concentration based on a partial dissolution of the matrix (see Ref. [318]). Refractory metals such as molybdenum, niobium, tungsten and tantalum powders are definitely difficult to dissolve or to keep in solution and require trace matrix separations to achieve a reliable volatilization of the analytes, e.g. in graphite furnace AAS. Here slurry sampling electrothermal AAS, as has been extensively investigated for these matrices by Krivan et al. (see e.g. Refs. [194, 294, 295]), is a worthwhile alternative, especially as contamination is prevented and calibration by addition of aqueous aliquots of the analytes is often possible.

AAS methods have also been widely proposed for the determination of the noble metals. In the case of flame AAS use of an oxygen–acetylene flame is often proposed. Graphite furnace AAS trace determinations (see e.g. Ref. [319]) can be made of Pt in environmental and biological samples as well as determinations in noble metals themselves.

Further industrial products that can be analyzed include glasses, ceramics, superconductors, cement, chemicals, pharmaceuticals, catalysts, semiconductors, paints and oils.

For determinations in glass, dissolution of the matrix by treatment with HF and subsequent analysis of the trace concentrate is certainly a feasable approach. In high-quality glasses used e.g. for the production of fibers, even very low concentrations of trace elements may be relevant, so hydrofluoric acid of very high purity has to be used for the sample decomposition. This can be obtained by sub-boiling procedures in plastic apparatus.

With ceramics, slurry sampling in the case of powders is of relevance. Here AAS methods are fairly useful even though the traces have individual thermochemical behaviors, and accordingly many of the analytical capabilities have to be compromised in the multielement determination methods when using the graphite furnace as an analyte vaporizer only, e.g. prior to ICP atomic spectrometry. The use of thermochemical reagents here can be particularly helpful, both for separating the evaporation of the matrix and analyte, as well as for influencing the degree of evaporation for the relevant trace element(s).

The determination of the main components in cement, such as Ca, with high precision is of importance, as is the determination of heavy metals such as thallium, however, because of environmental impacts. This can be performed successfully with AAS, provided the samples can be reliably brought into solution, as is now possible with microwave assisted dissolution.

Trace and ultratrace determinations are now very important for chemicals. For solid chemicals dissolution again may put limitations on the power of detection achievable due to contamination during the dissolution procedure. Graphite furnace atomic absorption together with plasma mass spectrometry are now of great importance for the analyzing acids used e.g. in the treatment of surfaces in microelectronic components.

For superconductor materials and precursors high-precision analysis is very important. To this aim flame atomic absorption can be applied, provided the precision is optimized, as is possible, for example, by using internal standardization and multichannel spectrometers. It should, however, be remembered that to control the stoichiometric composition, in addition the determination of non-metals such as N and O, which cannot be performed by atomic absorption spectrometry, is compulsory.

In pharmaceuticals it is mainly organic compounds that have to be determined. However, more and more elements such as Se and As now have to be determined as trace elements, whereas often substances such as MgO are relevant as major compounds, but then high precision is required.

Catalysts for the chemical industry have to be characterized with respect to their trace impurities and major components. Not only is their composition when they are used initially in chemical reactors important, but also their alteration in the course of time. As carbide forming elements such as V and Ti are often used, atomic absorption spectrometry could be problematic. This also applies to catalysts for exhaust gas detoxification in cars. Noble metals such as Pt, Pd and Rh are fixed on alumina supports often also containing cerium compounds. Both for the determination of the stoichiometry but also for the monitoring of the noble metal contents in used catalysts, AAS suffers from problems because of the need for sample dissolution as well as for the requirement to determine refractory oxide forming elements.

Semiconductors in particular have to be characterized with respect to their dopants. In the case of substances such as Si and GaAs, the determination of the elements B and P, for example, is very important for which AAS is only of limited use. For the metals at trace level the situation is different, with the problem that multi-element characterization is often required, which can be better served by plasma mass spectrometry.

Paint analysis for elements such as Fe in the case of TiO_2 based materials is often important. Also elements such as Cr and Hg may have to be determined.

For oil analysis both flame AAS as well as graphite furnace AAS are important. In the first case nebulizers tolerating suspended particles can be used and in the case of graphite furnace AAS direct solids sampling may be useful so as to avoid the need for sample dissolution as well as to allow coarse metal particles present in used engine oils to be dealt with.

Dissolution of environmentally relevant elements in slags and ores by fusion and subsequent determination by AAS can be applied.

In the analysis of high-purity substances, general matrix removal is often very important so as to pre-concentrate the elements to be determined. To this aim all separation techniques such as ion exchange, liquid–liquid extraction of a metal complex with organic solvents, fractionated crystallization, precipitation and co-precipitation as well as electrochemical methods may be used (for a systematic treatment, see Ref. [300]). These principles can also be applied in on-line systems, as is now possible with solid phase extraction. Here matrix elements or the analytes can be adsorbed as complexes onto the column and eluted for direct determination by AAS.

AAS is often used for analytical problems in the life sciences and environmental analyses. For the analysis of waste water, flame and furnace AAS are complementary to plasma emission spectrometry. The last facilitates the determination of a large number of elements per sample but is more expensive than flame AAS in terms of instrumentation and less sensitive than furnace AAS. Furnace AAS is very useful for the analysis of drinking water, but because of its monoelemental capability it is likely to be replaced more and more by plasma mass spectrometry, especially in large laboratories.

For the analysis of airborne dust a sample decomposition procedure, which takes into account that much silica material is present, together with ETAAS may be useful, but it again suffers from the monoelemental capability. Airborne dust can also be analyzed directly by slurry sampling and can even be directly sampled in the graphite tube. Integrated sampling analysis procedures especially for easily volatile elements such as Pb and Cd are feasible.

Many of the AAS procedures be it for water analysis, oil analysis, analysis of industrial products, have been described as DIN norms, now of great value to the analytical laboratory.

AAS is a useful tool for speciation work as well. This applies for combinations of species-specific extraction procedures and subsequent determination of the respective species by AAS. Just such a method was worked out for the case of Cr^{III} and Cr^{VI} compounds by Subramanian et al. [320]. Use is made of the fact that the Cr^{III} is slower than the Cr^{VI} to form complexes, as a result of complexation with 6 molecules of H_2O.

However, for speciation on-line coupling of gas or liquid chromatography with AAS is becoming more and more useful.

In furnace AAS, coupling with gas chromatography has been shown to be very useful for the determination of organolead [284] and organotin compounds [321]. By combining the cold-vapor technique with HPLC, a very sensitive method for the determination of Hg species at the sub-ng level becomes possible [322]. When applying cold-vapor AAS to the detection of mercury subsequent to the separation of the species by HPLC, which also enables thermally labile compounds to be separated, the organomercury compounds have to be destroyed to allow for the AAS determination. They can be destroyed by wet chemical oxidation with $H_2SO_4 - Cr_2O_7^{2-}$ or by photochemical oxidation. It is then possible to perform speciation of mercury in gas condensates easily, where the species can be separated by reversed phase HPLC [323].

The preferential reduction of the trivalent species in hydride generation can be used for the speciation of As, while the pentavalent species is first reduced by hydrogen after converting it into the trivalent species by a reduction with cystein. A similar way can be followed for SeIV and SeV! This can be done both in the case of conventional as well as of electrochemical hydride generation [324].

In addition, for speciation coupling of flow injection analysis and column chromatography with flame AAS and also a direct coupling of HPLC with flame AAS, as is possible with high-pressure nebulization, are most powerful. Here the Cr line in the visible region can be used, which makes the application of diode laser atomic absorption spectrometry possible [325]. This has been shown recently by the example of the determination of methylcyclopentadienyl manganese tricarbonyl.

4.8
Outlook

AAS with flames and furnaces is now a mature analytical approach for elemental determinations. However, its development has not yet come to an end. This applies to primary sources, where tunable diode lasers open new possibilities and even eliminates the requirement of using expensive spectrometers. It also applies to atom reservoirs, where new approaches such as further improved isothermal atomizers for ETAAS or the furnace in flame technique (see e.g. Ref. [326]) have now been introduced, but also to spectrometers where CCD based equipment eventually with smaller dimensions will bring innovation. Furthermore, it is clear that on-line coupling both for trace element–matrix separations and speciation will enable many analytical challenges to be more effectively tackeld.

Manufacturers

The most important manufacturers of atomic absorption spectrometry equipment worldwide now are:

Baird-Atomic, Bedford, MA, USA
Analytik Jena GmbH, Jena, Germany
Erdmann & Grün, Wetzlar, Germany
Hitachi, Tokyo, Japan
G.B.C., Australia
PerkinElmer, Norwalk, CT, USA
Unicam, Cambridge, UK
Varian Techtron, Mulgrave, Victoria, Australia
Shimadzu, Tokyo, Japan.

5
Atomic Emission Spectrometry

5.1
Principles

The history of atomic emission spectrometry (AES) goes back to Bunsen and Kirchhoff, who reported in 1860 on spectroscopic investigations of the alkali and alkali earth elements with the aid of their spectroscope [1]. The elements cesium and rubidium and later on thorium and indium were also discovered on the basis of their atomic emission spectra. From these early beginnings qualitative and quantitative aspects of atomic spectrometry were considered. The occurrence of atomic spectral lines was understood as uniequivocal proof of the presence of these elements in a mixture. Bunsen and Kirchhoff in addition, however, also estimated the amounts of sodium that had to be brought into the flame to give a detectable line emission and therewith gave the basis for quantitative analyses and trace determinations with atomic spectrometry.

By the beginning of the 20th century Hartley, Janssen and De Gramont had developed spectrochemical analysis further and flames, arcs and sparks became powerful analytical tools. To overcome the difficulties arising from unstable excitation conditions, in 1925 Gerlach introduced the principle of internal standard [327], where the ratio of the intensities of the signals of the analytes to the signals for a matrix or of an added element were made and the reference element was designated as the internal standard. This brought about the breakthrough of atomic spectroscopy for elemental analysis.

In atomic emission spectrometry a reproducible and representative amount of the sample is moved into a radiation source. The role of the radiation source in atomic emission spectrometry is of similar central importance as a dissolution procedure in wet chemical analysis. Therefore, continuous endeavors are always being made to improve old and to develop new radiation sources for atomic emission spectrometry, so as to be able to perform analyses with better figures of merit, as are required by the progress in science and technology. The role of the radiation source in AES does not end at the stage of free atom production as it does in AAS, but includes the excitation of levels from which transitions with high emission efficiency start and eventually also ionization, enabling ion lines to be used. Ideal sources for atomic emission spectrometry should therefore meet many requirements. The sources should have a very robust geometry, where the analyte is led through zones with hopefully a high excitation effciency, i.e. high electron temperatures, but low spectral background intensities (possibly deviations from thermal equilibrium). The source should have an inert gas atmosphere so as to keep interfering thermochemical reactions low. Also the influence of the analyte ionization and its contribution to the electron number densities in the radiation source should be low so as to keep matrix interferences low. The analyte dilution should be sufficient so as to obtain narrow lines, self-reversal also should be low and finally the instrument and operation costs in terms of power and gas consumption should also be low.

Qualitative analyses

These can be performed successfully with AES. Indeed, the unambiguous detection and identification of a single non-interfered atomic spectral line of an element is sufficient to testify to its presence in the radiation source and in the sample. The most intensive line under a set of given working conditions is known as the most sensitive line. These elemental lines are situated for the various elements in widely different spectral ranges and may differ from one radiation source to another, as a result of the excitation and ionization processes. Here the temperatures of the radiation sources are relevant, as the atom and ion lines of which the norm temperatures (see Section 1.4) are closest to the plasma temperatures will be the predominant ones. However, not only will the plasma temperatures but also the analyte dilutions will be important, so as to identify the most intensive spectral lines for a radiation source. Also the freedom from spectral interferences is important.

Despite the fact that spectral lines are very narrow (some pm) with respect to the whole spectral range, a large number of line coincidences can occur. Indeed, in the spectral range of 200–400 nm more than 200 000 lines for the elements of the Periodic Table have been tabulated and assigned to elements, and this is only a portion of the total number of spectral lines. Therefore, high-resolution spectrometers have to be used for the qualitative evaluation of the spectra but also use of more than one line is necessary to conclude the presence or absence of an element. Qualitative emission spectral analysis is easy when the spectra are recorded photographically or with a scanning monochromator. For the first case atlases are available, in which the spectra are reproduced and the most sensitive lines are indicated. To identify spectral lines often a photograph of the ion spectrum is produced (Fig. 88). Also books with spectral scans around the analysis lines with superimposed spectra from possible interferents have been published, e.g. for the case of ICP-OES (see that for example from Winge et al. [329]).

In addition, spectral line tables, in which the wavelengths of the spectral lines together with their excitation energy and a number indicating their relative intensity for a certain radiation source are tabulated, are very useful. They are available for different sources, such as arc and spark sources [330–332], but also in a much less complete form for newer radiation sources such as glow discharges [333] and inductively coupled plasmas [334].

Qualitative analysis by atomic spectrometric methods has now been given totally

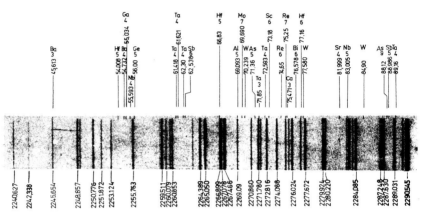

Fig. 88. Photographically recorded emission spectra from spectral atlas. (Reprinted with permission from Ref. [328].)

new possibilities, as simultaneous emission spectrometers where the whole analytically usable part of the spectrum is measured on a series of CCDs placed in the Rowland circle of a spectrometer are now available. This development could have been be foreseen when vidicon detectors became available, with which even for small spectral scans line identification became much easier, as shown for a rare earth mineral analyzed by ICP-OES (Fig. 89). With CCD spectrometers and the computer facilities allowing storage of many reference spectra and spectrum stripping, line identification becomes easy and unambiguous, as many lines per element can be checked. To a certain extent this can now also be achieved by rapid scanning spectrometers, where the spectral range of 200–800 nm can be scanned at high velocity, a high precision in random line positioning as well as peak intensitiy measurements becomes possible (Table 8) and no deterioration of spectral resolution as compared with slew-scanning high-resolution monochromators has to be taken into account, as shown by instrumental line widths [93]. This also allows spectral stripping in the case of stable sources as in ICP-OES, as shown for the case of Y in ZrO_2 e.g. in Ref. [93].

Quantitative atomic emission spectrometry

The intensity of an elemental atomic or ion line is used as the analytical signal in quantitative atomic emmision spectrometry. In fact the intensities are unequivi-

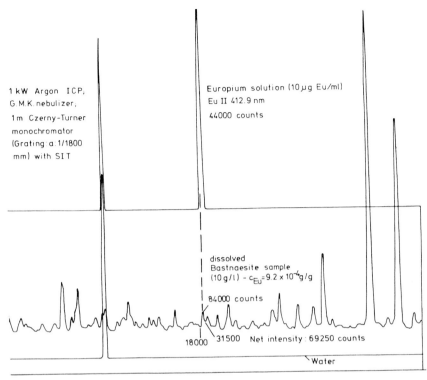

1 kW Argon ICP;
G.M.K. nebulizer;
1 m Czerny-Turner
monochromator
(Grating a. 1/1800
mm) with SIT

Europium solution (10 μg Eu/ml)
Eu II 412.9 nm
44000 counts

dissolved
Bastnaesite sample
(10 g/l) - $c_{Eu} = 9.2 \times 10^{-4}$ g/g

84000 counts

18000 31500 Net intensity : 69250 counts

Water

Fig. 89. Spectra of ICP-OES recorded with an SIT vidicon for the case of water, a europium solution and a dissolved bastnaesite sample containing europium [335].

Tab. 8. Relative standard deviations (RSD) of the intensities for selected Fe-lines as determined from five replicates of spectral scans acquired in the second order with a rapid scanning sequential monochromator for a solution containing 10 mg/L of Fe. (Reprinted with permission from Ref. [93].)

Line (nm)	Intensity (a.u.)	RSD (%)
239.927	26 200	4.0
238.076	37 800	3.5
238.204	725 000	3.2
238.324	23 700	1.8
238.863	114 000	2.6
239.562	512 000	2.6
239.924	132 000	3.1
240.488	277 000	4.4
240.666	101 000	2.0
241.052	153 000	1.4

cally related to the analyte number densities sampled. Indeed, the intensity of a spectral line is directly related to the number density in the source as:

$$I = [1/(2\pi)] \cdot h \cdot v \cdot [g_m/Z(T)] \cdot \exp(-E_m/kT) \cdot A \cdot n \tag{260}$$

where h is Planck's constant, v the line frequency, g_m the degeneracy of the excited level, $Z(T)$ the partition function for the level of ionization concerned, E_m the excitation energy, k Boltzmann's constant, T the excitation temperature, A the transition probability and n the analyte number density of the analyte in the ionization level concerned. Accordingly for an atom line $n = n_T - n_i$, with n_T being the notal number of analyte atoms per volume [(weight:atomic weight) × Avogadro number/ cm^3] and n_i the ion density. Furthermore:

$$n_i \cdot n_e/(n_T - n_i) = \{[(2 \cdot \pi \cdot m_e \cdot k \cdot T)^{5/2}]/h^3\}$$
$$\cdot (2 \cdot Z_i/(Z_a) \cdot [\exp(-E_i/k \cdot T)] \tag{261}$$

where n_e is the electron number density, m_e the mass of the electron, $Z_{i,a}$ the partition functions for atoms and ions, respectively, E_i the ionization energy and α_i the degree of ionization. N_T can be obtained by integrating over the source volume observed, making use of source modelling with respect to the analyte diffusion. The total number of analyte atoms in the source is related to this number in a rigid manner and to the influx of analyte through the sampling mechanism. The latter is, for example, the evaporation rate in arc emission work, the sample uptake rate in pneumatic nebulization with its efficiency taking into account transport losses, the sputter rate in glow discharge work, all of which can be calculated.

The number of photons entering the monochromator is determined by the space angle of observation and in its turn determines the flux at the entrance collimator (ϕ_{en}) as:

$$\phi_{en} = \tau \cdot I \cdot (\% \text{ of space angle}) \cdot (A_1 \cdot A_2)/(f_k \cdot \cos \beta) \tag{262}$$

where ϕ_{en} is the flux at the entrance collimator, τ is the optical transmittance of the entrance collimator, A_1 is the area of the entrance slit (slit width × slit height), A_2 is the area of the entrance collimator mirror and f_k its focal length, I the radiant density of the source over the 4π angle and β is the angle of the entrance collimator with the optical axis.

$$\phi_{out} = R_1 \cdot R_2 \cdot \phi_{in} \tag{263}$$

where R_1 and R_2 are the reflectivities of the grating and the exit collimator, respectively.

$$\phi_{Ph} = \phi_{out} \times \cos \theta \times s_{out}/s_{en} \tag{264}$$

where ϕ_{Ph} is the flux through the exit slit, s_{en} is the entrance and s_{out} the exit slit width and $\cos \theta$ the angle between the focal plane and the detector head. This photon flux ϕ_{Ph} is directly related to the signal of a photomultiplier through:

$$i_c = \phi_{Ph} \times Q.E. \times e \tag{265}$$

where i_c is the cathode current, $Q.E.$ the quantum efficiency and e the charge of the electron (in coulombs),

$$i_a = \theta(V) \cdot i_e \tag{266}$$

with i_a as the anode current and $\theta(V)$ the amplification over the dynode chain and:

$$V = R_{sel} \times i_a \tag{267}$$

$$\text{digital signal} = [\text{range}(A/D)]/V \cdot \max \cdot \text{number of digits} \tag{268}$$

with V the measured voltage for a selected resistor R_{sel} and an A/D convertor.

Accordingly, from the photomultiplier signal the number of analyte atoms brought from a given amount of sample into the radiation source can be calculated directly. As all constants in the calculation shown, however, are not known apriori, AES in practice is a relative method and a calibration has to be performed. The determination of the calibration function is an important part of the working procedure. The calibration function relates the intensity of a spectral line (I) to the concentration c of an element in the sample. From the work of Scheibe [336] and Lomakin [337] the following relationship between absolute intensities and elemental concentrations was proposed:

$$I = a \cdot c^b \tag{269}$$

where a and b are constants. The expression with absolute intensities is used virtually only in flame work or in plasma spectrometry, where very stable radiation sources are used. Normally the intensity ratio of an analyte line to a reference signal is used, as proposed by Gerlach. This leads to calibration functions of the form:

$$I/I_R = a' \cdot c^{b'} \tag{270}$$

The inverse function of the calibration function is the analytical evaluation function. It can be written as:

$$c = c_R (I/I_R)^{\eta} \tag{271}$$

or:

$$\log c = \log c_R + \eta \log(I/I_R) \tag{272}$$

The slope of the analytical evaluation curve in the bilogarithmical form in the case of trace analyses in most cases is 1. At higher concentrations η may be >1 as a result of the commencement of self-reversal. Particularly in trace analysis by AES the intensity of the spectral background I_U can be used as reference signal and then in Eq. (272), c_R should be replaced by c_U. The calibration constant c_U is the concentration at which the line and background intensities are equal (background equivalent concentration – BEC value) and can be calculated as:

$$c_U = [(I_X/I_U) \cdot 1/c]^{-1} \tag{273}$$

The ratio (I_X/I_U) and the calibration constants c_U for a given element and line depend on the radiation source and the working conditions selected but also on the spectral apparatus [338, 339]:

$$c_U = \frac{(dB_U/d\lambda)\Delta\lambda_L}{B_0} \cdot \frac{s_{eff} \cdot R_L}{s_0 \cdot R_0} \cdot \sqrt{[\pi/(\ln 16)]} \cdot \sqrt[4]{[(s_a/s_{eff})^4 + 1]}$$

$$c_U = \quad c_{U,\infty} \quad \cdot \quad A_1 \quad \cdot \quad A_2 \tag{274}$$

$c_{U,\infty}$ describes the influence of the radiation source, $(dB_U/d\lambda)$ is the spectral radiation density for the background intensity, B_0 is the radiant density for an analytical line at the concentration $c = 1$ and $\Delta\lambda_L$ is the physical width of the analysis line.

The second term (A_1) describes the influence of the spectral apparatus.

$$s_{eff} = (s_e^2 + s_0^2 + s_L^2 + s_z^2)^{1/2} \tag{275}$$

s_L is the spectral slit width corresponding to the physical width of the spectral line $(\Delta\lambda_L)$ and $R_L = \lambda/\Delta\lambda_L$, the resolution required to resolve the line width. s_z is the slit width corresponding to the optical aberrations in the spectral apparatus and is defined as:

$$s_z = \Delta\lambda_z \cdot dx/d\lambda \tag{276}$$

$\Delta\lambda_z$ can be determined by measuring the practical resolution of the monochromator from the deconvolution of the two components of the Hg 313.1 nm line doublet and subtraction of the contributions of the diffraction slit width $(s_0 = \lambda \cdot f/W)$ and the entrance slit width (s_e). The contribution of the natural width of the Hg lines can be neglected, as it is very low in the case of a hollow cathode lamp.

In the case of monochromators s_z is predominant, but in the case of a very high resolution it becomes less significant and $A_1 \rightarrow 1$.

The factor A_2 describes the influence of the profile of the analysis line and the influence of the effective measurement slit. In the case of photoelectric measurements the latter is the exit slit width of the spectrometer. This contribution is only relevant when $s_a/s_{eff} < 2$, which is virtually only the case with photographic measurements where the line profile is scanned with a very narrow densitometer slit. For photoelectric measurements $A_2 = s_a/s_{eff}$ as the exit slit width must be larger than the effective line width because of the thermal and mechanical stability of the system.

With the aid of these equations, detection limits and signal to background ratios that have been obtained with two different spectrometers can be compared.

For instance for a 0.9 m Czerny–Turner monochromator, with $R_0 = 216\,000$ (grating width: 90 mm and grating constant: 1/2400 mm), $s_e = 12$ µm ($\hat{s}_e = 4$), $s_a = 15$ µm ($\hat{s}_a = 5$), practical resolution: 0.007 nm ($R_z = 75\,000$); and for a 0.5 m Ebert monochromator: $R_0 = 134\,600$ (grating width: 60 mm and grating constant: 1/2442.4 mm), $s_e = 4$ µm ($\hat{s}_e = 1.5$), $s_a = 6$ µm ($\hat{s}_a = 2.3$), practical resolution: 0.01 nm ($R_z = 35\,000$), the ratio of the constants c_U and the detection limits for an analysis line at 300 nm and with a line width of $\Delta\lambda_L \cong 3$ pm in the case of the same excitation source and both spectral apparatus should be: c_L 0.9-m/c_L 0.5-m = 0.88 [180].

Detection limit

This is a very important figure of merit for an analytical method. As discussed in Section 2.1 in general, for an atomic emission spectrometric method it can be written as:

$$c_L = c_U \cdot (I_X/I_U) \tag{277}$$

I_X/I_U is the smallest line-to-background ratio that can still be measured. When the definition is based on a 99.86% probability, it is given by 3 times the relative standard deviation of the background intensities.

In the case of photographic or diode array registration the intensities of the line and the spectral background are recorded simultaneously, and:

$$\sigma = [\sigma^2(I_X + I_U) + \sigma^2(I_U)]^{1/2} \tag{278}$$

As the intensities of the line and the spectral background are nearly equal in the vicinity of the detection limit, their standard deviations will be equal and the relevant standard deviation reduces to:

$$\sigma = \sigma(I_U)\sqrt{2} \tag{279}$$

There is some difference in the standard deviations when line and background intensities are measured simultaneously, as in the case of photographic detection or with modern CCD spectrometers, or when they are measured sequentially as done in slew scanning systems. Indeed, in the first case the fluctuations of line intensities and background intensities for a considerable part are correlated, especially at low signal levels and thus partially cancel. This may lead to a considerable gain in power of detection.

The value of the standard deviation can be measured directly, but in the case of a photographic measurement it can be calculated from the standard deviation of the blackenings with the aid of Eq. (184). This shows that for trace analyses, emulsions with a high γ must be used. Accordingly, standard deviations $\sigma_r(I_U)$ of about 0.005 can be obtained. This assumes a smooth background and these values reflect only the graininess of the emulsion and do not include contributions from source fluctuations. In the case of photoelectric measurements with a photomultiplier, σ is the absolute standard deviation of the background intensity, measured at a wavelength adjacent to the spectral line or at a blank sample at the wavelength of the analytical line. It may consist of several contributions as:

$$\sigma^2(I_U) = \sigma_P^2 + \sigma_D^2 + \sigma_A^2 + \sigma_V^2 \tag{280}$$

σ_P is the photon noise of the source or the noise of the photoelectrons and $\sigma_P \approx \sqrt{n}$, when n is the number of photoelectrons considered; σ_D is the dark current noise of the photomultiplier and $\sigma_D \approx I_D$, when I_D is the dark current; σ_A is the flicker noise of the radiation source and $\sigma_A \approx \phi$, when ϕ is the radiant density; and σ_V is the amplifier noise.

σ_V in most cases can be neglected as compared with other sources of noise. In the case of very constant sources σ_A is small. Owing to the proportionality between σ_D and I_D it is advisable to choose a photomultiplier with a low dark current, so as to avoid detector noise limitations. Here limitations may remain when spectral lines at lower wavelengths are used. Indeed, then the spectral background intensities are usually low, so that the detector noise becomes predominant. This may specifically become the case when an Echelle spectrometer is used for the measurements. Owing to slit height limitations the photon flux onto the detector then becomes very small. In many cases this is balanced by the fact that Echelle spectrometers have a shorter focal length, because of which the overall radiant flux density in the focal plane is higher. Also in the case of CCD detector technology, limitations through the detector noise may occur. In particular, readout noise here may be limiting. This is because the detector is read many times so as to collect the charge accumulated during the numerous subsequent measurement cycles that are required to realize a sufficiently large linear dynamic range. In most other cases, the photon noise will be limiting for the level of detection.

When signals include a blank contribution, this will also have fluctuations, which limit the power of detection in practice, the influence of which can easily be shown. If I_X is the lowest measurable signal without blank contribution, I_X' the lowest analytical signal including a considerable blank value, I_U the background signal and I_{Bl} the blank signal, I_U, $(I_X + I_U)$, $(I_{Bl} + I_U)$ and $(I_X' + I_{Bl} + I_U)$ can be measured and:

$$(I_X' + I_{Bl} + I_U) - (I_{Bl} + I_U) = 3\sqrt{[\sigma^2(I_X' + I_{Bl} + I_U) + \sigma^2(I_{Bl} + I_U)]} \tag{281}$$

or

$$(I_X' + I_U) - I_U = 3\sqrt{[\sigma^2(I_X + I_U) + \sigma^2(I_U)]} \tag{282}$$

This leads to:

$$(I'_X/I_U) = (I_X/I_U)(1 + I_{Bl}/I_U)[\sigma_r(I_{Bl} + I_U)/\sigma_r(I_U)] \tag{283}$$

and the detection limit becomes [128]:

$$c'_L = c_U\{(I_X/I_U)(1 + I_{Bl}/I_U)[\sigma_r(I_{Bl} + I_U)/\sigma_r(I_U)]\}$$

or

$$c_L \cdot [\sigma_r(I_{Bl} + I_U)/\sigma_r(I_U)](1 + I_{Bl}/I_U) \tag{284}$$

As blank contributions themselves also scatter, which cannot be eliminated by blank subtraction, and as blank values scatter independently of the analytical signals, the presence of blank contributions always deteriorates the limit of detection achievable. Accordingly, to obtain the best limit of detection in AES the standard deviations of the background, the blank contributions and the c_U values should be minimized. The influence of the standard deviations of the line and background intensities are partly correlated. This has the consequence that the standard deviations of the net signals could be decreased when line and background intensities are measured simultaneously, as is possible with photographic emulsions and with the new array detectors that have simultaneous measurement capabilities at a large number of adjacent wavelengths. However, this is only true when the spectral background from the source and not the pixel-to-pixel noise is limiting. The c_U values are a function of the radiation source, the elements and the lines (reflected in $c_{U,\infty}$) and also of the spectrometer (through A_1 and A_2). In order to keep A_1 and A_2 as close to one as possible, a spectral apparatus with high resolving power (R_0 high), a low entrance slit ($s_e = 1.5s_0$) should be used and take $s_a = s_{eff}$. On the other hand, it should be confirmed that the spectral background intensities can still be measured. Indeed, the background intensities obtained are proportional to the entrance slit width and thus detector noise limitations could be encounterd when the slit widths become too narrow. Also thermal and mechanical stabilities become limiting in the case of narrow slit widths, by which, particularly in systems with on-peak integration instead of slew-scan procedures, the long-term stability can no longer be guaranteed.

Apart from the high power of detection, also the realization of the highest analytical accuracy is very important. This relates to the freedom of interferences. Whereas the interferences stemming from influences of the sample constituents on the sample introduction or on the volatilization, ionization and excitation in the radiation source differ widely from one source to another, most sources emit line-rich spectra and thus the risks for spectral interferences in AES are high. In the wavelength range 200–400 nm, as an example, only for arc and spark sources have more than 200 000 spectral lines yet been identified with respect to wavelength and element in the classical MIT Tables. Consequently, spectral interferences are much more severe than in AAS or AFS work.

Therefore, in AES it is also advisable to use high resolution spectrometers so as to minimize the risks of spectral interferences. A practical resolving power of 40 000, which guarantees that spectral lines at wavelengths of 300 nm and wavelength differences of 0.0075 nm can still be spectrally resolved, is advisable. This is particularly the case when trace determinations must be performed in matrices emitting line-rich spectra.

The nature of spectral interferences, may however, be different.

- Direct coincidence of lines: when the wavelength difference is less than the spectral bandwidth of the spectral apparatus.
- Line wings: spectral lines of matrix elements may have a width extending to up to several pm, as is known e.g. from the calcium lines in water samples from the early work on ICP-OES [341]. Here an estimation of the spectral interference for analytical lines situated on the wings has to be made by weighting the intensity contributions at well-defined wavelength differences away from the line center.
- Contributions from band emission spectra, especially in the region of intensive band systems (CN at 370–385 nm, N_2^+ at 390–400 nm, NH at 340–350 nm, OH at 310–320 nm). Also here weighted corrections, as in the case of matrix line wings may be necessary.
- Stray radiation contributions usually have a more even spectral dependence and in a number of cases can be corrected for by measuring the stray radiation intensity on one side of the analytical line only.

Knowledge of the atomic spectra is also very important so as to be able to select interference-free analysis lines for a given element in a well-defined matrix at a certain concentration level. To do this, wavelength atlases or spectral cards for the different sources can be used, as they have been published for arcs and sparks, glow discharges and inductively coupled plasma atomic emission spectrometry (see earlier). In the case of ICP-OES, for example, an atlas with spectral scans around a large number of prominent analytical lines [329] is available, as well as tables with normalized intensities and critical concentrations for atomic emission spectrometers with different spectral bandwidths for a large number of these measured ICP line intensities, and also for intensities calculated from arc and spark tables [334]. The problem of the selection of interference-free lines in any case is much more complex than in AAS or AFS work.

5.2
Atomic emission spectrometers

As all elements present in the radiation source emit their spectrum at the same time, from the principles of AES it is clear that it is a multielement method and is very suitable for the determination of many elements under the same working conditions. Apart from simultaneous determinations, so-called sequential analyses can also be carried out, provided the analytical signals are constant. Sequential and simultaneous multielement spectrometers both have their own possibilities and limitations.

Sequential spectrometers

In most cases these include Czerny–Turner or Ebert monochromators with which the lines can be rapidly selected and measured one after another. Owing to the line-rich emission spectra the focal lengths are often up to 1 m and a grating with a width of up to 100 mm and a constant of 1/2400 mm is used. These instruments require very accurate presetting of the wavelength, which is difficult on a random-access basis. Indeed, the halfwidth of a spectral line corresponds to a grating angle of 10^{-4} degrees. Therefore, work can be performed in a scanning mode and with stepwise integration at different locations across the line profile or advanced techniques such as optical encoders and direct grating drives can be used. Systems employing fixed optics with a multislit exit mask, a moveable photomultiplier and fine wavelength adjustment with the aid of computer-controlled displacement of the entrance slit are also used. Sequential spectrometers are flexible as any elemental line and any background wavelength within the spectral range (usually from 200 to up to 600 nm) can be measured, however, they are slow when many elements must be determined. They are of particular use in the case of stable radiation sources such as plasma sources at atmopheric pressure or glow discharges operated under steady state sputtering conditions.

Spectral background intensity measurements can be accomplished by slew-scan procedures for the grating or also for the entrance slit. However, the spectral background and the line intensities in the case of monochannel instruments are not measured simultaneously, which is actually possible with array detectors.

An interesting device for rapid scanning of spectra (the whole analytical range from 165 to 750 nm in 2 min) with a high resolution (7 pm line widths measured) and with individual integration times of 2.5 ms is the IMAGE system [93]. The high dynamic range is realized through rapid changes in the photomultiplier voltage while keeping the currents within a narrow range. The precision of wavelength access is so good that spectrum subtraction readily becomes possible as shown in the case of ICP-OES for the determination of Zr in a solution containing Fe (Fig. 90).

Simultaneous spectrometers

Today these virtually always make use of photoelectric radiation detection, whereas earlier, spectrographs with photoplate recording were in wide use, especially for qualitative and semi-quantitative analyses. Simultaneous photoelectric spectrometers usually have a Paschen–Runge optical mounting. They have many exit slits and are particularly suitable for rapid determinations of many elements in a constant and well-understood sample matrix and the achievable precision is high. However, the analysis program is fixed and accordingly simultaneous spectrometers are not suitable for the analysis of random samples. Owing to the stability require-

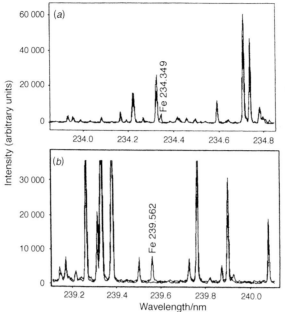

Fig. 90. ICP atomic emission spectra for a 10 g/L Zr solution with and without 2 mg/L Fe in the vicinity of the Fe II 234.349 nm (a) and Fe II 239.562 nm (b) lines obtained with the IMAGE approach. (Reprinted with permission from Ref. [342].)

ments larger exit slits are often used, through which thermal drifts can be overcome and lower spectral resolution is obtained. In applications of trace analysis in particular, background correction is often required. This can be achieved by computer-controlled displacement of the entrance slit or by a rotation of a quartz refractor plate behind the entrance slit (Fig. 91). Here slew-scan procedures also enable scans from different samples to be superimposed, such as solutions of real samples, the dissolution reagents required and water e.g. in ICP-OES. In the case of copper a positive error could be attributed to a coincidence with a molecular band, probably from the B–O species (Fig. 92) [343].

As a result of the decrease in the price of CCDs, providing up to 20 CCDs is now feasable, covering the whole spectral range from the VUV to the visible, even including the alkali element lines with the aid of a supplementary grating, as shown in Fig. 93 for a commercially available system. Such a system makes line identification very easy, as has already been shown by earlier work where a SIT vidicon was coupled to a conventional Czerny–Turner monochromator. By spectrum comparisons, analytical lines of the rare earths in a complex dissolved mineralogical sample in ICP-OES could easily be found. This is done by comparing back with stored spectra from pure analyte solutions, and at the same time background correction becomes easy as every wavelength is digitally accessible, and through appropriate software even the simulation of background spectra and their subtraction is possible. When the whole analytical range becomes accessible multiline calibration becomes easy, also in post-measurement evaluation when the spectra are saved. This

Fig. 91. Background correction with a computer-controlled simultaneous ICP-OES spectrometer.

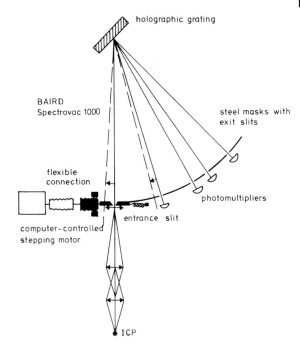

holographic grating

BAIRD
Spectrovac 1000

steel masks with exit slits

flexible connection

photomultipliers

computer-controlled stepping motor

entrance slit

ICP

may be very useful for improving or controlling the analytical accuracy limitations resulting from spectral interferences.

A further advantage of a multi-CCD system is the linear dispersion and the one-dimensional spectral dispersion. This both facilitates comparison of recorded and saved spectra and enables a high radiation throughput as the whole height of the

Fig. 92. Spectral interferences in ICP-OES as recorded in wavelength scans obtained by displacement of the entrance slit. 3 kW Ar-N$_2$ ICP, Baird 1000 1 m spec-trometer, sample: 0.5534 mg of air-borne dust sample HBW 2 per mL (–), 19 mg/mL of HBO$_3$, 19 mg/mL of HF, 4.9 N HNO$_3$ (– –); the same after the decomposition (–·–·); water (····), line: Cu I 324.754 nm. (Reprinted with permission from Ref. [343].)

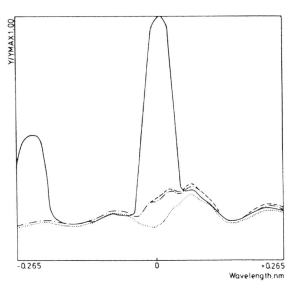

Y/YMAX 1.00

-0.265

0

+0.265

Wavelength,nm

Fig. 93. ICP-OES spectrometer CIROS–Paschen–Runge circle with detector alignement. (Courtesy of Spectro Analytical Instruments.)

entrance slit can be used. However, there certainly are line interference problems, as with 20 CCDs of 1024 pixels each the spectral resolution for a 1 m Paschen–Runge instrument with a grating constant of 1/2400 mm may become too low in a number of cases.

Echelle spectrometers

In addition, Echelle spectrometers are often used [50]. By combination of an order-sorter and an Echelle grating either in parallel or in crossed-dispersion mode, high practical resolution (up to 300 000) can be realized with an instrument of fairly low focal length (down to 0.5 m) (Fig. 94). Therefore, the stability as well as the luminosity are high. By using an exit slit mask with a high number of preadjusted slits, highly flexible and rapid multielement determinations are possible.

As high-quality solid-state detectors are now available (Table 9) relatively compact high-resolution electronic spectrometers can be realized [344]. According to the grating equation, the wavelength difference $\Delta\lambda(k)$ between two subsequent orders with the same diffraction angle β, which also gives the spectral length of order k is

(A)

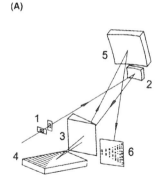

(B)

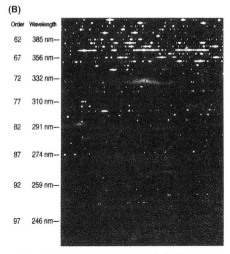

Fig. 94. Echelle spectrograph with internal order separation in a tetrahedral mounting. (1): Entrance slits, (2): spherical collimator mirror, (3): prism, (4): Echelle grating, (5): spherical camera mirror, (6): focal plane (A). Part of an iron spectrum recorded with an Echelle spectrograph with collimator and camera focal length of 500 mm, original size of the part: 20 × 15 mm² (B). (Reprinted with permission from Ref. [344].)

given by:

$$\Delta\lambda(k) = \lambda(k) - \lambda(k+1) = \lambda(k)/k \tag{285}$$

where $\lambda(k)$ is the central wavelength of the order k. To achieve the splitting of the spectral range of interest into as many sub-spectra as are necessary, the number k has to be according to Eq. 285 and the grating equation:

$$d \gg \lambda \tag{286}$$

For a groove distance e.g. of $d = 13.33$ μm (75 grooves per mm) the wavelength range between 190 and 850 nm is split into approximately 100 orders ($k = 125–28$), which superimpose for $\alpha = 68°$ within a diffraction angle ranging from $\beta_{min} = 56.5°$ to $\beta_{max} = 63.8°$.

Another specialty of the Echelle grating relates to the angular dispersion, normally given by:

$$d\beta/d\lambda = k/(d \cdot \cos\beta) \tag{287}$$

and for the case of autocollimation of an Echelle grating ($\alpha = \beta = \Theta_B$, the Blaze angle:

$$d\beta/d\lambda = 2 \tan\Theta_B \tag{288}$$

Tab. 9. Performance of selected charge-coupled devices. (Reprinted with permission from Ref. [344].)

Detector	Type	Pixel area/μm^2	Number of pixels (area/mm^2)	Max. quantum (10% limits)	Full well capacity/e$^-$	Read-out noise (e$^-$ rms)
Kodak KAF-1000	CCD	24.0 × 24.0	1024 × 1024 (24.6 × 24.6)	35%, 400–1080 nm	500 000	<15 500 kHz readout
Kodak KAF-6300	CCD	9.0 × 9.0	3027 × 2048 (27.65 × 18.43)	45%, 400–1080 nm	85 000	9–10 500 kHz readout
EEV 05-30	CCD	22.5 × 22.5	1242 × 1152 (27.5 × 25.9)	45%, 400–1080 nm	500 000	4–6 50 kHz readout
SITeCCD-10245B	CCD	24.0 × 24.0	1024 × 1024 (24.6 × 24.6)	70%, 190–1050 nm	400 000	5–7 50 kHz readout
Reticon RA2000J	CCD	13.5 × 13.5	2048 × 2048 (27.6 × 27.6)	50%, 200–1050 nm	100 000	3 50 kHz readout
DALSA1A-D9-5120	CCD	12.0 × 12.0	5120 × 5120 (61.44 × 61.44)	30%	130 000	80 50 MHz readout
CIDTEC CID 38 PPR	CID	28.0 × 28.0	512 × 512 (14.34 × 14.34)	40–60%, 200–1100 nm	1 500 000	250 50 kHz readout 25 100 readouts (non-destructive)

Thus Echelle gratings have a constant and much larger relative angular dispersion than classical diffraction gratings and the spectral resolution $d\lambda/\lambda$ of an Echelle spectrometer is almost constant and at least 2–8 times higher than the corresponding resolution of a classical grating spectrometer. The orders, however, always have to be separated, for which prisms are used, preferably positioned transversally. The detector size in the case of a two-dimensional detector such as a CCD and the spectral coverage required necessitates optical mapping of generally at least 50–150 diffraction orders on to a plane of typically 25×25 mm^2. The distance between different orders of an Echelle spectrum is given by the dispersive element used for order separation and is not constant. Taking the distances between two neighboring orders at the upper and lower ends of the wavelength range, their difference increases in the ratio: $\lambda_{max}/\lambda_{min}$. The corresponding slit height h for a minimum order distance is given empirically by:

$$H < [B/(m \cdot \sqrt{\lambda_{max}/\lambda_{min}})] - b \qquad (289)$$

where B is the width of the detector perpendicular to the diffraction orders, $m = k(\lambda_{min}) - k(\lambda_{max})$ is the total number of diffraction orders and b is the pixel width, denoting that the minimum distance of 1 pixel between the orders has to be guaranteed. For an Echelle spectrum with $m = 100$, recorded by an array detector with 1000×1000 pixels and 25×25 mm^2 total area, h becomes <93 µm.

Echelle spectrometers can also be made small, when only part of the spectrum is to be used, while still having a considerably high spectral resolution. In Echelle systems, dedicated parts of the spectrum are often measured with a separate built-in spectrometer. This is e.g. the case in ICP-OES for the spectral range where the most sensitive alkali lines are found [345].

As the most sensitive lines of a number of elements such as S, P, C and As are at VUV wavelengths, access to this spectral range is required, which necessitates the elimination of oxygen from the entire optical path. This can be achieved by continuous evacuation of the spectrometer, purging with nitrogen or argon or by the use of sealed gas-filled spectrometers. Accordingly sensitive lines of Br, Cl, I, N, P and S become available for the ICP (see Refs. [346, 347] for N and Heitland [348] for the other elements) (Table 10). Then, however, the path between the ICP and

Tab. 10. VUV atomic emission lines for the halogens, phosphorus and sulfur in ICP-OES. [348].

Cl	134.724
Br	163.340
I	161.761
P	178.290
S	180.734
S	182.037

the spectrometer must be free of air as well. This can be easily achieved with a spark, or in the case of the ICP by using an axially viewed ICP with the appropriate interface or by using a special torch [347].

Often wavelength measurements are made at low wavelengths in the second order, so as to increase the spectral resolution and to increase the wavelength coverage up to the longer region, where the alkali elements have their most sensitive lines. Then solar-blind photomultipliers are used to filter out long-wavelength radiation of the first order.

5.3
Flame emission

The oldest of the spectroscopic radiation sources, a flame, has a low temperature (see Section 4.3.1) but therefore good spatial and temporal stability. It easily takes up wet aerosols produced by pneumatic nebulization. Flame atomic emission spectrometry [265] is still a most sensitive technique for the determination of the alkali elements, as e.g. is applied for serum analysis. With the aid of hot flames such as the nitrous oxide–acetylene flame, a number of elements can be determined, however, not down to low concentrations [349]. Moreover, interferences arising from the formation of stable compounds are high. Further spectral interferences can also occur. They are due to the emission of intense rotation–vibration band spectra, including the OH (310–330 nm), NH (around 340 nm), N_2^+ bands (around 390 nm), C_2 bands (Swan bands around 450 nm, etc.) [20]. Also analyte bands may occur. The S_2 bands and the CS bands around 390 nm [350] can even be used for the determination of these elements while performing element-specific detection in gas chromatography. However, SiO and other bands may hamper analyses considerably.

5.4
Arcs and sparks

Arc and spark AES have found wide use for the analysis of solids and are still important methods for routine analysis, especially when many elements have to be determined in numerous samples.

5.4.1
Arc emission spectrometry

5.4.1.1 Arc characteristics
Electrical arc sources for atomic emission spectrometry use currents of between 5 and 30 A, and the burning voltage between two electrodes of different polarity in the case of an arc gap of a few mm then ranges between 20 and 60 V. Dc arcs can

be ignited by bringing the electrodes together, which causes some ohmic heating. Under these conditions dc arcs have a normal (decreasing) current–voltage characteristic due to the high rate of material ablation by thermal evaporation. In a dc arc the areas between the cathode fall near to the cathode layer, the positive column and the anode fall area can be distinguished.

The cathode delivers a high number of electrons, carrying the greater part of the current, by glow emission. This electron current heats up the anode to its sublimation or boiling temperature (up to 4200 K in the case of graphite) and the cathode reaches temperatures of ca. 3500 K as a result of atom and ion bombardment. An arc attaches to a limited number of small burning areas. Owing to the high temperatures of the electrodes containing or constituting the sample, sample volatilization is mainly the result of thermal evaporation. Therefore the anions present may influence the volatilization of metals considerably, as the thermochemical behavior of the compounds determines the volatilization of an element. This can lead to high matrix influences but at the same time can be used to obtain selective volatilization, removing spectral interferences due to matrix element lines.

The plasma temperature in the carbon arc is of the order of 6000 K and it can be assumed to be in local thermal equilibrium. According to the temperatures obtained, it could be expected that in arc emission spectrometry mainly the atom lines will be the most sensitive lines, when considering their norm temperatures. Arcs are usually used for survey trace analysis but also for the analysis of pure substances when the highest power of detection is required. However, they may be hampered by poor precision (RSDs of 30% and higher).

5.4.1.2 Dc arc spectrometry

The sample is placed in or on an electrode with a suitable geometry and volatilizes as a result of the high electrode temperature. Electrodes made of graphite and with a beaker form are often used. In the case of refractory matrices anodic evaporation is used and the positive column can be observed, or for very sensitive methods also the cathode layer where the volatilized sample ions are enriched (cathode layer effect).

Solid samples

These must be powdered, and sample amounts of up to 10 mg (powder or chips) are eventually brought into the electrode after mixing with graphite powder and briquetting. Thermochemical aids such as NH_4Cl can be added, so as to create similar volatilization conditions for all samples. In addition, spectrochemical buffers such as $CsCl$ may be added to control the arc temperature or the electron pressure. Easily ionized elements normally have a very strong influence on the line intensities, for several reasons. Firstly, the ionization of easily ionized elements removes energy from the plasma and cools it, through which the electrical characteristics are changed and the intensities of lines are influenced through the Boltzmann function. Secondly, the ionization of an easily ionized element increases the elec-

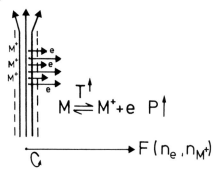

Fig. 95. Ambipolar diffusion in spectro-chemical sources.

tron number densities in the plasma, which delays the ionization of the analytes and thus influences the atom to ion line ratios. Thirdly, when elements ionize, the pressure in the plasma, which is proportional to the particle number densities, increases and the plasma increases its volume. This is still increased as the electrons diffuse more easily to the outer plasma zones than the positive ions, providing an additional electric field (Fig. 95), a mechanism known as ambipolar diffusion.

The strong influence of changes in the ionization through different matrices thus will strongly influence the geometry of arc plasmas as well, which explains the high interferences of easily ionized elements. In most cases internal standards are used. Rules for the selection of suitable lines for internal standardization have been set up and can be derived from the formulae giving the intensities of atomic emission lines. It is clear that a suitable line pair should make use of either two atom or two ion lines, the ionization potentials of the two elements should be fairly similar, as should the excitation energies as well as the transition probabilities of the lines selected. Also their wavelengths should be fairly close so as to avoid differences in the spectral response of the detector.

The plasma is operated in an atmosphere of a noble gas or often also in O_2 or CO_2, so as to create favorable dissociation and excitation conditions or to avoid intense bands such as the CN bands. Universal procedures for the analysis of oxide powders by dc arc analysis are well known and are still in use (see e.g. Ref. [351]).

Dc arcs are also of use for the analysis of metals, as is still done for copper with the so-called globule arc. Here distillation of volatile elements from the less volatile matrix is applied.

Stabilized dc arcs

Such arcs can be obtained by sheathing gas flows or by wall-stabilization, as e.g. described by Riemann [352]. Magnetic fields can be used to provide rotating arcs, which have a better precision. Here a magnetic field should be applied transversally to the direction of the arc, and the rotation frequency is a function of the magnetic field strength as well as of the arc current, as described in classical dc arc papers from Todorovic et al. (see e.g. Ref. [353]).

Dry solution analysis

This can be performed successfully both on metal electrodes (e.g. copper) or with graphite electrodes which have been previously impregnated with a solution of Apiezon (e.g. in benzene). In the latter case carriers such as CsCl may be useful so as to increase the line to background ratios as a result of their influence on the current voltage characteristics [354]. The copper arc method has found use in the determination of the rare earths in solutions, which was historically important in radiochemistry related topics [355–357].

Dc arc spectrography is still a most powerful method for trace determinations in solids even with difficult matrices such as U_3O_8. Here the detection limits for many elements are down in the sub-$\mu g/g$ range [358]. It is still in use for survey analysis, especially in the case of ores, minerals and geological samples. In work with unipolar arcs, re-ignition of the arc is often facilitated by providing an hf discharge over the dc arc, by which the arc channel is kept electrically conductive.

Dc arc work has been given new impetus since the availability of CCD spectrometers, allowing simultaneous detection of all spectral lines photoelectrically, which could be expected to revive dc arc work for general survey analyses (see e.g. Ref. [359]).

5.4.1.3 Ac arc spectrometry

Ac arcs have an alternating polarity at the electrodes. To reignite the arc an hf discharge is often used, which is superimposed onto the ac discharge gap. Because thermal effects are lower and the burning spot changes more frequently in the ac arc, the reproducibility is better than in the case of dc arcs. Here RSDs of the order of 5–10% may be obtained.

5.4.2
Spark emission spectrometry

Spark emission spectrometry is a standard method for the direct analysis of metallic samples and is of great use for production as well as for product control in the metallurgical industry (for a treatment of this subject see Ref. [360]).

5.4.2.1 Sparks

Sparks are short electrical capacitor discharges between at least two electrodes. The spark current is delivered by a capacitor which is charged to a voltage U_C by an extra circuit (Fig. 96A). When the voltage is delivered directly by an ac current an ac spark is obtained. The voltage can also be delivered after transforming to a high voltage and rectifying, in which case dc spark discharges occur. In both instances the voltage decreases within a very short time from U_c to U_B, which is the burning voltage. During the formation of a spark, a spark channel is first built and then a vapor cloud is produced at the cathode. This plasma may have a temperature of 10 000 K or more whereas in the spark channel up to 40 000 K may be reached. This plasma is in local thermal equilibrium but its lifetime is short and the after-

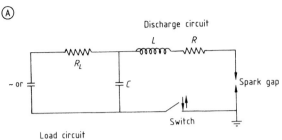

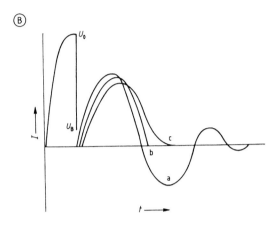

Fig. 96. Circuitry in spark generator for optical emission spectrometry (A) and types of spark discharges (B), a: oscillating spark, b: critically damped spark and c: over-critically damped spark.

glow is soon reached. Here the temperature and the background emission are low but due to the longer lifetime of atomic and ionic excited states line emission may still be considerable.

In a discharge circuit with capacity C, an inductance L and a resistance R the maximum discharge current is:

$$I_0 = (U_0 - U_B)\sqrt{(C/L)} \approx U_0\sqrt{(c/L)} \qquad (290)$$

The spark frequency within the spark train is:

$$f = 1/(2\pi)\sqrt{[1/LC - R^2/(4L^2)]} \qquad (291)$$

or when R is small:

$$f = 1/(2\pi)\sqrt{(1/LC)} \qquad (292)$$

The capacitor may store energy of:

$$Q = 1/2U_C^2 \cdot C \qquad (293)$$

of which 20–60% is dissipated in the spark and the rest in the auxiliary spark or as heat. Depending on R, L and C the spark has different properties, e.g. when L increases the spark becomes harder.

As is shown in Fig. 96B, there can also be oscillating sparks ($R < 2\sqrt{L/C}$) (a), critically damped ($R = 2\sqrt{L/C}$) (b) or overcritically damped sparks ($R > 2\sqrt{L/C}$) (c).

High voltage sparks are of $10 < U_C < 20$ kV and are frequently used. Their time cycle is regulated mechanically or electrically. Medium voltage sparks ($0.5 < U_C < 1.5$ kV) are now used for metal analysis. They require an hf discharge for ignition and are usually unipolar discharges in argon. Accordingly, band emission is low, the most sensitive lines of elements such as C, P, S, and B can be excited and the VUV radiation is not absorbed. Often, a high energy is used to prespark, by which the sample melts locally, is freed from structural effects and becomes homogeneous. Frequencies of up to 500 Hz are often utilized, by which high energy can be dissipated and a high rate of material ablation is obtained. Often a unidirectional spark is used, which is possible by rectification in the spark circuit. Then no material ablation at the counter electrode is to be expected and e.g. the use of tungsten counter electrodes is possible without suffering from contributions of the line-rich tungsten spectra to the spectral background.

5.4.2.2 Analytical features

Spark emission spectrometry enables a rapid and simultaneous determination of many elements in metals, including C, S, B as well as P and even gases such as N and O. Therefore, spark emission atomic emission spectrometry is very complementary to x-ray spectrometry for analyses in metallurgical laboratories [361]. Analysis times including presparking (5 s) may be down to 10–20 s and the method is of great use for production control. Special sampling devices are used to sample liquid metals, e.g. from the blast furnace or convertor and particular care must be taken that samples are homogeneous and do not include cavities. The detection limits obtained in metal analysis differ from one matrix to another but for many elements they are in the µg/g range (Table 11). Thus the power of detection of spark analysis is lower than with dc arc analysis.

The analytical precision is high. By using a matrix line as an internal standard line, RSDs at the 1% level and even lower are obtained.

Matrix interferences are high and relate to the matrix composition but also to the metallurgical structure of the sample. Therefore, a wide diversity of standard samples are used and also matrix correction procedures are applied, as discussed previously (see Section 4.2.1). More appropriate selection of the analytical zone of the spark is now possible with the aid of optical fibers, which could allow the number of matrix interferences to decrease. Indeed, the energy addition zone and the signal generation zone can thus be separated, by which changes in sample composition have less influence on the energy uptake in the spark, the signals become less affected and calibration becomes easier.

The acquisition of the spectral information from each single spark separately is now possible with advanced measurement systems, using integration in a small capacitor and rapid computer-controlled readout combined with storage in a multichannel analyzer. With the aid of statistical analysis, sparks with outlier signals can be rejected and the precision improved accordingly. Furthermore, pulse differential height analysis enables discrimination between a number of elements such

Tab. 11. Detection limits (µg/g) for steels in spark and glow discharge OES and glow discharge MS.

Element	Spark OES[a]	GD-OES[b]	GD-MS[c]
Al	0.5	0.1	
B	1	0.3	
Cr	3	0.05	
Cu	0.5	0.3	
Mg	2	0.9	
Mn	3	0.2	
Mo	1	0.8	
Nb	2	0.6	
Ni	3	0.1	
Pb	2		0.004
Si	3	0.4	
Ti	1	0.6	
V	1	1	
Zr	2	1.5	

[a] Spark discharge in argon using polychromator [360];
[b] Grimm-type glow discharge with 1 m Czerny–Turner monochromator [362];
[c] Grimm-type glow discharge and quadrupole mass spectrometer [363].

as Al and Ti, between dissolved metals and excreted oxides. Indeed, the first will give low signals and the last high intensities as they stem from locally high concentrations [364]. Also by measurements in the afterglow of single sparks or by the application of cross excitation in spark AES (see e.g. Ref. [365]), there is still room for improvement in the power of detection.

As a number of elements have their most sensitive lines at VUV wavelengths, such as N, O, P, S, Cl and H, access to the VUV region is very important, as is possible through the use of sealed spectrometers with air-tight flushed spark stands. For elements that are highly relevant metallurgically, such as C, O, H and N, efforts to lower the blank values due to the environment gases are still required to be able to replace classical C, H, N, O analyzers based on sample combustion.

Spark AES has also been used for oil analysis. Here the oil is sampled by a rotating carbon electrode. Also electrically non-conducting powders can be analyzed after mixing them with graphite powder and briquetting.

5.5
Plasma source AES

Plasma sources developed from flames and are now very important tools for multielement determinations in liquid samples and in solids following sample dissolution. In order to overcome the disadvantages related to the low temperatures of chemical flames but to develop sources with a similarly good temporal stability and versatility with respect to the sample introduction, efforts were directed towards electrically generated discharges called plasma sources. At atmospheric pressure these discharges have a temperature of at least 5000 K and provided their geometry is suitably optimized, they allow the efficient take up of sample aerosols. Moreover, in most cases they are operated in a noble gas thus chemical reactions between the analyte and working gas are avoided.

Among these so-called plasma sources in particular dc plasma jets, inductively coupled plasmas (ICP) and microwave discharges have become important sources for atomic emission spectrometry, mainly for the analysis of solutions (see Ref. [28]).

5.5.1
Dc plasma-jet AES

Dc plasma jets were first described as useful devices for solution analysis by Korolev and Vainshtein [366] and by Margoshes and Scribner [367]. The sample liquids were brought into an aerosol form by pneumatic nebulization and arc plasmas operated in argon and with a temperature of about 5000 K were used.

5.5.1.1 Types of plasma jets
So-called current-carrying and current-free plasma jets can be distinguished. With the first type cooling as a result of changes in aerosol introduction has only a slight influence. Indeed, the current decrease resulting from a lower number density of current-carriers can be compensated for by an increase in the burning voltage and this stabilizes the energy supply. The discharge, however, is very sensitive to changes in the electron number density, as caused by varying concentrations of easily ionized elements. When they increase the burning voltage breaks down and with it the energy supply, which again results in drastic temperature decreases.

With the so-called current-free or "transferred" plasma, the observation zone is situated outside the current-carrying zone. A source such as this can e.g. be realized by the use of a supplementary gas flow directed perpendicular to the direction of the arc current and by the observation zone being in the tail-flame. In this observation zone no current is flowing. This type of plasma reacts significantly on cooling as no power can be delivered to compensate for temperature drops. Therefore, it is fairly insensitive to the addition of easily ionized elements. They do not cause a temperature drop but only shift the ionization equilibrium and give rise to ambipolar diffusion, as discussed previously.

The dc plasma jet described for example by Margoshes and Scribner [367] is a current-carrying plasma. This also applies to the disk stabilized arc according to Riemann [352], where the form of the plasma is stabilized by using several disks with radial gas introduction keeping the plasma form stable under different sample loads and the influence of the plasma composition on the plasma properties low. However, the plasma described by Kranz [368] is a transferred plasma.

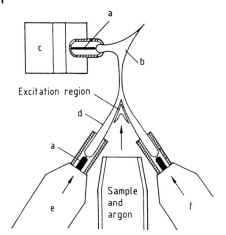

Fig. 97. Three-electrode plasma jet. (a): electrodes; (b): tail flame; (c): cathode block; (d): plasma column; (e, f): argon + anode block front (Reprinted with permission from Ref. [369].)

5.5.1.2 Three-electrode plasma jet

The three-electrode dc plasma jet developed by Spectrometrics Inc. (Fig. 97) [369] uses two graphite anodes protected by sheathing gas flows and both having a separate power supply (2×10 A, 60 V). There is a common shielded cathode made of tungsten. The sample solution is normally nebulized by a cross-flow nebulizer and the aerosol is aspirated by the gas flows around the anodes. The observation zone is small (0.5×0.5 mm^2) but very intensive and located just below the point where the two positive columns join. The optimum combination for this plasma is with a high-resolution Echelle monochromator with crossed dispersion, as the latter has an entrance slit height of around 0.1 mm. It also has the advantages of a current-free plasma, as the observation zone is just outside the current-carrying part of the discharge but still close enough to it to be compensated for rapid cooling. The plasma temperature is of the order of 6000 K, $n_e = 10^{14}$ cm^3 [370], and the plasma is almost in LTE.

The detection limits for most elements are of the order of 5–100 ng/mL. For elements with very sensitive atomic lines such as As, B and P, the detection limits are slightly lower than in ICP-AES [371]. The high level of detection is also certainly related to the high resolution of the Echelle spectrometer used. Different concentrations of alkali elements, however, cause higher matrix effects than in ICP-AES and may even necessitate the use of spectrochemical buffers. The analytical precision achievable is high and RSDs below 1% can be reached. The system can cope with high salt contents (>100 g/L) and has found considerable use e.g. for water analysis [372], and especially for the analysis of seawater, brines and even oils.

The multielement capacity is limited, as the optimal excitation zones for most elements differ slightly, and thus as a result of the small observation zones considerable losses in the power of detection occur when using compromise conditions. Owing to the high resolution of the Echelle spectrometer, DCP-AES is also very useful for the analysis of materials with line-rich emission spectra, such as high-alloy steels, tungsten, molybdenum and other refractory metals. It has also entered into speciation work, where applications such as the determination of metalloproteins by coupling DCP-OES with chromatography have been described [373].

5.5.2
Inductively coupled plasma AES

Inductively coupled plasmas (ICPs) in their present form go back to the work of
Reed [374], who used these sources for crystal growth in the 1960s. They were introduced into spectrochemical analysis, by Greenfield et al. in 1964 [375] and by
Wendt and Fassel [376], both of whom used them in their present form as sources
for atomic emission spectrometry.

5.5.2.1 The inductively coupled plasma
In this type of plasma the energy of a high-frequency generator (in the frequency
range of around 6–100 MHz) is transferred to a gas flow at atmospheric pressure
(mostly argon) in a quartz tube system with the aid of a working coil (Fig. 98). The
electrons take up energy and collide with atoms, by which a plasma with a temperature of up to 6000 K is formed. At a suitable gas flow, geometry of the torch and
frequency, a ring-shaped toroidal plasma is formed, where the sample aerosol
passes centrally through the hot plasma. The burner consists of three concentric
quartz tubes. The aerosol is led with its carrier gas through the central tube. Between the outer and the intermediate tube a gas flow is introduced tangentially. It
takes up the high-frequency energy but also prevents the torch from melting.

A torch of the type according to Fassel has a diameter of ca. 18 mm and can be
operated at 0.6–2 kW and with 10–20 L/min of argon. In so-called mini-torches gas
consumption can be down to about 6 L/min, through the use of a special gas inlet

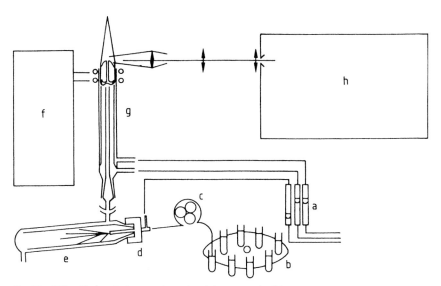

Fig. 98. ICP-optical emission spectrometer. (a) gas supply, (b)
sampler, (c) peristallic pump, (d) nebulizer, (e) nebulization
chamber, (f) hf generator, (g) ICP torch, (h) spectral apparatus,
data acquisition, data processing.

or by further reduction of the torch dimensions [377]. It has been reported that ICPs can even be operated at powers of down to 200 W and with an argon consumption as low as 1 L/min [378]. However, this plasma is no longer a robust source that is capable of taking up wet aerosols of any aqueous sample solution [379]. Greenfield used a larger torch (outer diameter of up to 25 mm), in which an ICP can be operated at higher power and where also argon, oxygen or even air can be used as the outer gas flows (up to 40 L/min). Here an intermediate argon flow (about 8 L/min) is used. With the Fassel torch an intermediate gas flow (about 1 L/min) is useful when analyzing solutions of high salt contents or organic solutions, so as to prevent salt or carbon deposition, respectively.

The operation of a plasma based on air has also been tried specifically for environmental monitoring purposes [380, 381]. The plasma can only be succesfully ignited using argon and then slowly changing the gas from argon to air or nitrogen. These discharges are operated at 1.5–2 kW, and with respect to smoothness of the spectral background and power of detection are much less powerful than the argon ICP. As in addition they are difficult to ignite and to keep stable, they have not achieved a breakthrough as yet.

The ICP is almost in local thermal equilibrium. Indeed, the excitation temperatures (from atomic line intensity ratios) are about 6000 K [382] and the rotation temperatures (from the rotation lines in the OH bands) are 4000–6000 K (see Refs. [383, 384]). From the broadening of the H_β-line, an electron number density of 10^{16} cm^{-3} is obtained, whereas from the intensity ratio of an ion and an atom line of the same element the electron number density found is 10^{14} cm^{-3}. It has also been reported that measured line intensity ratios of ion to atom lines are higher by a factor of 100 than those calculated for a temperature of 6000 K and the electron number density found is 10^{16} cm^{-3}. This indicates the existence of over-ionization. This can be understood from the excitation processes taking place. They include the following.

- Excitation through electrons or processes involving radiation:
  $Ar + e \rightarrow Ar^*$, Ar^m (metastable levels: $E \sim 11.7$ eV), and
  Ar^+ ($E_i = 15.7$ eV)
  $Ar + h\nu \rightarrow Ar^+$ (radiation trapping)
  $M + e \rightarrow M^*$, $M^{+*} + e$ (electron impact)
  $M^+ + e \rightarrow M + h\nu$ (radiative recombination)
- Excitation through argon species:
  $Ar + M \rightarrow Ar + M^*$
  $Ar^m + M \rightarrow Ar + M^+$ (Penning effect)
  $Ar^+ + M \rightarrow Ar + M^{+*}$ (charge transfer)

Accordingly, an over-population of the argon metastable levels would explain both the over-ionization as well as the high electron number density in the ICP. Indeed, it could be accepted that argon metastables act both as ionizing species and at the same time are easily ionized [385]. This could explain the fairly low interferences caused by easily ionized elements and the fact that ion lines are excited very efficiently, despite the fact that their norm temperatures are much higher than the plasma temperatures. However, the discrepancies shown are not encountered to such an extent when the temperatures from the Saha equation are used in the calculations [386]. Nevertheless, it can be accepted that a number of processes are predominant in well-defined zones of the plasma, as indicated by spatially-resolved measurements of various plasma parameters (see e.g. Ref. [387]). New detailed knowledge can be achieved by using Thomson scattering, with which both electron number densities and electron temperatures can be determined irrespective of whether the plasma is in local thermal equilibrium or not, and this can be achieved with high spatial resolution. Several investigations on the contribution of different excitation mechanisms, the influence of easily ionized elements on the processes in the different zones of the plasma and the influence of working parameters have been published by Hieftje and coworkers [388–390].

5.5.2.2 **Instrumentation**

ICP generators are R-L-C-circuits producing a high-frequency current at high power (Fig. 98). The end-stage can be a triode, but nowadays transistor technology can also be used. The hf energy can be led to the remote plasma stand with the aid of a coaxial cable. This requires the use of a matching unit so as to tune the plasma optimally before and after ignition with the generator. However, in other systems the working coil can be provided as an integral part of the oscillator. Generators with a fixed frequency as well as generators where the frequency can change with the load are also used. In the first case, a change in the load changes the ratio of the forward power and the power reflected to the generator and necessitates retuning, whereas in the second case a small frequency shift occurs. Both types of generators may perform equally well provided the power supply to the plasma, which usually amounts to about 50% of the energy dissipated in the generator, remains constant. The stability of the power and the constancy of the gas flows are both essential in order to obtain the highest analytical precision.

ICP generators with frequencies ranging from 27 to 100 MHz have been used and in general 27.12 and 40 MHz generators are now available. As a result of detailed investigations by Mermet and coworkers only slight differences in the performance in terms of detection limits and matrix interferences have been measured (see e.g. Ref. [391]).

The ICP can be viewed side-on as well as end-on for atomic emission spectrometry. Whereas in the beginning side-on observation only was used, end-on observation was later found to have a number of advantages: (i) the optical pathway is longer, which may remove detector noise limitations especially when working at low UV and VUV wavelengths; (ii) it is easy to realize a system that is transparent at VUV wavelengths, enabling determinations to be made of elements such as the halogens, phosphorus, sulfur but also many metals at lines free of spectral interferences and with high line-to-background intensity ratios [392]. Therefore, limitations with respect to the linear dynamic range and higher easily ionized element interferences as a consequence of the larger influence of geometry changes of the plasma may occur [393].

In plasma spectrometry the sample must be brought as a fine aerosol into the plasma, which as a rule is achieved by pneumatic nebulization. The aerosol should enter the hot plasma zones with a low injection velocity so as to realize residence times in the ms range that are required for an efficient atomization and excitation. Therefore, the carrier gas flow should not exceed 1–2 L/min. With advanced pneumatic nebulizers for ICP spectrometry an aerosol with a mean droplet diameter in the 5 µm range can be obtained with an efficiency of a few % [139, 140] for a gas flow of 1–2 L/min and a sample uptake rate of 1–2 mL/min. The concentric nebulizer made of glass (Meinhard nebulizer) is self-aspirating, the cross-flow nebulizer may have capillaries made of different materials and is pump-fed but is less sensitive to clogging in the case of solutions containing high salt contents. The Babington-type nebulizer can be used successfully for slurry nebulizations as was first shown by Ebdon and Cave [116]. Later it was used in ICP-AES for many types of powder analyses including also ceramic powders [117]. The fritted disk nebulizer has not found wide use. Apart from those mentioned, new approaches for ICP-AES lie in the use of the jet-impact nebulizer [142], direct injection nebulization [121] and high-pressure nebulization [143] (see Section 3.1).

The high efficiency nebulizer [110] works at high pressure delivering a high nebulization efficiency at very low sample uptake, which is ideally suited for coupling ICP-AES or mass spectrometry with chromatography for speciation purposes. High efficiency is also obtained with the so-called oscillating nebulizer, as diagnosed when using different advanced aerosol study techniques, as reviewed by McLean et al. [122]. In plasma spectrometry the idea of micro total analysis systems (µTAS) incorporating sample preparation, trace–matrix separation and determination in one system has also become increasingly important. The miniaturization of all steps beginning with those of sample pretreatement allows it to work with the most minute amounts of sample as e.g. required in biomedical research, to use lowest amounts of reagent and to minimize chemical waste problems. To this aim the use of high-efficiency nebulizers working with minute amounts of sample is very suitable. So far work has been done with capillary electrophoresis, which can easily be housed in microstructured systems on a chip, as shown e.g. by Bings et al. [394].

In pneumatic nebulization for ICP-AES continuous sample feeding requires a sample aspiration time of about 30 s, so as to attain a stationary signal, a measurement time of around 5–10 s and a rinsing time again of 30 s at minimum. However, discrete sampling is also possible with injection systems known from flame AAS [125, 126] and by flow injection. Work with sample aliquots of down to 10 µL then becomes possible, which e.g. is particularly useful when it comes to working with microsamples [140], or for the analysis of solutions containing high salt contents [395].

For ICP-AES both sequential and simultaneous as well as combined instruments are used. In sequential spectrometers special attention is given to the speed of the wavelength access and in simultaneous spectrometers to the provision of background correction facilities. In combined instruments a number of frequently used channels are fixed and with a moving detector or an integrated monochromator access to further lines is possible. Echelle instruments gained interest as they allow the highest resolution at low focal length to be realized and as they can be successfully combined with advanced detectors such as CCDs. Paschen–Runge spectrometers with a large number of CCDs covering the whole analytically usable spectral range, certainly for ICP-AES, allow full use to be made of the spectral information at all wavelengths simultaneously.

5.5.2.3 Analytical performance

Power of detection

The nebulizer gas flow, the power, and the observation height are important parameters to be optimized so as to get the highest power of detection for an element (single element optimization) or a number of elements (compromise conditions). Such optimizations can be done by single-factor studies or Simplex optimization (see e.g. Refs. [396, 397]). The nebulizer gas influences the aerosol droplet size and nebulizer efficiency but also the plasma temperature. Accordingly, for each element and line there will be a fairly sharp optimum for the aerosol gas flow where the line-to-background ratio is maximum. The power mainly determines the plasma volume and will be optimum for so-called soft lines (atom lines and ion lines of elements with low ionization potential) at rather low values (1–1.2 kW), whereas for hard lines (lines of elements with high excitation potential and highly energetic ion lines) as well as in the case of organic solutions the optimum value is higher (1.5–1.7 kW). The optimum observation height will be a compromise between the analyte number densities (highest in the lower zones) and the completeness of atomization and excitation (a few mm above the working coil). Only elements such as the alkali metals are measured in the tail flame. Usually an observation window (e.g. 2×2 mm^2) that is not too small is selected so as to secure sufficiently high photon fluxes through the spectrometer in order to avoid detector noise limitations. As the analytical lines in ICP-AES have a physical width of 1 to up to 20 pm and the spectra are line rich the use of a high-resolution spectrometer is required.

The detection limits for most elements are at the 0.1 (Mg, Ca,...) to 50 ng/mL level (As, Th,...). In particular, for elements which have refractory oxides (such as Cr, Nb, Ta, Be, rare earths,...), a fairly low ionization potential and sensitive ion lines, the detection limits are lower than in AAS. On the contrary, the detection limits for elements with high excitation potentials, such as Cd, Zn and Tl, are lower in AAS than in ICP-AES. The detection limits for B, P, As,..., which have more sensitive atom lines, are somewhat lower in DCP-AES than in ICP-AES. For P, S, N, O and F, the most sensitive lines are at VUV wavelengths and here with vacuum spectrometers and VUV transparent pathways and optics, values of 450 ng/mL (Br), 200 ng/mL (Cl), 55 ng/mL (I), 3.9 ng/mL (P) and 3.5 ng/mL (S) can be obtained [348].

The detection limits obtained by radial and axial observation were found to be up to one order of magnitude lower in the axial as compared with the radial viewing, as reported by Ivaldi and Tyson [393]. These results were obtained with an Echelle spectrometer using CCD detection. The results were found to be partly due to better signal to background ratios but also to lower RSD values. The latter may be understood from the better stabilization of the plasma.

Interferences

Most interferences in ICP-AES are additive interferences and relate to coincidences of spectral background structures with the elemental lines used. They are minimized by the use of high-resolution spectrometers and background correction procedures. Indeed, in the spectral wavelength range of 200–400 nm more than 200 000 atomic emission lines of the elements have been listed and in addition molecular bands also occur. Accordingly, coincidences will often occur within the spectral line widths themselves. In order to facilitate the selection of interference-free spectral lines in ICP-AES, Boumans worked out a system where the interferences for about 900 prominent lines [334] are given for spectrometers with a certain spectral slit width in terms of critical concentration ratios. Further line tables for ICP-AES [398] and atlases [329] are available, although incomplete. In ICP-AES multiplicative interferences also occur. They may have several causes, namely:

- Nebulization effects: As discussed earlier, differences in the physical properties of the different sample and calibration solutions lead to variations in the aerosol droplet size and thus also in the efficiency of the nebulizer and the sample introduction. This effect is strongest in the case of free sample aspiration and relatively low nebulizer gas flow and can be minimized (see Section 3.1).
- Easily ionized elements have a complex influence: Firstly they may cause a decrease in excitation temperature, as energy is consumed for ionization. Further, they may shift the ionization equilibrium for partially ionized elements as they influence the electron number densities through their easy ionization. They may also cause changes in plasma volume as a result of ambipolar diffusion or eventually quench exciting species such as metastables [399]. After correcting for the influence of easily ionized elements on the spectral background, signal enhancements or suppressions are negligible up to concentrations of 500 µg/mL of Na, K, Ca or Mg. At larger differences in concentration matrix-matching of the standards should be applied.

Ivaldi and Tyson [393] reported that interferences caused by NaCl were found to be nearly twice as high in the case of axial viewing than in radial viewing and also atom to ion line intensity ratios were influenced more severely in the case of axial viewing.

Analytical precision

As the sample introduction into the plasma is very stable, RSDs in ICP-AES are about 1% and the limiting noise is proportional to this, including flicker noise and nebulizer noise.

However, as shown in the noise spectra (Fig. 9), there is a frequency dependent component at about 200 Hz, this being due to the gas flow dynamics, as has been mentioned by Belchamber and Horlick [400]. The flicker noise present increases when particles are entering the ICP that can no longer be completely evaporated. This is e.g. the case when blowing powders into the ICP.

At wavelengths below 250 nm, however, detector noise limitations may occur especially with high-resolution spectrometers. For integration times exceeding the 1 s level, the full precision is normally achieved. In the case of CCD detection, readout noise may also become important.

With the aid of an internal standard, fluctuations in sample delivery in particular and up to a certain extent changes in excitation conditions can also be compensated for. To appreciate the full benefit of internal standardization, the intensities of the analytical line and the internal standard line, and in trace determinations eventually also of the spectral background near the analytical line, should be measured simultaneously, which can be easily done in the case of a CCD spectrometer. The full analytical precision of ICP-AES in fact can be thus obtained. This can actually be demonstrated by the determination of the main components, as is required in stoichiometric determinations of advanced ceramics. After optimizing the sample dissolution procedure for the case of silicon–boron–carbo–nitride, which can be made speeded up by microwave-assisted decomposition with acids [401] silicon at the 60% level could be determined with a sequential spectrometer with an absolute error of analysis (standard deviation inluding replicated analyses as well as decompositions) of 1% or better. In the case of high-temperature superconductors, the precision at the 0.4–3% (m:m) level for Pb–Bi–Sr–Ca–Cu–O superconductors can be even higher, which is understandable from the less stringent dissolution conditions required and the thermochemical properties of the related elements, which are also less prone to volatilization losses and contamination from the laboratory environment. With the aid of analyses with an ICP-CCD based spectrometer, formulae can be determined as: $Pb_{0.404\pm0.001}Bi_{1.795\pm0.007}Sr_{1.95\pm0.02}Ca_{2.11\pm0.02}Cu_{3.04\pm0.02}O_{10.8+0.1}$ [402]. In many cases this results in relative errors of 1% and lower. This is possible by simultaneous measurements of line and background intensities, leaving more room for improvements by internal standardization. This, however, was found to be very critical with respect to line pair selection.

It can thus be seen that ICP-AES could now replace spectrophotometric and titrimetric methods completely with respect to the precision achievable, also when considering the main components, as in addition it has been shown to be suitable for advanced working materials [403].

The principles of the selection of line pairs in the case of ICP-AES have been described well and in detail by Fassel et al. [404].

For an intensity ratio of two atomic emission lines from two different elements:

$$I_{qp(Xo)}/I_{sr(Ro)} = \{[n_{(Xo)} \cdot A_{qp} \cdot v_{qp} \cdot g_q \cdot Q_{Ro}]/[n_{(Ro)} \cdot A_{sr} \cdot v_{sr} \cdot g_s \cdot Q_{(Ro)}]\}$$
$$\cdot \{\exp[(E_s - E_q)/kT]\} \tag{294}$$

Fluctuations in number densities of both elements X (analyte) and R (reference element) are cancelled. In order to keep the influence of fluctuations in the excitation temperature low, lines with similar excitation energies should be used and also their partition functions should be of the same order of magnitude, rendering the choice difficult. Two atom or two ion lines should also be selected and in order to correct for changes in sample delivery, the ionization has to be taken into account:

$$n_{(X+)}/n_{(R+)} = [n_{(X)} \cdot K_{(Xo \rightarrow X+)} \cdot (n_e^2 + K_{(Ro \rightarrow R+)} \cdot n_e + K_{(Ro \rightarrow R+)} \cdot K_{(R+ \rightarrow R++)})]/$$

$$[n_{(R)} \cdot K_{(Ro \rightarrow R+)} \cdot (n_e^2 + K_{(Xo \rightarrow X+)} \cdot n_e + K_{(Xo \rightarrow X+)} \cdot K_{(X+ \rightarrow X2+)})$$

$$(295)$$

with:

$$K_{(Xo \rightarrow X+)} = [n_{(X+)} \cdot n_e]/n$$

$$= [(2 \cdot \pi \cdot m_e \cdot k \cdot T)^{3/2}]/(h^3)\} \cdot [2 \cdot Q_{(X+)}/Q_{(Xo)}]\exp[-E_{i(Xo)}/(k \cdot T)]$$

$$(296)$$

$$K_{(X+ \rightarrow X2+)} = [n_{(X2+)} \cdot n_e]/n$$

$$= \{[(2 \cdot \pi \cdot m_e \cdot k \cdot T)^{3/2}]/(h^3)\} \cdot [2 \cdot Q_{(X2+)}/Q_{(X+)}]$$

$$\cdot \exp[-E_{i(X+)}/(k \cdot T)]$$

$$(297)$$

and similarly for R.

These equations show that the analyte and reference elements should have ionization energies of the same order of magnitude. Changes occurring similarly for analyte and reference element number densities can be fairly well compensated for by selecting I_{qp}/I_{sr} as the analytical signal, however, this applies less for changes in the temperatures. Departures from LTE in the ICP should be taken into account and also that analytical lines are selected for which the norm temperatures are far above the plasma temperatures. This is still more stringently controlled in the case of multielement determinations, where the selection of different reference elements for a number of analytes may actually be required.

The possibilities of simultaneous background correction in the case of a two-channel spectrometer have been described in early ICP papers by Meyers and Tracy [405]. In the case of brass alloys with internal standardization, RSDs below 0.1% can be obtained with simultaneous ICP-AES spectrometers for copper and zinc, where ICP-AES comes near to the precision achievable with x-ray fluorescence spectrometry for metal analysis [406].

Multielement capacity

The multielement capacity of both sequential and simultaneous ICP-AES is high as is the linear dynamic range (up to 5 orders of magnitude). It is restricted on the lower side by the limits of determination and at the higher concentrations by the beginning of self-reversal. This only starts at analyte concentrations of about 20 000 µg/mL, which is due to the fact that analyte atoms hardly occur in the cooler plasma zones. The linear range for axial viewing and radial viewing of the plasma may differ somewhat. For axial viewing as the optical pathway is longer, the self-absorption and risks of self-reversal increase. The latter in particular is more likely when the analyte has to pass through a longer but cooler zone. This can be avoided both by skimming the ICP as well as by the use of a gas flow entering perpendicular to the ICP axis and blowing the analyte away from the cooler zones. These differences in linear dynamic range were also reported by Ivaldi and Tyson [393].

Special techniques

The analytical performance of ICP-AES can be considerably improved by using alternative techniques for sample introduction.

Organic solutions Solutions such as in MIBK, xylene or kerosene can also be nebulized pneumatically and determinations in these solvents can be performed directly with ICP-AES (see e.g. Ref. [407]). The addition of oxygen or the use of higher power is helpful so as to avoid carbon deposits. It has also been suggested that the use of burners with sharp edges and improved flow dynamics are very helpful at preventing carbon deposits [408]. Oils can be analyzed after dilution 1:10 with xylene (c_L: 0.5–1 µg/g). Although xylene has been used as a diluent, however, MIBK and other ketones may be more problematic especially at a lower powers of operation. Heavy metals can be extracted from waste waters as APDTC complexes with organic solvents (such as MIBK, xylene or CCl_4) and thus separated from the accompanying easily ionized elements that cause interferences [407].

Advanced pneumatic nebulizers There has been increasing interest in these nebulizers for ICP-AES.

Maximum dissolved solids nebulizers [111] are usually cross-flow nebulizers operated at a high pressure. They are often made from PTFE, including the capillaries, and thus can be used for work with HF containing solutions. In this case a nebulization chamber made of PTFE and an ICP torch with an internal tube made of Al_2O_3 ceramics or BN should be used.

High-efficiency nebulizers are also cross-flow nebulizers, usually operated at very high pressures and are available from several manufacturers for ICP work. Fritted disk nebulizers have not made a great breakthrough for practical ICP work. More applications can be found for the Babington nebulizers when it comes to slurry nebulization for fine powders, based again on Ebdon's work. The Légère nebulizer and the grid nebulizer are other types of Babington nebulizers that have been developed and have been shown to offer good efficiencies at relatively low gas flows. Systems delivering monodisperse aerosols are very worthwhile for the study of evaporation processes and because of their relationship to the power of detection and interference effects in ICP-AES. Towards this aim, the MDMI (monodisperse dried microparticulate injector) has been used by Olesik (see e.g. Ref. [409]). It is a system that combines a droplet generator and an oven, reproducibly generating mono-disperse droplets of controllable size and velocity on demand.

Ultrasonic nebulization   This has been applied since the early work on ICP-AES [151]. Both nebulizer types where the sample liquid flows over the nebulizer transducer crystal and types where the ultrasonic radiation (at 1 MHz frequency) is focussed through a membrane on the standing sample solution have been used. When applying aerosol desolvation the power of detection of ICP-AES can be improved by a factor of 10 by using ultrasonic nebulization. This specifically applies to elements such as As, Cd and Pb, which are of environmental interest. However, because of the limitations discussed in Section 3.2, the approach is of particular use in the case of dilute analytes such as in water analysis [150]. Additional fine detailed development, however, is regularly carried out, as with ICP-AES the process is crucial for elements such as Cd, As and Pb for which threshold values in fresh water samples can just still be measured reliably with this type of sample introduction. Such a development is the microultrasonic nebulizer (μUSN) operated with argon carrier gas, as described by Tarr et al. [410].

High pressure nebulization   As has been described by Berndt, this has been shown to provide both a gain in power of detection as well as good short- and long-term stability in ICP-AES. It is interesting to see the gain in power of detection when the process is performed at high temperature [145]. A further development is its on-line coupling with sample decomposition of suspended material, which is important for water and for biological samples [130].

Thermospray

Since their introduction in the 1980s thermospray systems have been shown to increase the aerosol generation efficiencies and accordingly the power of detection. In ICP-AES they have been further developed into systems where a gain in the power of detection and high accuracy can be achieved for real samples as compared with conventional pneumatic nebulization [411].

It should be kept in mind that both thermospray nebulization and high-pressure nebulization [143] successfully allow the analyte introduction efficiency to be increased and thus also the power of detection, however, again only when aerosol desolvation is applied. They are especially interesting for speciation by on-line coupling of ICP-AES and HPLC, as shown later.

Electrothermal evaporation and direct sample insertion   This procedures allow the sampling efficiency to increase so much that work with microsamples becomes possible and at the same time the power of detection for the concentration can be improved further. ETV-ICP-AES has been applied with graphite cups [189], graphite furnaces [186] but also with tungsten filaments [412]. It has been applied not only with dry solution residues but also with dry suspension residues. The latter approach has been developed extensively by Krivan and coworkers and applied to the analysis of refractory powders (e.g. AlN and Al_2O_3 [413]). In the case of Al_2O_3 powders, it was shown by Wende [34] that volatilization aids were very helpful at increasing the

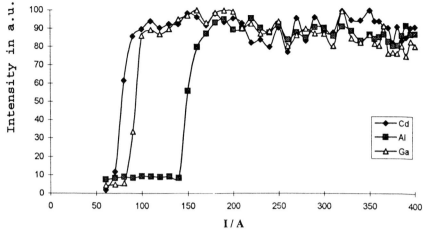

Fig. 99. Volatilization of dry solution aliquots from a graphite boat evaporation system coupled to a simultaneous ICP optical emission spectrometer at different heating currents [34].

number of elements that can be volatilized from the matrix. It was also possible to avoid volatilization of the matrix in a number of cases. The use of AgCl as a thermo-chemical reagent, even in the case of relatively coarse powders weighed into a sampling boat, resulted in a fairly good volatilization of Ca, Fe, Na and Si from the matrix, however, volatilization occurs according to the boiling points of the analytes (Fig. 99).

Direct sample insertion allows the direct analysis of used oils [189] and of micro-amounts of sediments [193] as well as the determination of volatile elements in refractory matrices [191]. Difficulties lie in the calibration and in the signal acquisition. The latter necessitates a simultaneous and time-resolved measurement of the transient signals for the line and background intensities to be made in trace analysis. This may become easier when applying CCD detection with the appropriate software. Also in the case of direct sample insertion the use of thermo-chemical reagents has been found to be useful [414]. An extremely sensitive technique lies in its combination with laser ablation [415]. A very high degree of sample introduction efficiency is reached, so the power of detection may be the highest possible in plasma atomic emission spectrometry.

Hydride generation   As compared with pneumatic nebulization, hydride generation with flow cells [154] allows the detection limits for As, Se, etc. to be decreased by 1.5 orders of magnitude down to the ng/mL level. Thus a hydride generator is a useful accessory to ICP-AES for water analysis. However, the limitations due to possible systematic errors from heavy metals and from analytes being in the non-mineralized form remain. Working at high acid concentrations, the use of complexing agents to mask the heavy metals causing the interferences (tartrate in the case of copper for example) and mineralization by treatment with sulfuric acid until fuming to decompose organoarsenicals is helpful. Electrochemical hydride generation avoids blank contributions from the high amounts of acids that have to be used as well as from the sodium borohydride and the NaOH required to stabilize its solutions. This was first proposed by Liu et al. [416] for AAS work and was optimized by Hueber and Winefordner [417]. Comparative studies of conventional and electrochemical hydride generation with ICP-AES have also been performed and showed that the detection limit of 0.8 ng/mL for conventional hydride generation can certainly be improved by electrochemical hydride generation.

Gas and liquid chromatography   Coupled with ICP-AES these are useful approaches for speciation work, as shown by the early work of Cox et al. [418] for Cr^{III}/Cr^{VI} speciation, where Al_2O_3 filled columns and elution with acids and NaOH were used. Anion exchangers could also be used. New impetus has arisen from the interfacing of thermospray and high-pressure nebulization to ICP-AES. ICP-AES remains of interest for the element-specific detection in the speciation of silicon compounds and is a good alternative to ICP-MS, where considerable spectral interferences hamper the power of detection that can be obtained.

Direct solids sampling   This is of special interest, despite the fact that ICP-AES is mainly of use for the analysis of liquids. In a number of cases it enables the same precision and accuracy to be obtained as in work with solutions but without the need for time-consuming sample dissolution involving analyte dilution.

For compact solids arc and spark ablation are a viable approach for metals [100, 214]. Aerosols with particle sizes at the μm level [216] and detection limits at the μg/g level are obtained [213]. Owing to the separate ablation and excitation stages, matrix influences are particularly low, as shown for aluminum [100] and for steel samples [213]. In the first case, only for supereutectic silicon concentrations were matrix effects obtained (Fig. 100). For low-alloyed samples straight calibration curves are obtained and in the case of high-alloyed steels, even samples with widely different Cr or Ni contents are on the same calibration curves, which, are in fact slightly curved.

Laser ablation can be used with similar figures of merit [419], also for electrically non-conducting samples and with lateral resolution (up to around 10 μm). The aerosol produced with a Nd:YAG laser has a particle size in the μm range. With steel samples, no selective volatilization was found to occur, which contrasted sharply with the case of brass. However, in the latter case spark ablation was also found to result in a high selective volatilization [225]. A very stable material ablation was possible with the so-called LINA (laser induced argon spark) spark system [297], where through a rotation of the focussing lens the plasma is moved over the sample surface. With ICP-AES, detection limits in steels were found to be in the μg/g range, even when measuring the spectral background with a high-purity steel sample (Table 12) and, as shown for a standard reference sample, fairly good analytical accuracy was obtained. It was found that by using the intensity of an Fe line as the internal standard signal the precision could be significantly improved.

5.5.2.4 Applications
ICP-AES is now a routine analytical method, which is of specific use when a large number of elements have to be determined in many samples, whether they are

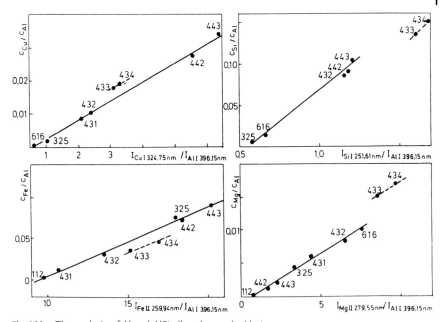

Fig. 100. The analysis of Al and AlSi alloys by spark ablation coupled with ICP-OES. 3 kW argon-nitrogen ICP, Spectrovac 1000 and 1 kV spark at 25 Hz. (Reprinted with permission from Ref. [100].)

Tab. 12. Detection limits obtained with LINA spark coupled with sequential ICP-AES [217].

Sample matrix	Element/line (nm)	Concentration range (%)	Limit of detection (%)
Low-alloyed steel	Ni II 231.604	0.59–4.45	0.002
	Cr II 267.716	0.102–2.95	0.004
	Mo II 281.615	0.09–1.00	0.007
	Mn II 293.306	0.047–0.95	0.002
	Si I 212.412	0.033–1.46	0.006
Brass	Fe II 259.940	0.022–0.73	0.004
	Ni II 231.604	0.009–0.36	0.004
	Sn II 189.989	0.19–0.92	0.018
	Pb II 220.353	0.49–3.01	0.095
Aluminum	Fe II 259.940	0.0079–0.98	0.004
	Cu I 324.754	0.021–2.8	0.035
	Si I 212.412	0.20–9.22	0.024

solutions or dissolved solids. Therefore, in many cases it is complementary to AAS as the power of detection of furnace AAS cannot be rivaled for most elements. ICP-AES is of special interest for the analysis of geological samples, environmental analysis, clinical analysis, food analysis, metal analysis, chemicals analysis and in the certification of reference materials.

In the analysis of geological samples, sample dissolution with respect to nebulization effects but also to ionization interferences is important [399]. The analysis of rare earth minerals without [420] and with removal of the matrix has been reported.

The analysis of environmentally-relevant samples is a major field of application. Based on the work of Garbarino and Taylor [421], a method has been proposed by the US EPA (Environmental Protection Agency) [422] and later by DIN [423] for waste water analysis. The latter, standardized procedure describes the sample decomposition, the analytical range for 22 elements and frequent interferences of ICP-AES in waste water analysis. For the analysis of natural waters, hydride generation [424], preconcentration based on liquid–liquid extraction of the dithiocarbamate complexes [425], adsorption of trace elements onto activated carbon [426] and also co-precipitation [e.g. with $In(OH)_2$] [427], etc. have been reported and special emphasis has been given to speciation (as given in the Refs. in [428]) and on-line preconcentration [134].

In clinical analysis, Ca, Fe, Cu, Mg, Na and K can be determined directly even in microsized serum samples [128]. Also electrothermal evaporation and direct sample insertion are particularly useful from this point of view. Generally, however, the power of detection of ICP-AES is too low for most analytical problems encountered in the life sciences.

Metals analysis in particular should be mentioned in the field of industrial products. Apart from solution analysis also direct metal analysis by spark ablation is very useful [100]. Through the availablity of CCD spectrometers the analytical precision that can be achieved is so high, that even for major components errors as low as or even lower than in classical chemical methods can be achieved. This is documented by the analysis results for stoichiometric determinations of high-temperature superconductors of the Y–Ba–Cu–O type (Fig. 101) [402]. It also applies for ceramics of the Si–B–N–C type, where sample decomposition by various methods including microwave-assisted dissolution, dissolution by fusion and decomposition with acids at high-pressure combined with sequential ICP-AES successfully allowed the stoichiometry to be controlled [401]. In addition, the analysis of refractory powders by slurry nebulization for Al_2O_3 [429], SiC [430] and even for ZrO_2 [132] powder samples or analyses subsequent to well-investigated sample decomposition methods, also making use of microwave-assisted dissolution, [431] are very powerful methods.

5.5.3
Low-power high-frequency plasmas

High-frequency discharges at low power and with capacitive power coupling, known as the stabilized capacitively coupled plasma (SCP), as described by Gross et al. [432]

Concentration aimed at:

$$Pb_{0.40}Bi_{1.80}Sr_{2.00}Ca_{2.10}Cu_{3.00}O_{10.0+x}$$

Analysis by sequential ICP-OES:

$$Pb_{0.405\pm0.007}Bi_{1.78\pm0.03}Sr_{1.97\pm0.01}Ca_{2.08\pm0.02}Cu_{3.07\pm0.05}O_{10.7\pm0.1}$$

Analysis by simultaneous CCD-based ICP-OES (Optima 3000):

$$Pb_{0.404\pm0.001}Bi_{1.795\pm0.007}Sr_{1.95\pm0.02}Ca_{2.11\pm0.02}Cu_{3.04\pm0.02}O_{10.8\pm0.1}$$

Fig. 101. Precision achievable in the analysis of high-temperature superconducting powders by sequential and simultaneous ICP-OES [402].

for element-specific detection in chromatography may, in addition to ICP-AES, be a useful alternative especially when advanced sample introduction techniques are used.

The plasma is sustained in a water-cooled quartz tube with an internal diameter of 1 mm and is powered through two annular electrodes around the capillary and within a distance of about 10 mm. The plasma has a length of about 20 mm and can be operated with helium or argon at a power of 100 W at maximum and a frequency of 27.12 MHz. The plasma itself is very well established between the electrodes and is fairly robust.

For sample introduction flow-cell hydride generation can be used without the need to remove the excess of hydrogen from the reaction gases, giving a detection limit of 40 ng/mL for As. Also the discontinuous hydride generation technique using NaBH$_4$ pellets can be used and then an absolute detection limit for As of 5 ng is obtained. These values are better by a factor of 2 than with a surfatron MIP as the atomic emission source but by a factor of 5 higher than in ICP-AES. Alternatively pneumatic nebulization can be applied provided a desolvation of the aerosol is performed and the figures of merit are then similar to those of an MIP according to Beenakker (detection limits in the 1–10 ng/mL range) [433].

With electrothermal evaporation from a tungsten filament and quartz fiber optics, detection limits are at the 50–100 pg level for many elements, except for Fe which is subjected to spectral interferences from tungsten lines. This was established from the use of different working gases and especially from experiments with the addition of H$_2$ to the argon. In the case of coupling with graphite furnace atomization Cu, Mg and Fe can be determined in serum samples without dilution for Fe and Cu and with a 1:100 fold dilution for Mg [434].

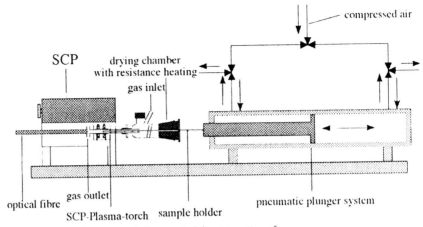

Fig. 102. Set-up for pneumatically operated direct insertion of dry solution residues into an SCP as a source for atomic emission spectrometry. (Reprinted with permission from Ref. [435].)

When working with the SCP, the plasma is so well established between the electrodes, that it is possible to insert a sample into the plasma, without moving it away. This opens up the possibility of drying sample aliquots on a quartz rod and then etching them from the rod by inserting it into the plasma (Fig. 102) [435]. Both with Ar and with He, for Pb, Mg, Cd and Cu detection limits in the 100–600 pg range are obtained for 20 μL samples, however, with considerable interferences from easily ionized elements.

5.5.4
Microwave plasmas

Microwave plasmas are operated at 1–5 GHz. They are produced in a magnetron. The electrons emitted by a glowing cathode are led through a cavity with a series of radially arranged resonance chambers to which an ultra-hf (UHF) field is applied. Strong UHF currents are produced by resonance, which can be coupled out by the anode and transported by a coaxial cable. The microwave current is led to a waveguide. Only when the waveguide has dimensions below certain critical values can microwaves be transported. The transport efficiency can be regulated with the aid of moveable walls and screws. A standing wave is produced in a cavity and the microwave energy is coupled into the resonator by a loop or with an antenna. Both $\lambda/2$, $\lambda/4$ cavities, etc. have been proposed [436] and allow tuning with respect to a favorable ratio of forward to reflected power. Microwave energy can also be transported through strip lines of suitable dimensions formed by metal vapor deposition or galvanically deposited on electrically non-conducting substrates. These strip lines should be protected against heating through reflected power losses by cooling with a fan, a cooling-water circuit or a Peltier cooling element. They can have dimensions of below 1 mm in width, several 0.1 mm in thickness and varying length (up to 10 cm) and are very useful for attemps to miniaturize microwave resonant structures.

Capacitively coupled microwave plasma

This type of single-electrode discharge goes back to the work of Mavrodineanu and Hughes [437]. Here the microwaves are led to a pin-shaped electrode and the surrounding burner housing acts as the counter electrode. A bush-form of discharge burns at the top of the electrode and the sample aerosol enters concentrically with this pin (Fig. 103). The plasma, which can be operated with 2–4 L/min argon, helium or nitrogen (at 400–800 W), is not in LTE (T_{ex} = 5000–8000 K and T_g = 3000–5700 K, n_e = 10^{14}) (see the radially and axially resolved measurements of Bings et al. [438]) and it is a current-carrying plasma. As the sample aerosol enters with a fairly low efficiency into a hot zone where the background emission also will be high, the detection limits especially for elements with thermally stable oxides will be poor (see e.g. Ref. [439]), for example, Na 589 nm: 0.05; Be 234.9 nm: 0.06; Mg 285.2 nm: 0.04; Al 396.1 nm: 0.16 µg/mL. In addition, alkali elements cause high matrix effects; e.g. for 500 µg/mL they amount to 100% in the case of Mg. Therefore, spectrochemical buffers are often used, such as CsCl. Because of the possibility of working with He, the source is also useful for the speciation of widely diverse elements including P, S and the halogens, as shown by Hanamura et al. [440].

A considerable advantage of the CMP is that it can also easily be operated with air only. Even in the case of organic solutions, such as diluted oils, trace determinations are then possible [441]. The high stability of an air operated CMP also allows optical emission spectrometry to be used with a CMP for real-time air quality monitoring. For environmentally relevant elements such as Fe, Cu, Pb and Cd, detection limits in air are obtained (Table 13), which are, for example, below the thresholds set by legislation in Germany [442]. It was also found that high water loadings, as are often present after the washing out of stack gases, or CO_2, as is always present in stack gases of waste incineration, significantly influence but do not drastically deteriorate the detection limits. It also should be possible to decrease the dimensions of a CMP considerably without deteriorating the analytical performance. In addition, wall stabilization of the plasma by additional gas flows will certainly improve the robustness of the CMP still further and improve both power of detection as well as decrease the sensitivity towards variations in easily ionized element concentrations.

Microwave induced plasmas

These electrodeless discharges (MIPs) are operated in a noble gas at fairly low power (mostly below 200 W). They have become important as sources for atomic

(A)

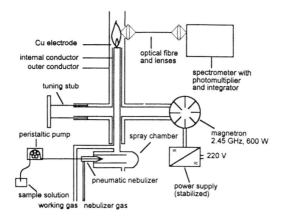

(B)

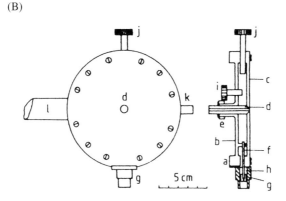

(C)

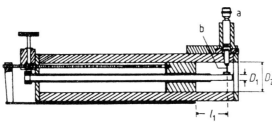

Fig. 103. Capacitively coupled microwave plasma (A) [442] and microwave induced plasma discharges (B, C). (B): MIP in a TM$_{010}$ resonator according to Beenakker; (a): cylindrical wall, (b): fixed bottom, (c): removable lid, (d): discharge tube, (e): holder, (f): coupling loop, (g): PTFE insulator, (h): microwave connector, (i, j): tuning screws, (k): fine tuning stub, (l): holder (reprinted with permission from Ref. [442, 443] and related literature). (C): Surfatron type MIP; original version: 20 mm $< l_1 <$ 250 mm, D_1: 11 mm and D_2: 40 mm, cut-off frequency: 2 GHz (reprinted with permission from Ref. [453]).

Tab. 13. Detection limits (μg/m^3) obtained with OES using an air CMP with different water loading [442]. Maximally allowed concentrations of metals in flue gases in Germany: Σ Cd, Tl, Hg $<$50 μg/m^3 and Σ As, Co, Cr, Cu, Mn, Ni, Pb, Sb, Sn and V $<$500 μg/m^3.

Element/line (nm)	Water loading (g/m^3)		
	32 $\pm$ 4	50 $\pm$ 5	222 $\pm$ 13
Cd I 326.10	29		2000
Co I 345.35			100
Cr I 425.43	0.6	0.5	13
Fe I 371.99	0.6	1.2	51
Mg I 285.21	0.4		25
Mg II 279.55	3.1		254
Ni I 361.93		4.3	130
Pb I 405.78	7.2	4.3	250

spectrometry since Beenakker [443] succeeded in operating a stable plasma at atmospheric pressure with less than 1 L/min argon or helium and a power below 100 W (Fig. 103). Here, the so-called TM$_{010}$ resonator is used. It is a cylindrical resonant cavity (diameter: ca. 10 cm) where the power enters through a loop or an antenna and the filament plasma is contained in a centrally located capillary. At higher powers toroidal argon [444] and diffuse helium [445] discharges can also be obtained. The first of these requires a wet aerosol to be led into the plasma and careful centering of a 4 mm internal diameter quartz tube in the cavity. At a power of 200–300 W a plasma can also be sustained with nitrogen and under the name of MINDAP has been described to be a powerful excitation source [446]. Even at higher power the same resonator can be used and plasmas in helium with high excitation efficiency have been obtained, as investigated by Wu and Carnahan [447].

The MIP is generally not in LTE and metastable argon or helium species are assumed to contribute to the excitation. Moussanda et al. [448] studied the spectroscopic properties of various MIP discharges and observed that the rotational temperatures are about 2000 K. MIPs are very useful for coupling with electrothermal evaporation, where detection limits in the upper pg range can be obtained. This can be realized both with graphite furnace atomization [180] and tungsten filament atomizers [178]. The excitation of desolvated aerosols is also possible. Wet aerosols can only be taken up with the toroidal argon MIP. Accordingly, this allows element-specific detection in liquid chromatography [449]. The detection limits are rather low, which is understandable from the low spectral background intensities in the center of the plasma (Table 14) [450].

Helium MIPs are excellent for element-specific detection in gas chromatography, as has been commercially realized [451]. In this way, not only are the halogens and other elements relevant in pesticide residue analysis but also organolead and organotin compounds determined down to low concentrations. This makes MIP-AES very useful for speciation work [321, 452]. It has been shown that the delocalized helium MIP gave low detection limits for elements with high excitation potentials

Tab. 14. Detection limits (both on a 3σ basis) obatined with OES using a toroidal MIP and pneumatic nebulization of solutions without desolvation as compared with those using a filament-type MIP and aerosol desolvation obtained by Beenakker et al. (Reprinted with permission from Ref. [450].)

Element (nm)	Detection limit (μg/mL)	
	toroidal MIP	filament MIP
Co I 240.73	0.05	0.36
Cr I 425.43	—	0.21
Cr I 357.86	0.01	—
Cu I 324.75	0.003	0.015
Ga I 417.20	0.003	0.015
Mg I 285.12	0.002	0.008
Mg II 279.55	0.002	0.009
Mn I 403.08	0.015	0.08
Mn II 257.61	0.006	0.05
Sr II 407.77	0.002	0.008

such as Cd and P, however, it could also only be operated with dry sample vapors or aerosols.

The MIP is also useful for the excitation of volatile hydride forming elements after stripping the hydrides from the excess of hydrogen. With different trapping techniques detection limits down to the sub-ng level can easily be obtained [156]. A further useful type of MIP is the so-called surfatron described by Hubert et al. [453] (Fig. 103). Here the power enters through an antenna plunging perpendicularly to near the plasma capillary and a plasma is formed by microwave coupling through a slit and power propagation along the surface of the discharge. As compared with the MIP obtained in a TM_{010} resonator according to Beenakker, this plasma is more stable and allows more variability in sampling, as e.g. shown by direct comparative studies with the same tungsten filament atomizer [454].

Also in a TE_{010} resonator, an argon MIP can be operated at a power below 100 W. The discharge has been found to be very stable and can be coupled with electrochemical hydride generation without the need to remove the excess of hydrogen before releasing the reaction gases into the MIP [158]. The detection limits were slightly higher than in the argon or helium surfatron (2.4 versus 1 ng/mL for As) and by a factor of 2 lower than in flow-cell hydride generation using $NaBH_4$. When trapping hydrides in a hot graphite furnace detection limits for As, Sb and Se were found to be 0.08, 0.12 and 0.14 ng/mL in the case of a helium surfatron MIP and determinations in real samples such as chinese tea could be performed [455].

The MIP torch described by Jin et al. [456] (Fig. 104) allows operation with both argon and helium discharges and so the use of hydride generation or electrothermal evaporation is possible. The high excitation efficiency in the MIP torch (MPT) and the high robustness are understandable from the high electron tem-

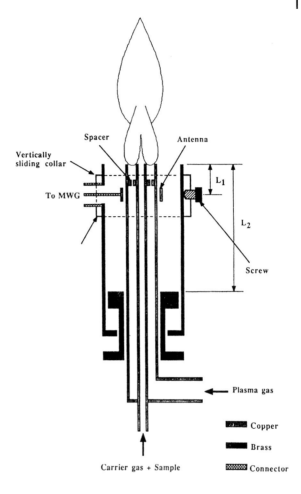

Fig. 104. Microwave plasma torch according to Jin et al. (Reprinted with permission from Ref. [456].)

peratures (16 000–18 000 K) and electron number densities $(2–4 \times 10^{20}$ m$^{-3})$, as determined from Thomson scattering [457] as well as from the relatively high gas temperatures of 3600 K, as approximated by the radially resolved rotation temperatures measured by Engel et al. [458].

After careful optimization and use of a high-efficiency Légère nebulizer, aerosols from aqueous solutions and acetonitrile containing solutions can be introduced into the plasma and the detection limits for elements with sensitive atom lines such Cr are still at the 100 ng/mL level [459]. In the case of flow-cell hydride generation the detection limits found for As, Sb and Se without removing the excess of hydrogen from the reaction gases were at the 2–5 ng/mL level, being no more than a factor of 5 higher than in ICP-AES [460]. Owing to the relatively high gas temperatures, an MIP in an MPT can also be used successfully for evaporating and exciting spark ablated aerosols. The differences of the detection limits obtained

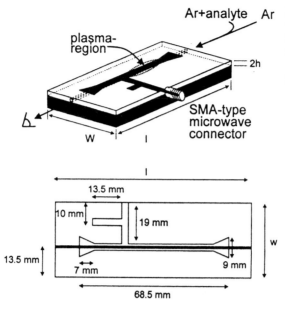

Ar+analyte   Ar

plasma-
region

2h

SMA-type
microwave
connector

△

W   I

Fig. 105. Resonant structure
for MSP miniaturized MIP
inside a quartz wand powered
through microstrips.
(Reprinted with permission
from Ref. [463].)

I

13.5 mm

10 mm   19 mm

13.5 mm   9 mm
7 mm

w

68.5 mm

stripline width: 3 mm

from those of spark ablation ICP-AES for steels, after considering the differences in spectrometers used, were less than an order of magnitude [212].

Through modelling of the electric fields and the influences of the load the dimensions of the MPT can be optimized so that the plasma can be ignited and operated without retuning, making the torch construction much simpler [461].

One of the latest approaches in microwave induced plasmas consists of the realization of microwave discharges in microstructured resonant cavities using microstrip lines for power transmission (an MSP). Such a strucure was achieved by Bilgic et al. [462] (Fig. 105). In a 1 mm diameter channel made in quartz plates a plasma is sustained and powered through microstrip lines produced electrochemically and by metal vapor deposition. The structure is able to withstand a 20 W argon plasma for hours, as cooling is provided by a copper plate forming the grounded electrode. The gas temperatures measured in the plasma were at the 600 K level and the excitation temperatures at 4000 K. When coupled with mercury cold vapor generation devices, a detection limit of 50 pg/mL of Hg was obtained and high analytical short- and long-term precisions were obtained [463]. In a modified version it turned out that it was also possible to obtain a stable discharge with helium at 5–30 W and at 1–10 W in argon at gas flows of 50–1000 mL/min [464]. In the case of helium, chlorine could be successfully excited, showing the potential of the developments especially in combination with miniaturized CCD spectrometers for process control and element-specific detection in gas chromatography, and also in the so-called μTAS (micro total analysis systems).

5.6
Glow discharge AES

Glow discharges have long been recognized as unique sources for atomic emission spectrometry. Their features relate to the possibility of analyte introduction by sputtering as well as to the advantages of non-LTE plasmas, allowing the discrete excitation of terms even including those of high energy. Their analytical properties are related to the geometric and electrical properties of the glow discharges. These discharges have a high burning voltage (up to several kV) and a fairly low current density as compared with arc sources (around 0.1 A/cm^2 as compared with several 10 A per mm^2 for arc sources). The plasma is mainly delocalized and covers the whole electrodes uniformly (up to several cm^2), whereas an arc plasma is very constricted in dimensions (as little as <1 mm^2). This results in a very stable and low-noise operation.

The fact that glow discharges have much lower atom number densities than atmospheric pressure plasmas is responsible for the fact that the measurement of fundamental parameters such as electron number densities, electron temperatures, etc. by techniques such as Thomson scattering is much more difficult than in the case of atmospheric pressure plasma discharges, and that this is only now becoming a field of active research. Moreover, the lower collision frequency, which causes large departures from local thermal equilibrium, is responsible for the fact that many more processes are significantly involved in the excitation mechanisms than in the case of atmospheric pressure plasmas.

Modelling work in this field started later and the first reliable models for the sample volatilization and excitation are only now becoming available as are calculations of the emission spectra, as performed by Bogaerts et al. (see e.g. Ref. [465]). The species assumed to be present in an argon glow discharge plasma include electrons, argon (Ar$^+$, Ar$_2^+$ and Ar^{2+}) ions, fast argon atoms, argon atoms in various excited levels, sputtered material atoms and ions, also in the ground state and in excited levels. The formation of these species and their energies are calculated using a combination of Monte Carlo models (for the electrons, for the Ar$^+$ ions, fast argon atoms and material atoms in the sheath adjacent to the cathode and for the thermalization of material ions), fluid models (for the slow electrons and argon ions) and collisional–radiative models (for the excited argon and material levels). These models are coupled together due to the interaction processes between the species. The last model takes into account that in the case of an argon discharge, for example, the following processes are involved in the populating and de-populating processes of 64 relevant argon levels (see Ref. [466]:

- electron, fast argon ion, fast and thermal argon atom impact excitation and de-excitation between species at all possible levels;
- electron, fast argon ion, fast and thermal argon atom impact ionization from all the levels;
- three-body electron–ion recombination at all levels, where the third body is an electron, a fast argon ion, a fast or a thermal argon atom;

- radiative electron–ion recombination at all levels;
- radiative decay between all levels;
- associative ionization for the levels with excitation energy above 14.71 eV, this being the ionization energy of Ar_2^+, i.e. $Ar^* + Ar^\circ \rightarrow Ar_2^+ + e^-$.

The four 4s levels play a key role in analytical glow discharges (e.g. for Penning ionization of the sputtered atoms) and they cannot easily be depopulated by radiative decay (due to forbidden transitions for the metastable levels, and to radiation trapping for the resonant levels). Therefore some additional loss processes are incorporated for these levels, in order to describe them with more accuracy:

- Penning ionization of sputtered atoms;
- 4s–4s collisions leading to the ionization of one of the atoms, or leading to the formation of Ar_2^+ (associative ionization);
- two-body collisions with argon ground state atoms, yielding "collision induced emission": $Ar^*(4s) + Ar^\circ \rightarrow Ar^\circ + Ar^\circ + h\nu$;
- three-body collisions with argon ground state atoms, leading to the formation of Ar_2^+: $Ar^*(4s) + Ar^\circ + Ar^\circ \rightarrow Ar_2^+ + Ar^\circ$; and
- diffusion and subsequent de-excitation at the cell walls.

Finally, the radiative decay to the ground state is affected by so-called radiative trapping. Indeed, the ground state argon atoms will re-absorb a fraction of the emitted radiation, so that only a small fraction (defined by the escape factor, of the order of 10^{-4} for pressures of about 1 mbar) can actually escape. Therefore the Einstein probability for radiative decay must be multiplied by this escape factor to obtain the net decay rate. Because all these processes are collisional or radiative, this type of model is called a "collisional–radiative model". All the cross-sections, rate coefficients, transition probabilities etc. have been presented in papers by Bogaerts et al. Other data required to calculate the population and loss rates, such as densities, fluxes and energies of electrons, argon ions and atoms, must be taken from the results of models describing these plasma species. In this way the "collisional–radiative" model is coupled with the other models, and solved iteratively until convergence is achieved.

The features of glow discharges can be realized in the hollow cathode and related sources, in glow discharges with flat cathodes with dc or rf power and in special sources such as the so-called gas-sampling glow discharges.

5.6.1
Hollow cathodes for AES

The energetically important parts of the discharge (cathode layer, dark space and negative glow) as well as the sample are inside the cathode cavity. The volatilization results from cathodic sputtering and/or thermal evaporation. This depends on the fact whether the whole cathode with its outer and inner wall is subjected to sputtering without any cooling (hot hollow cathode), the outer wall is shielded by a quartz enclosure without applying further cooling (transitional hollow cathode) or if the outer wall of the cathode is shielded and directly cooled (cooled hollow cathode).

The characteristic depends on the discharge gas (at a few mbar of argon: $i = 0.2–2$ A, $V = 1–2$ kV; at 10–20 mbar of helium: $i = 0.2–2$ A, $V < 1$ kV) but also on the cathode mounting. Indeed, in a cooled cathode the characteristic is normal and the analyte volatilizes by cathodic sputtering only, whereas in the hot hollow cathode thermal evaporation also takes place, by which the characteristic especially at high currents may become normal. In this case thermal effects could even lead to a strong selective volatilization, which can be made use of analytically. The latter was shown to occur in the case of brass, as it could be demonstrated by electron probe micrographs of partly molten brass samples, the outer layers of which are less rich in Zn [467].

The analyte is, in any case, excited in the negative glow contained in the hollow cathode only. Owing to the high residence times in the excitation zone and the high line-to-background ratios, resulting from the non-LTE character of the plasma (gas temperatures of 600–1000 °C [21] and much higher electron temperatures) the hollow cathode is the radiation source with the lowest absolute detection limits. In this sense it has found wide use for analysis of dry solution residues, e.g. in the case of the rare earth elements, or subsequent to matrix separation in the analysis of high-purity substances (see Refs. in [468]). The Doppler widths of the lines are low and accordingly this source has even been used for isotopic analysis (determination of U^{235}/U^{238}). To date it is still employed for the determination of volatile elements in a non-volatile matrix such as As or Sb, in high-temperature alloys [469], where detection limits of below 1 µg/g can be obtained.

Hollow cathode glow discharges operated in helium are also of great use for the detection of the halogens, which can then be detected in gas chromatographic effluents, with detection limits in the pg/s range for Cl and Br and a linear response of 1.5 decades [470]. Another interesting application is the detection of trace elements in air and airborne dust. This can be done off-line, where when using graphite electrodes as collectors, detection limits for Pb and Cd in the µg/m³ range can be determined [471]. Through the direct introduction of air and its particulates through the side wall of a hollow cathode, total toxic metal loads and with end-on introduction the gaseous load only can actually be differentiated [472]. A further treatment of the hollow cathode lamp as an atomic emission spectrometric source is given in Ref. [473].

5.6.2
Furnace emission spectrometry

Similar to hollow cathodes electrically heated graphite furnaces placed in a low-pressure environment can also be used for sample volatilization. The analyte released can then be excited in a discharge between the furnace and a remote anode. This source is known as FANES (furnace atomic non-resonance emission spectrometry) and was introduced by Falk et al. [474]. Because of the separation of volatilization and excitation its absolute detection limits are in the pg range. For real samples, however, the volatilization and excitation interferences may be considerable.

5.6.3
Dc glow discharges with a flat cathode

With this source compact electrically-conducting samples can be analyzed directly. They must be flat, and are then taken as the cathode and ablated by cathodic sputtering. The sputtered material is excited in the negative glow of the discharge, which is usually operated at a few mbar of argon. As the sample is ablated layer-by-layer both bulk and in-depth profiling analyses are possible.

Glow discharge sources

A useful device was first described by Grimm in 1968 [475]. Here the anode is a tube (diameter: 8–10 mm) of which the distance from its outer side to the cathode block and the sample is less than the main free path length of the electrons (around 0.1 mm). Furthermore, the anode–cathode interspace is kept at a lower pressure than the discharge lamp. According to the Paschen curve, no discharge can take place at these locations and the glow discharge is restricted to the cathode surface (Fig. 106). In later versions, a lamp with a floating restrictor and a remote anode, in which the second vacuum is superfluous, has also been used.

The Grimm lamp has an abnormal characteristic, the current is selected to be at 40–200 mA and the burning voltage is usually below 1 kV. Both in krypton as well as in neon the burning voltage is higher than in argon, but it also depends on the

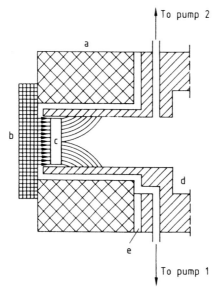

Fig. 106. Glow discharge lamp according to Grimm. (a): cathode block; (b): sample; (c): negative glow; (d): quartz window; (e): insulator (Reprinted with permission from Ref. [232].)

cathode material. The ablated material is deposited at the edge of the burning crater, limiting the discharge time to a few minutes, and inside the anode tube, making a regular rinsing stage necessary. In AES the glow discharge is viewed end on.

Characteristics

Sputtering　After ignition of the discharge a burn-in time is required to achieve a
sputtering equilibrium. The surface layers have to be removed first, but once equilibrium is reached the discharge penetrates with constant velocity into the sample.
Preburning times are usually up to 30 s for metals (at 90 W in argon, zinc: 6 s; brass:
3–5 s; steel: 20 s; and Al: 40 s) but depend on the sputtering conditions (shorter at
high voltage, low pressure, etc.). Therefore, in bulk analysis a high-energy preburn at
low pressure and high voltage and current is often used, with durations below around
10 s so as to shorten the total time of analysis.

The burning crater has a structure depending on the material structure or inclusions, and the electrical field may induce a small curvature, especially with the
classical Grimm lamp [476]. In the glow discharge with a floating restrictor this
curvature was found to be less pronounced. The form of the sputtering profiles can
also actually be calculated from models and this in good agreement with experimentally recorded sputter profiles [239].

The material volatilization is of the order of a few mg/min and increases in the
order: C < Al < Fe < steel < Cu < brass < Zn. It also depends on the gas used,
which is normally a noble gas. Helium is not used, as due to its small mass sputtering is too limited. The sputtering rate increases in the order: neon < argon <
krypton. It is proportional to $1/\sqrt{p}$ [240], when p is the gas pressure. Furthermore,
the sputtering rates of alloys can be predicted to a certain extent from the values for
the pure metals [232].

Excitation　Collision with high-energy electrons are important for the excitation, as
are also other processes that have already been discussed such as charge transfer or
excitation by metastables. The electron temperatures are 5000 K (slow electrons for
recombination) and >10 000 K (high-energy electrons for excitation) and gas temperatures are below 1000 K, a result of which is that the line widths are narrow. The
importance of processes such as charge transfer has been shown for the case of copper lines by Steers et al. [477]. Here the use of Fourier transform spectrometry was
very useful so that many lines and their intensities could be recorded. Because of the
high resolution, which could then be obtained, self-reversal was found to occur in the
case of resonance lines in the Grimm-type glow discharge [40]. Steers et al showed
that the use of cross excitation with microwaves was very helpful, so as to supplementary excite the ground-state atoms released from the sample and remaining
at high number densities in front of the negative glow, which causes self-reversal
[478].

Because of the low spectral background intensities the limits of detection from
the side of the source will be low and for metals will be down to the µg/g range in a
classical Grimm lamp [479]. Also elements such as B, P, S, C, As, ... can be determined. The plasma is optically thin, but for resonance lines of the matrix elements,

self-reversal may occur. Altogether, however, the linear dynamic range will be high as compared with arc or spark sources. As a result of the stable nature of the discharge the noise is low. There is hardly any flicker noise and only some frequency dependent components from the vacuum and the line occur on purely white noise. Thus, RSDs obtained without an internal standard can easily be below 1%. However, owing to the low radiant densities shot noise limitations may occur. Because of its stable nature the glow discharge source can be coupled with Fourier transform spectrometry. In such a case e.g. a detection limit down to 30 µg/g can be obtained for molybdenum in steel samples. At the same time the highest spectral resolution can be obtained [40].

As the sample volatilization is due to cathodic sputtering only, matrix interferences as a result of the thermochemical properties of the elements do not occur. This has been shown impressively in early comparative studies of glow discharge atomic spectrometry and spark emission spectrometry with aluminum samples (Fig. 107) [480]. It must be stated, however, that with advanced sparks, where through the use of fiber optics only those parts of the spark plasma are observed that are not involved in sample ablation, matrix interferences in the case of spark emission spectrometry are also lower. The analysis of similar alloys with different metallographic structures by glow discharge atomic spectrometry can often be carried out with one calibration.

Applications

Bulk analysis of metals   Glow discharge AES has found wide use in the bulk analysis of metals. This applies less to production control, as the preburn time required is long as compared with sparks, than it does to product control. The easy calibration due to the abscence of volatilization interferences and low spectral interferences as well as to the high linear dynamic range are advantageous. The method is particularly useful for elements having high sputtering rates such as copper. Samples have to be available as flat and vacuum-tight disks (thickness: 10–30 mm), but with special holders threads and metal sheets can also be analyzed. The samples must be polished and free from oxide layers prior to analysis.

Depth-profile analysis   As the sample is ablated layer-by-layer with a penetration rate of 1–3 µm/min depth-profile analysis is also possible. The intensities of the analyte lines are measured as a function of time. However, the sputtering rates of alloys with varying compositions must be known so as to be able to convert the time scale into a depth scale. The intensities must be related to concentrations, which can be done by using theoretical models and sputtering constants, as proposed by Bengtson et al. [481]. The power of detection can be quite good and the depth-resolution is of the order of 5 nm, when elemental concentrations >0.1% must be monitored. Depth-profile analysis is now a major area of application of glow discharge emission spectrometry in the laboratories of the metal industry [see e.g. (Fig. 108)].

The depth of the crater can also be measured from the radiation leaving from the glow discharge emission source in the case of transparent samples ablated by rf glow discharges (see Section 5.6.4). Indeed, part of the emitted light is reflected at

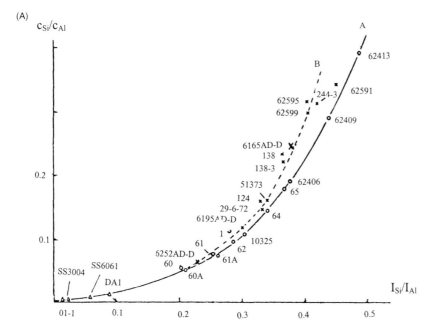

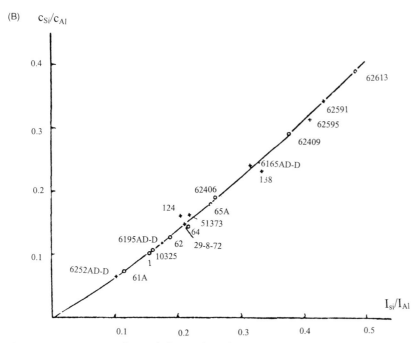

Fig. 107. Determination of Si in Al alloys with medium voltage spark OES (A) and OES using a Grimm-type glow discharge (B). [480].

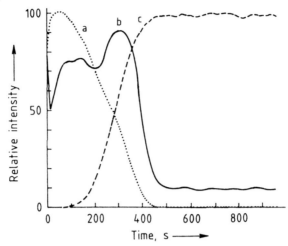

Fig. 108. Depth-profile through a galvanized steel sheet surface. The Zn coating is ≈18 μm thick, and the Al concentration in the bulk is 0.049%. The voltage *U* increases from 510 V in the Zn to 740 V in the bulk steel. (a): Zw; (b): Al; (c): Fe (Reprinted with permission from Ref. [481].)

the sample surface and modulates the line intensity and in the case of transparent layers a technique based on the interference patterns has been developed for measuring layer thicknesses [482].

Electrically non-conducting powders These can also be analyzed after mixing with Cu powder and briquetting into pellets, as first shown by El Alfy et al. [210] for geological samples and ores. The performance depends greatly on the particle size of the powders, but detection limits in the μg/g range are possible for the light elements (Al, Be, B, . . .).

5.6.4
Rf glow discharges

Apart from dc also rf glow discharges are very useful sources for atomic spectrometry. As a result of the bias potential built up in the vicinity of the sample, nonconducting samples such as ceramics can also be ablated directly and analyzed by atomic emission spectrometry, as shown by Winchester et al. [483] (Fig. 109). In rf glow discharges the power can be coupled into the plasma through the sample, which has the disadvantage that sample thickness has to be corrected for in a number of cases, however, approaches using a sideways coupling of the power into the plasma have also been developed [484]. In this case the same source can be operated with rf and dc power.

Extensive measurements of electrical characteristics and plasma parameters for rf glow discharges have been performed by Marcus et al., e.g. with Langmuir probes and they revealed higher electron number densities [485] than in dc glow discharges, however, comparisons are difficult because of constructional differences. Potential applications of rf glow discharges include analyses of multilayers as well as studies with coated glasses and layered ceramics.

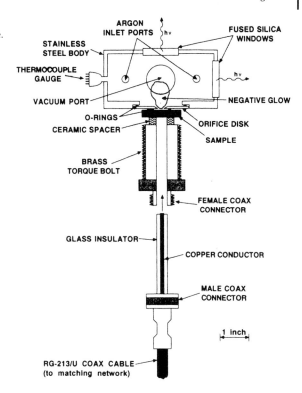

Fig. 109. Rf glow discharge atomization/excitation source. (Reprinted with permission from Ref. [483].)

5.6.5
New developments

Since it was found that a great part of the analyte released by sputtering consists of ground-state atoms, cross excitation by dc discharges [486], rf discharges [487] and microwave discharges [362] has been investigated. The microwave assisted glow discharge, as represented in Fig. 110, has been studied extensively by Leis et al. [362]. It was found that the supplementary excitation in the case of steel samples resulted not only in a gain in net intensities and thus in a decrease of shot noise limitations, but also in line-to-background intensity ratios with a consistent precision and accordingly a gain in the power of detection. This gain was of the order of a factor of 10, as shown for steels (Table 11).

Jet-enhanced sputtering can be incorporated into glow discharge sources so as to increase the material ablation. This goes back to Gough [296] and is very useful when using glow discharges as atom reservoirs for atomic absorption spectrometry. Indeed, the atom number density can then be increased. Commercial instrumentation for direct atomic spectrometric analysis of metal samples using jet-assisted glow discharges as atom reservoirs are available and the features for steel analysis have been shown by Ohls [298]. Reliable analyses for minor constituents are possible. In atomic emission work, it was found that the jet-assisted glow discharge

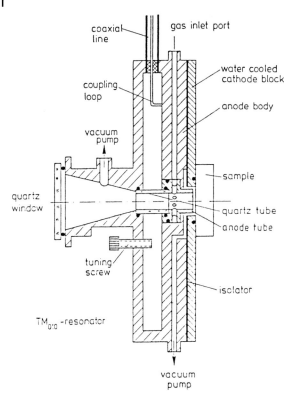

coaxial line

gas inlet port

coupling loop

vacuum pump

quartz window

tuning screw

$TM_{0\cdot0}$ -resonator

water cooled cathode block

anode body

sample

quartz tube

anode tube

isolator

vacuum pump

Fig. 110. Glow discharge lamp with integrated microwave cavity for atomic emission spectrometry. (Reprinted with permission from Ref. [362].)

has higher signal-to-background ratios for non-resonant lines of elements with high excitation energies such as Zn, as expected from the increase in ablation rates (from about 1 to 5 mg/min for copper samples). Through measurements on brass, it could be shown that this is not true for copper resonance lines, which can be understood from the increased self-reversal. Also, selective sputtering for Cu and Zn as a result of redeposition was found to be possible [238].

Magnetic fields were found to influence the sputtering rates in glow discharges as they change the geometry of the glow discharge plasma. This was found to be the case both with dc and rf discharges. With dc discharges and copper samples sputtering rates were found to increase from 100 to 200 µg/min for a magnetic field of 10 kG, 1–2 mm thick samples and 20–60 W at 2–4 mbar. The changes in the form of the plasma were found to relate back to the sputter profiles [245]. In the case of rf discharges and glass or ceramic samples, ring-shaped plasmas as well as plasmas with the highest density in the center were also obtained, alongwith the appropriate sputter profiles [246].

Furthermore, it was found that simple types of dc discharges were also capable of exciting vapors and even molecular species such as volatile hydrides [488].

In the case of a helium discharge sustained in a Grimm-type glow discharge lamp, vapors of halogenated hydrocarbons can be introduced successfully through the cathode, both when using hollow cathode as well as with flat cathode geometries. As a result of the introduction of CH_2Cl_2 vapors the current at the constant burning voltage drops considerably [489], which can be understood from the energy required for the sample decomposition. The highest line-to-background ratios are obtained, however, around the sampling orifice, despite the fact that it would be expected that the analyte immediately spreads out in all directions after entering the discharge chamber.

From ratios of C:Cl line intensities, it is possible to determine empirical formulae, which makes the method very valuable for the identification of compounds when applying element-specific detection in gas chromatography. McLuckey et al. [490] introduced gas sampling glow discharges as soft ionization sources for organic molecules in mass spectrometry. Pereiro et al. [491] showed that they were also prominent sources for atomic emission spectrometry. They determined the halogens present in halogenated hydrocarbon vapors in helium and thus demonstrated the potential as element-specific detectors for gas chromatography. It has been shown that gas-sampling glow discharges with flat cathodes can be operated with argon, helium and with neon and can also be used to excite vapors produced by hydride generation (Fig. 111) [488]. The current–voltage characteristics of such systems are fairly flat and detection limits for arsenic down to 20 ng/mL can be obtained. Spatially resolved measurements also showed that the sample introduction location remains clearly visible. The source is also a viable ion source for mass spectrometry (see Section 6.2.2).

5.7
Laser sources

High-powered lasers have shown to be useful sources for the direct ablation of solids. In atomic emission spectrometry, ruby and Nd:YAG lasers have been used since the 1970s for solids ablation. When laser radiation interacts with a solid a laser plume is formed. This is a dense plasma containing both atomized material and small solid particles, which have evaporated and or been ejected due to atom and ion bombardment from the sample. The processes occurring and the figures of merit in terms of ablation rate, crater diameter (around 10 μm) and depth at various energies as well as the expansion velocities of this plume were described back in the 1980s [223]. The plume is optically thick and as an emission source seemed hardly of interest. By applying cross-excitation with a spark or with microwave energy, however, the power of detection can be considerably improved, as described in the respective textbooks (see e.g. Ref. [221]). It was found that the absolute detection limits were down to 10^{-15} g, but the RSDs were no better than 30%. The method, however, can be applied to the identification of mineral inclusions and of microdomains in metallurgical analysis, forensic samples and archeology.

There has been renewed interest in the method, mainly due to the availability of improved Nd:YAG laser systems. In addition, different types of detectors, such as microchannel plates coupled to photodiodes and CCDs, in combination with mul-

(A)

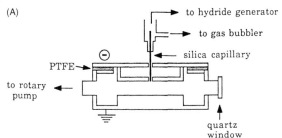

Fig. 111. Gas sampling glow discharge (A) and continuous-flow hydride generation system (dimensions in mm) (B), (a): reaction column, (b): fritted glass disk, (c): gas–.liquid separator and (d): condenser (Reprinted with permission from Ref. [488].)

(B)

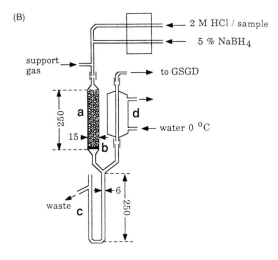

tichannel analyzers make it possible for an analytical line and an internal standard line to be recorded simultaneously, by which the analytical precision can be considerably improved. By optimizing the ablation conditions and the spectral observation, detection limits obtained using the laser plume as a source for atomic emission spectrometry are in the 50–100 µg/g range and RSDs are in the region of 1% as shown by the determination of Si and Mg in low-alloyed steels [224, 227]. This necessitates the use of slightly reduced pressure, so that the atom vapor cloud is no longer optically very thick and the background emission intensities become lower. In the case of laser ablation of brass samples at normal pressure and direct emission spectrometry at the laser plume, selective volatilization of Zn was still encounterd, irrespective of the laser wavelength used, as shown by Gagean and Mermet [225].

New advanced fs-lasers were recently found to produce less buffer-gas plasma above the sample surface and especially at reduced pressure, and thus allow more accurate determinations of Cu and Zn in brass samples to be performed than with ns pulses [492].

Manufacturers

Equipment for atomic emission spectrometry is now available from various manufacturers. Several types of sources [arc, spark, ICP, MIP, glow discharge (GD), etc.] are offered as indicated for the respective firms, of which most are listed below.

- A. F. Analysentechnik, Tübingen, Germany: MIP
- ARL, Fisons, Ecublens, Switzerland: arc, spark, ICP, GD
- Baird Atomic, Bedford, MA, USA: arc, spark, ICP
- GBC, Australia: ICP
- Hilger, UK: arc, spark, ICP, GD
- Hitachi, Japan: ICP
- Instruments S. A., Longjumeau, France: arc, spark, ICP, GD
- Labtam, Australia: arc, spark, ICP
- Leco, St. Joseph, MI, USA: GD, ICP
- Leeman Labs, USA: ICP
- Linn, Hersfeld, Germany: ICP
- PerkinElmer, Norwalk, CT, USA: ICP
- Shimadzu, Japan: arc, spark, ICP
- Spectro Analytical, Kleve, Germany: arc, GD, spark, ICP
- Thermo Jarrell Ash, USA: arc, spark, ICP, GD
- Varian, Australia: ICP

6

Plasma Mass Spectrometry

The plasma sources developed for atomic emission spectrometry have also been shown to be very suitable ion sources for mass spectrometry. This is particularly true for electrical discharges at pressures in the 1–5 mbar range and sources at atmospheric pressure since the powerful vacuum systems became available, with which the pressure difference between the mass spectrometer (of the order of 10^{-5} mbar) and the source can be bridged.

Elemental mass spectrometry, however, goes back to the use of high-vacuum arcs and sparks, with which ultratrace and survey analyses of metal samples could be performed (for an overview, see e.g. Refs. in [68, 69]). Spark source mass spectrography with high-resolution sector field mass spectrometers, is still very useful for a survey characterization of electrically-conducting solids down to the ng/g level. The spectra can be recorded on photographic plates, which are a permanent document and at least enable semi-quantitative analyses to be made. At the ng/g level this approach is suitable for the quality control of materials required in microelectronics. The technique has become very useful since computer-controlled densitometers have been available, which automatically record the blackenings of the elemental lines on the photoplates and convert them into logarithmic intensities, these being proportional to the logarithmic concentrations.

Spark source mass spectrometry requires expensive sector-field mass spectrometers and despite the possibility of automated read-out of the spectra, highly skilled laboratory personnel for data evaluation are also required. This situation arises from the fact that high-vacuum sparks are very powerful sources producing vast amounts of multiply charged ions making the spectra line rich. The analytical precision achievable in spark mass spectrometry is low, however, special precautions such as the use of rotating electrodes have been found to be helpful. Spark source mass spectrometry has been used for the analysis of compact metals, from which electrodes could be made by turning-off or by fixing drillings in electrode holders. In the case of metals of low melting point, such as gallium, liquid nitrogen cooled electrodes of the gallium were used. For electrically conducting as well as non-conducting powders electrodes were pressed from mixtures with gold powder, which is ductile and leads to electrodes that are electrically and heat conductive. The method is very powerful for dry solution residues as a result of its sensitivity and has been used for impurity detection in liquid aliquots obtained from trace–matrix separations in the analysis of high-purity materials.

In the course of the late 1970s new mass spectrometric methods, which made use of the plasma sources known from optical atomic spectrometry came into use. They will be treated in detail and consist in particular of ICP mass spectrometry (ICP-MS) and glow discharge mass spectrometry (GD-MS), which have contributed to a considerable portion of the progress that has been made in elemental analysis as compared with spark source mass spectrometry.

6.1
ICP mass spectrometry

Plasma mass spectrometry developed on the basis of the experience gathered with the extraction of ions from flames, as was done for diagnostic purposes in the mid-1960s [493].

Douglas and French [494], towards the end of the 1970s, used an atmospheric pressure microwave plasma, into which they introduced dry aerosols, as an ion source for mass spectrometry and reported very low detection limits as compared with other methods of solution analysis. In 1980 Houk et al. [495] first showed the analytical possibilities of ICP-MS and described the sampling of ions through a water-cooled aperture (of 0.6–1 mm) into a vacuum chamber with a pressure of a few mbar, maintained by a powerful roughing pump. From this chamber a second sampling takes place, through a second aperture of the same dimensions, into the high vacuum of a quadrupole mass spectrometer. This development became possible as a result of the improvement in stability and the spectral resolution of 1 dalton that is achievable with quadrupole mass filters, as well as by the further development of vacuum technology.

As the method has the same excellent possibilities for sample introduction as ICP-AES but in addition enables much lower detection limits to be achieved, covers more elements and allows the determination of isotope concentrations, ICP-MS rapidly developed into an established analytical method [496].

6.1.1
Instrumentation

In ICP-MS (Fig. 112) the ions formed in the ICP are extracted with the aid of a conical water-cooled sampler into the first vacuum stage where a pressure of a few mbar is maintained. A supersonic beam is formed and a number of collision processes take place as well as an adiabatic expansion. A fraction is sampled from this beam through the conical skimmer placed a few cm away from the sampler. Behind the skimmer, ion lenses focus the ion beam now entering a vacuum of 10^{-5}. This was originally done with the aid of oil diffusion pumps or cryopumps, respectively, but very quickly all manufacturers switched to turbomolecular pumps backed by roughing pumps.

The sampler and the skimmer are usually made of stainless steel and are both conically shaped with different cone angles. The sampler can also be made of copper, which has a better heat conductivity. In the case of HF containing solutions,

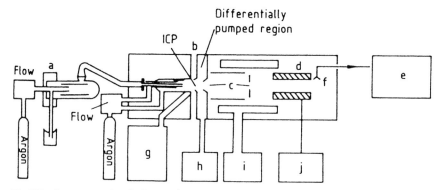

Fig. 112. Instrumentation for low-resolution ICP-MS.
(a): nebulizer; (b): sampler; (c): ion optics; (d): quadrupole;
(e): electronics; (f): detector; (g): rf-generator; (h): roughing
pump; (i): turbomolecular pump: (j): quadrupole rf generator.

platinum samplers can also be used. This is particularly worthwhile for the analysis
of geological samples subsequent to wet chemical dissolution and removal of the
silicates. The distance between the sampler and skimmer is critical with respect to
the maximum power of detection and minimal ionization interferences. This also
applies to the power transmission to the rf coil, where considerable differences
were found for coils powered centrally and coils powered at one of the ends [497].

The processes in the intermediate stage together with their influence on the ion
trajectories in the interface and also behind the second aperture (skimmer) are very
important for the transmission of ions and for related matrix interferences, this
being the topic of fundamental diagnostic studies (see e.g. Refs. in [496]). Vaughan
and Horlick [498] used the commercial program MacSimion to model the einzel
lens and the Bessel box input lens system of an inductively coupled plasma mass
spectrometer and compared the results obtained for different lens voltage settings
with experimental data.

Today's commercially available instrumentation fall into three groups, namely
quadrupole based instruments, high-resolution instruments and time-of-flight systems. For each system different ion extraction conditions must be selected. In a
quadrupole, as a result of the high transmission, no high extraction voltage is
required, which is totally different for a sector field instrument. Here either the
spectrometer or the plasma must be at a high potential. With quadrupoles a mass
resolution of 1 dalton over the whole spectral range means the method suffers
from isobaric interferences especially at masses below 80 dalton. In the case of
sector field instruments, the resolution may be up to 5000 while in commercial
instruments sufficient transmission can still be maintained. As the spectral background intensities are low, detection limits are lower, not only as a result of the
absence of a number of interferences but also because of better signal-to-background ratios [499]. For time-of-flight ICP-MS, both othogonal as well as linear
end-on arrangements are used. The instruments available use a reflectron [74].

Duty cycles are still fairly low as ion packages in a time scale of µs are sampled but spectrum development and acquisition still takes orders of magnitude longer.

Despite high throughput, the power of detection has as yet not been fully exploited. Dual-channel mass spectrometers, where the ion beam is split inside the mass spectrometer [87] are very useful for isotope ratio work. However, they still use different detectors. Mattauch–Herzog arrangements with a flat focal plane have also been introduced for plasma mass spectrometry work. They are ideally suited for work with array detectors, which are certainly very useful for isotope ratio work, as within a narrow mass range isotope signals are actually measured simultaneously, together with their respective background signals [500]. Such detectors are available as microchannel plates coupled through a phosphor to photodiode array detectors.

6.1.2
Analytical features

The analytical features of ICP-MS are related to the production of ions in analytical ICPs as well as to the highly sensitive ion detection and the nature of the mass spectra themselves.

On the basis of the Saha–Eggert equation, the degrees of ionization for the chemical elements can be calculated for a plasma temperature of 6000 K, an electron number density of 10^{15} and partition functions according to Ref. [5] (Table 15) [501]. They range from over 99% for sodium to 10^{-5} for chlorine. The argon ICP thus has good prospects for the determination of metals but for non-metals has a number of limitations as a result of these considerations.

ICP mass spectra

In low-resolution ICP-MS the resolution in the spectra at best is 1 dalton and cluster ions are found particularly in the lower mass range signals, which in a number of cases cause spectral interferences with analyte ions. Cluster ions may be formed from different types of compounds present, namely:

Tab. 15. Degree of ionization of some elements in an analytical ICP (temperature: 6000 K, n_e: 10^{15}, electronic partition functions according from Ref. [5]. (Reprinted with permission from Ref. [501].)

Element	Ionization energy (eV)	Degree of ionization (%)
Ba	5.21	9.386
Cl	13.01	1.7×10^{-5}
Fe	7.87	0.231
Cu	7.724	0.637
Na	5.138	0.996
S	10.357	9.7×10^{-3}
Y	6.51	0.867

- solvents and acids: H^+, OH^+, H_2O^+,..., NO^+, NO^{2+},...., Cl^+,..., SO^+, SO_2^+, SO_3H^+ (when residual H_2SO_4 is present in the measurement solution);
- radicals of gases in the surrounding atmosphere: O_2^+, CO^+, CO_2^+, N_2^+, NH^+, NO^+,...;
- reaction products of the above mentioned species with argon: ArO^+, $ArOH^+$,..., $ArCl^+$, Ar_2^+,....

These cluster ions give strong signals in the mass range up to 80 dalton and may hamper the determination of the light elements as a result of spectral interferences. At higher masses, the singly charged ions of the heavier elements are found, as are doubly charged ions of the light elements. They occur particularly for elements with low ionization energies.

In the mass spectra signals for compounds of the analyte with various other species also occur, such as MO^+, MCl^+, MOH^+, MOH_2^+,.... These ions may be formed by the dissociation of nitrates, sulfates or phosphates in the plasma or they may result from reactions between analyte ions and solvent residuals or oxygen in the plasma or eventually in the interface between the plasma and the mass spectrometer inlet.

In early work it was recognized that the selection of the sampling zone in the plasma and the aerosol carrier gas flow may strongly influence the occurrence of cluster ions and doubly charged analyte ions [502, 503]. This is clearly demonstrated by the results for lanthanum, where even doubly and triply charged ions may be expected in the argon plasma. This necessitates specific optimization of the power and especially of the carrier-gas flow, which influence both temperature and residence times of the sampled substance in the plasma. This is documented e.g. for the intensities of the signal for the ArO^+ and the Cu^+ ions in Fig. 113 [504].

Signals of the different isotopes are obtained for each element. Their intensity ratios correspond to the isotopic abundances in the sample. Use can be made of this fact for calibration by isotopic dilution with stable isotopes and in tracer experiments. The isotopic abundances of the elements, however, are also useful for the recognition of spectral interferences. The spectra in ICP-MS are less line rich than in ICP-AES. With quadrupole mass spectrometers, however, the resolving power of the spectrometers is also lower than in ICP atomic emission spectrometers. Accordingly, cluster ions will often cause spectral interferences and appropriate correction procedures must be worked out. With high-resolution sector field instruments (Fig. 114) which have a resolution of 5000 or more, a number of these interferences will not occur. This is clearly shown by calculations of the minimum resolution that should be obtained so as to avoid the most frequent interferences occurring in plasma mass spectrometry with argon inductively coupled plasmas (Table 16) [499]. The fact that prominent spectral interferences in the analysis of real samples such as dissolved ceramic powders can be avoided is reflected by the signal scans for Cr and Fe, signals commonly interfered with by cluster ion signals, as seen in Fig. 115 [504].

In ICP-MS the background intensities are usually low. They are mainly due to the dark current of the detector and to signals produced by ions scattered inside the

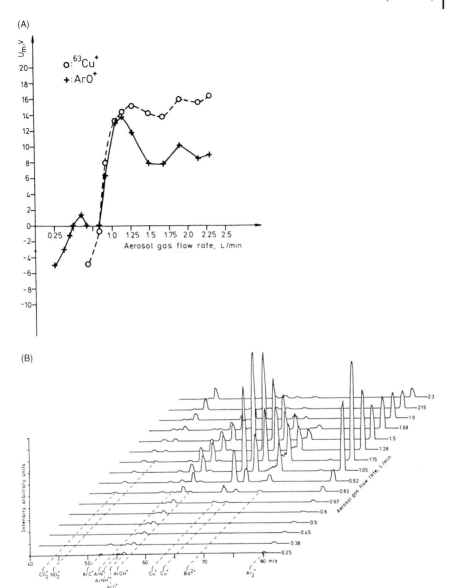

Fig. 113. Maximum values of ion energies for two ion species (A) and mass spectra in the mass regio 42–82 dalton in ICP-MS (B) when sampling ions at different gas flow rates for a 1 µg/mL solution of Cu at radial distance 0 mm and 1.5 kW, standard solution. (Reprinted with permission from Ref. [497].)

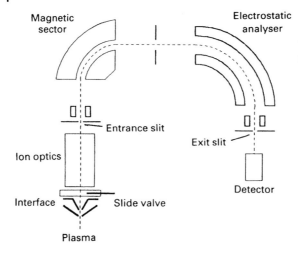

Magnetic sector

Electrostatic analyser

Entrance slit

Exit slit

Ion optics

Interface   Slide valve

Detector

Plasma

Fig. 114. High-resolution ICP mass spectrometer. (Reprinted with permission from Ref. [504].)

mass spectrometer as a result of collisions with residual gas species or reflection. In order to shield for the direct UV radiation, a beam stop is often supplied in the mass spectrometer. Accordingly, in ICP-MS the ICP only contributes slightly to the background, which contrasts with ICP-AES where the background continuum arises from interactions between electrons and ions in the plasma, from molecular bands, from wings of broad matrix lines and from stray radiation. Therefore, calibration by standard additions is often very successful in ICP-MS, as the systematic errors originate more from signal enhancements or depressions than from spectral background interferences.

Tab. 16. Minimally required resolving power (*R*) in order to avoid spectral interferences in ICP-MS (extracted from Ref. [499]).

Isotope	Polyatomic ion	R required
^{27}Al	$^{12}C^{15}N$, $^{13}C^{14}N^+$	1454
^{51}V	$^{12}C^{14}N^1H^+$	918
^{52}Cr	$^{40}Ar^{12}C^+$	2375
	$^{35}Cl^{16}O^1H^+$	1671
^{55}Mn	$^{39}K^{16}O^+$	2670
	$^{37}Cl^{18}O^+$	2034
^{56}Fe	$^{40}Ar^{16}O^+$	2502
	$^{40}Ca^{16}O^+$	2479
^{60}Ni	$^{44}Ca^{16}O^+$	3057
	$^{23}Na^{37}Cl^+$	2409
^{63}Cu	$^{40}Ar^{23}Na^+$	2790
	$^{31}P^{16}O_2^+$	1851
^{64}Zn	$^{32}S^{16}O_2^+$	1951
^{75}As	$^{40}Ar^{35}Cl^+$	7775
^{77}Se	$^{40}Ar^{37}Cl^+$	9181

Fig. 115. High-resolution ICP mass spectra from elements which suffer from Cl-induced spectral interferences in 0.4 M HCl: (a): 20 ng/mL of V, resolution 5000, (b): 20 ng/mL of Cr, resolution 5800 and (c): 100 ng/mL of As, resolution 7500. (Reprinted with permission from Ref. [504].)

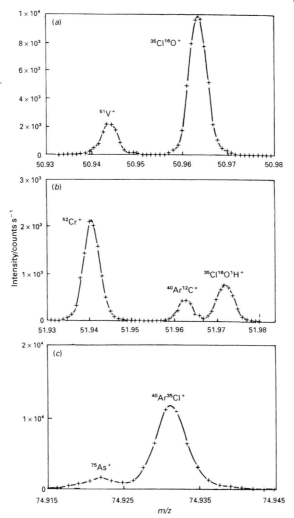

Optimization

For the optimization of ICP-MS with respect to the highest power of detection, minimal spectral interferences and signal enhancements or depressions, as well as highest precision, the most important parameters are the power of the ICP, its gas flows (especially the nebulizer gas), the burner geometry and the position of the sampler as well as the ion optics parameters. These parameters determine the ion yield and the transmission and accordingly also the intensities of the analyte and interference signals. At increasing nebulizer gas flow the droplet size decreases (see Section 3.1), and thus the analyte introduction efficiency goes up, however, this is at the expense of the residence time in the plasma, the plasma temperature

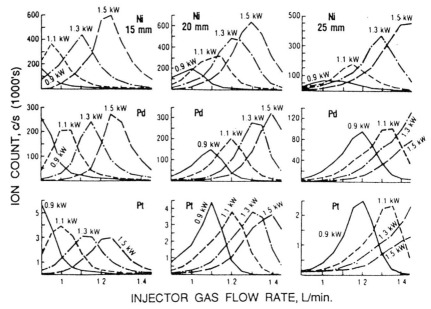

Fig. 116. Dependence of Ni$^+$, Pd$^+$ and Pt$^+$ signals on injector gas flow rate for a range of forward powers and sampling depths of 15, 20 and 25 mm. (Reprinted with permission from Ref. [505].)

and thus also of the ionization, as shown by optimization studies (Fig. 116) [505]. The optimization of the carrier gas flow, the ion sampling location and the power is more critical in ICP-MS than in ICP-AES. Indeed, in the latter case, the observation window is often of the order of 2×2 mm^2 and thus considerably larger than in ICP-MS (below 1 mm^2). This leads to sharper optimization maxima in ICP-MS than in ICP-AES and accordingly larger differences between results for single element optimization and results at compromise conditions, respectively.

Furthermore, changes of the nebulizer gas flow also influence the formation and the breakdown of cluster ions, requiring optimization with respect to minimum spectral interferences as well.

The carrier gas flows influence the ion energies, as shown for ^{63}Cu$^+$ and the ArO$^+$ in Ref. [497] and also the geometry of the aerosol channel. Normally the aerosol gas flow is between 0.5 and 1.5 L/min. It must be optimized together with the power, which influences the plasma volume and therewith the kinetics of the different processes taking place, and the position of the sampler. By changing the voltages at the different ion lenses the transmission for a given ion can be optimized, enabling the optimization of its detection limit and the minimization of interferences. In multielement determinations a compromise must always be made.

For the easily ionized elements working at so-called cool plasma conditions has been shown to be very successful. From the calculation of the degrees of ionization it can be seen that here mainly singly ionized species are formed instead of multiply ionized species (see detection limits listed in Ref. [499]. For Na, K, Ca and Al detection limits in the sub-ppt range then become achievable.

Power of detection

In order to obtain the optimum power of detection, the analyte density in the plasma, the ionization and the ion transmission must be maximized. The power will thus be between 0.6 and 2 kW and the sampler at about 10–15 mm above the tip of the injector. The detection limits, obtained at single element optimum conditions differ considerably from those at compromise conditions, but are still significantly lower than in ICP-AES (Table 17). In general it can be said that for the heavy elements, which have complex atomic term schemes and accordingly very

Tab. 17. Detection limits in ICP-OES and ICP-MS.

Element	Detection limit		
	ICP-OES (ng/mL) [329]	ICP-MS (ng/mL) [506]	HR-ICP-MS (ng/L) [499]
Ag	7	0.03	0.4[a]
Al	20	0.2	0.4[a]
Au	20	0.06	0.8
B	5	0.04	5.4
Cd	3	0.06	0.5[a]
Ce	50	0.05	—
Co	6	0.01	0.14[a]
Cr	6	0.3	0.24[a]
Fe	5	—	0.9[a]
Ge	50	0.02	0.7
Hg	20	0.02	—
La	10	0.05	—
Li	80	0.1	0.012[a]
Mg	0.1	0.7	0.2[a]
Mn	1	0.1	0.14[a]
Ni	10	—	0.3[a]
Pb	40	0.05	0.12[a]
Se	70	0.8	—
Te	40	0.09	—
Th	60	0.02	—
Ti	4	—	0.4
U	250	0.03	—
W	30	0.05	0.15
Zn	2	0.2	0.2[a]

[a] Cold plasma conditions.

line-rich spectra but with low intensity lines in ICP-AES, the detection limits in ICP-MS are much lower than in ICP-AES. For most elements the detection limits are very similar, except for the elements for which spectral interferences are a limiting factor. This applies to As ($^{75}As^+$ with $^{40}Ar^{35}Cl^+$), Se ($^{80}Se^+$ with $^{40}Ar^{40}Ar^+$) and Fe ($^{56}Fe^+$ with $^{40}Ar^{16}O^+$). The detection limits for elements with high ionization potential are usually very poor because of the limited formation of positive ions. They may, however, be lower when they are detected as negative ions (c_L for Cl^+: 5 and for Cl^-: 1 ng/mL).

The acids present in the measurement solution and the material of which the sampler is made (Ni, Cu, ...) may influence the occurrence of spectral interferences considerably and accordingly the detection limits for a number of elements. This is particularly important when analyzing solids subsequent to sample dissolution by treatment with acids. Here the measurement solution should hopefully not contain chlorine, phosphate or sulfate ions. If they do, it is advantageous to remove them by precipitation or fuming off and taking up the analytes in dilute nitric acid. The detection limits for ICP-MS in the case of solids thus suffer from the necessary sample dilution. In this case the sample concentration in the measurement solution is often limited to 500 µg/mL, as for solutions of Al_2O_3 or SiC powders [507]. This is due to the risk of sample depositions blocking the sampler and contrasts with ICP-AES where for these materials sample concentrations up to 5 or even 50 mg/mL are possible. For elements such as B, Mg, Fe, etc. the power of detection of ICP-AES with respect to the solid samples is accordingly higher than in ICP-MS.

The detection limits also depend to a great extent on the type of mass spectrometer used. The values for quadrupole mass spectrometers are by orders of magnitude higher than in the case of high-resolution mass spectrometry. This applies particularly for strongly interfered elements such as Fe [499]. With time-of-flight ICP-MS, the detection limits in the case of the light elements are up to one order of magnitude better than in quadrupole based ICP-MS. However, for the heavy elements they are up to a factor of five worse [508].

Precision and memory effects

The constancy of the nebulizer gas flow is of prime importance for the precision achievable in ICP-MS. After stabilizing the nebulizer gas flows, relative standard deviations can be below 1%. They can be improved still further by internal standardization [509]. The tolerable salt concentrations (1–5 g/L) are much lower than in ICP-AES, because of the risks of sampler clogging and depend on the respective salts. The memory effects may become limiting and in the case of a high matrix load rinsing times of 1–2 min are required.

A considerable portion of the noise originates from the nebulizer and the noise level can generally be decreased by cooling the spray chamber. The latter is shown to decrease the white noise somewhat as well as some $1/f$ contributions and also the standard deviations for the blanks [41]. This leads directly to an improvement in the detection limits.

Interferences

Signal enhancements and depressions resulting from the matrix relate to the nebulization, influences on the ionization in the ICP and to changes in the plasma geometry as well as in the ion energies. Changes of the nebulizer gas flow influence the nebulization effects (see Section 3.1), however, they also lead to changes of the aerosol channel geometry and the plasma temperature and hence also the interferences of easily ionized elements [510]. Space charge effects have also been shown to play an important role in the suppression of signals as a result of alkali metals [511]. Up to a certain amount these effects can be adequately corrected for by using an internal standard, for which the selection is only possible empirically. However, the guidelines are that elements with similar masses and ionization potentials should be chosen. Thus, it is often advisable to select more internal standards to cover elements in the whole analytical mass range, such as Sc for the low mass elements and Rh for the heavier elements. The use of internal standards was shown in early work to be very successful for serum samples [509]. The matrix effects can be eliminated to such a degree that calibration with aqueous solutions in the case of diluted serum samples is possible.

In mass spectrometry signals are obtained for each isotope present. With the low mass resolution of quadrupole mass spectrometers (≈ 1 dalton), this leads to a number of isobaric interferences, which can be corrected for with appropriate software. This type of interference depends only slightly on the working conditions, which is not the case for spectral interferences resulting from doubly charged ions, background species or cluster ions. The background species at low masses [512] cause considerable spectral interferences e.g. for $^{28}Si^+$ (with $^{14}N^{14}N^+$), $^{31}P^+$ (with $^{16}N^{16}OH^+$), $^{80}Se^+$ (with $^{40}Ar^{40}Ar^+$). Species such as $^{40}Ar^{16}O^+$ not only interfere with $^{56}Fe^+$, but due to all isotopic combinations and any hydrides present, also with other transition metal isotopes (^{52}Cr, ^{53}Cr, ^{54}Cr, ^{54}Fe, ^{55}Mn, ^{56}Fe, ^{57}Fe, ^{58}Ni, ^{58}Fe and ^{59}Co). In the case of HCl, additional interferences for further transitional metal isotopes will result from Cl^+, ClO^+, ClN^+ and $ArCl^+$ species. Under fixed working conditions these interferences do not change very much with the matrix composition and can be corrected for by subtraction. However, they limit the power of detection. The interactions of signals from doubly charged ions and cluster ions change considerably with the power, the sampler position and the nebulizer gas flow. These interferences are particularly important for elements which have relatively low ionization potentials or that form thermally stable oxides (e.g. Ba, Sr, Mg, Ti). This has been shown by measurements of the signals of singly charged (M^+), and doubly charged (M^{2+}) ions as well as of metal oxides (MO^+) and hydroxides (MOH^+), e.g. in the case of Ba [503, 505].

Considerable progress with respect to spectral interferences in the case of quadrupole mass spectrometry has been possible through the use of collision- and reaction-induced dissociation of cluster species. This work goes back to the research of Douglas and French [513] and Barinaga et al. [514]. For collision-induced dissociation, it is advantageous to provide for hexapole or octapole arrangements between the skimmer and the mass spectrometer itself, and to fill these arrangements with the appropriate gases. For collision-induced dissociation, helium as well as xenon, have been successfully used and for reaction-induced dissociation, e.g., hydrogen, nitrogen, oxygen and ammonia, as shown for the case of H_2 in Ref. [515], and it has been used in the determination of S [516], B, Li, etc.

A reduction of the interferences resulting both from spectral overlap and ionization can be realized in a number of cases by removal of acids and matrix residuals, as shown by on-line removal of these species by separations on solid phases, which has now been tested in the case of exchanges made at modified nebulizer surfaces [517].

Isotopic dilution analysis

Dilution with stable isotopes offers the possibility of performing tracer experiments but also of circumventing systematic errors. The principle [518] can be applied for every element which has at least two stable or longlife isotopes. For its application the analyte with a known isotopic composition but which differs from that of the sample is added to a known amount of sample, and mixed thoroughly. The isotopic abundance ratio R then is given by:

$$R = [N_S \cdot h_S(1) + N_A \cdot h_A(1)]/[N_S \cdot h_S(2) + N_A \cdot h_A(2)] \tag{298}$$

N_S is the number of atoms of the element to be determined in the sample and N_A the number of atoms in the amount of standard added. h_S and h_A are the abundances of the isotope (1) and (2) in sample and standard added. Thus, N_S or the absolute mass G_S is given by:

$$G_S = G_A \cdot [h_A(1) - R \cdot h_A(2)]/[R \cdot h_S(2) - h_S(1)] \tag{299}$$

R results from the signals of the isotopes, $h_S(1)$ and $h_S(2)$ as a rule are the natural abundances. G_A is the absolute amount of standard added, $h_A(1)$ and $h_A(2)$ are known from the isotopic composition. Isotope dilution in ICP-MS has been applied in studies on Pb (see e.g. Ref. [519]). Also tracer experiments for Fe in biological systems (see e.g. [520]) have been described. The precision achievable in the determination of isotope ratios for abundances which differ by a factor of less than 10 is in the lower percent range.

For the determination of isotope ratios, the precision of TOF-ICP-MS has been studied in a preliminary comparison with other mass spectrometer systems [521]. Typical isotope ratio precisions of 0.05% were obtained, thus overtaking sector field mass spectrometry with sequential detection, for which values of 0.1–0.3% for $^{63}Cu/^{65}Cu$ in Antarctic snow samples have been reported [522]. Similar results were obtained by Becker et al. [523] for Mg and Ca in biological samples (0.4–0.5%). In principle, the features of TOF-ICP-MS may be superior to those of sequential sector field or quadrupole mass spectrometry, however, true parallel detection of the signals as is possible with multicollector systems may be the definitive solution, as shown by Hirata et al. [524]. Here the use of detectors which allow true parallel measurement of the signals within the relevant mass range, just as the CCDs do for optical atomic spectrometry, may be the ultimate solution and bring about the final breakthrough for ICP-MS isotope ratio measurements as is required in isotope dilution mass spectrometry.

Alternative methods for sample introduction

Apart from continuous pneumatic nebulization, all sample introduction techniques known from ICP-AES have been used and are of use for ICP-MS. Similar to ICP-AES, the analysis of organic solutions is somewhat more difficult [525].

The addition of oxygen was found to be helpful when nebulizing effluents from HPLC containing organic eluents such as acetonitrile. This was useful when using ICP-MS for on-line detection in speciation as well as in trace–matrix separations. Here, however, it is useful to use desolvation, even in the case of low consumption, high efficiency nebulizers, such as the HEN or DIHEN. This can be done efficiently with membrane desolvation using Nafion membranes [148].

The use of ultrasonic nebulization just as in ICP-AES allows the sampling efficiency to be increased. The high water loading of the plasma has to be avoided, as is possible with desolvation, not only to limit the cooling of the ICP but also to keep the formation of cluster ions and the related spectral interferences low. These are complex, as for example, a change in water loading also influences the pressure in the intermediate stage [526]. The use of high-pressure nebulization in ICP-MS has similar advantages and is suitable for coupling ICP-MS with HPLC. The set-up that is of use in speciation work is the same as the one used for on-line trace–matrix separations [527]. With the formation of volatile hydrides, the detection limits for elements such as As, Se and Sb can also be improved. As shown in Ref. [528], improvements for Pb were also obtained. They are due to improved analyte sampling efficiency but also to the decrease in cluster ion formation resulting from the introduction of a water-free analyte, which also applies to the cold vapor technique in the case of Hg.

Electrothermal vaporization (ETV) in addition to its features for the analysis of microsamples, in ICP-MS has the additional advantage of introducing a dry analyte vapor into the plasma. Hence, it has been found to be useful for elements for which the detection limits are high as a result of spectral interferences with cluster ions. In the case of ^{56}Fe, which is subject to interference by ^{40}ArO$^+$, Park et al. [529] showed that the detection limit could be improved considerably by ETV. For similar reasons the direct insertion of samples in ICP-MS leads to the highest absolute power of detection (detection limits in the pg range and lower [530, 531]).

Transient signals arise from electrothermal vaporization, where accordingly the number of elements that can be determined for one vaporization event is very limited, unless drastic decreases in analytical precision and power of detection are accepted. In this respect TOF-ICP-MS offers some advantages. Here simultaneous measurements of different isotopes are possible during one evaporation event without losses in analytical performance. It is even possible to monitor the separation of the volatilization of interferents from the evaporation of the analytes in real-time, which is very helpful for removing both spectral and ionization interferences, as shown by Mahoney et al. [80].

Despite ICP-MS being mainly a method for the analysis of liquids and solids subsequent to dissolution, techniques for direct solids sampling have also been used. They are required particularly when the samples are difficult to bring into solution or in addition are electrically non-conducting and thus difficult to be analyzed with glow discharge or spark techniques. For the case of powders, such as coal, slurry nebulization with a Babington nebulizer can be applied in ICP-MS as well [532]. ETV offers good possibilities not only for powders but even for granulates also. With a novel furnace into which graphite boats can be introduced and where halogenated carbons can be used as volatilization aids, different volatile halogenide forming elements can be successfully evaporated from Al_2O_3.

When applying slurry sampling under continuous ultrasonic treatment of the slurry, accurate results for powdered biological substances [533], and also for WC [534] and Al_2O_3 powders down to the sub-μg/g level [535] can be obtained. For the direct analysis of metals spark ablation can be applied and the detection limits are in the ng/g range, as shown for steels [536]. When analyzing metals [537] as well as electrically non-conducting samples, laser ablation combined with ICP-MS is very useful. A Nd:YAG laser with a repetition rate of 1–10 pulses/s and an energy of around 0.1 J has been used. For ceramic materials such as SiC, the ablated sample amounts are of the order of 1 ng and the detection limits down to the 0.1 μg/g level. When using lasers with different wavelengths, material ablation as well as the minimum crater diameters achievable may well vary, both of which also vary with the gas atmosphere (argon or helium), as studied by Guenther and Heinrich [538]. Despite the availability of advanced lasers for certain sample types a number of questions remain. For the analysis of Cu–Zn alloys, the calibration behaves non-linearly, which can be explained by a change in the mass ablation with the composition, and also when using different lasers with different pulse lengths and wavelengths [539]. Laser ablation is now so controllable that, in the case of multi-collector ICP-MS, isotopic analyses for individual grains of minerals can be performed with a precision in the isotopic ratio of better than 0.02% (2σ level) [524]. Laser ablation can also be used successfully for samples which are difficult to analyze directly by any other method, e.g. plastics and glass samples. The use of individual laser spikes will be especially interesting in the case of TOF-ICP-MS, as then it will be possible to perform real multielement determinations from the material cloud generated by a single laser shot.

6.1.3
Applications

ICP-MS is especially promising for the areas where ICP-AES is applied but where further improvement in the power of detection is required. This is the case in trace analysis for geological samples and specifically for hydrogeological samples as well as for trace determinations in metals, in biology and medicine as well as in environmental analysis.

Geological samples

In the case of geological samples ICP-MS is of interest where multielement determinations are required and ICP-AES cannot be used because of the lack of power of detection or spectral interferences. This applies to the determination of low concentrations of the rare earth elements [506, 540]. Clogging and corrosion of the sampler may be critical and requires rinsing and working with solutions having concentrations below 0.1%. Hydrogeological samples, as described by Garbarino and Taylor [541], can be analyzed very accurately for Ni, Cu, Sr, Cd, Ba, Tl and Pb by isotope dilution ICP-MS. For a series of trace elements ETV was found to be useful so as to reduce spectral interferences. This applies particularly to volatile elements such as Tl [542] but in the case of a metal filament vaporizer also to Pt, Pd, Ru and Ir [543]. For the investigation of inclusions in minerals, laser ablation ICP-MS is very powerful. It can even be used for the analysis of liquid inclusions in minerals, which provide important information for geologists [544].

Metals and ceramics

Trace determinations down to the sub-µg/g level are possible in metals and ceramics as the analyte concentrations may be up to 5 g/L. ICP-MS therefore is really an alternative to ICP-AES for the analysis of metals with line-rich spectra. This has been shown in the case of high-temperature alloys [545]. However, matrix interferences also finally limit the power of detection and matrix removal is useful to make further improvements to the power of detection and calibration. All classical principles of trace–matrix separation are very helpful in this respect. In the case of Al_2O_3 powders subsequent to sample dissolution, the Al can be separated off by fixing the APDC complexes of a number of analytes on a solid phase and by releasing them into the ICP after matrix removal [131]. For SiC the acid residues remaining after decomposition of the powders with fuming sulfuric acid can be removed by fuming off and taking up the residue with dilute nitric acid, by which the detection limits can often be improved by one order of magnitude [546]. Using high-resolution ICP-MS the detection limits for Al_2O_3 ceramics are found to be of the same order of magnitude as in the case of quadrupole ICP-MS coupled on-line to matrix removal. In the latter case, however, Cr could not be determined down to low concentrations, as a result of the interference of the $^{52}Cr^+$ signal with the ArC^+ signal [504].

For the direct analysis of steels by ICP-MS, Jiang and Houk [547] used arc ablation and reported detection limits at the 0.1 µg/g level and calibration curves being linear to concentrations of 0.1%. Arrowsmith and coworkers showed that both in metals as well as in ceramic samples direct analyses could be performed with laser ablation coupled to ICP-MS [537, 548, 549]. Laser ablation ICP-MS is now commercially available and of great use for survey analysis of solids and also for inclusions.

Biological and medical samples

For biological and medical samples, ICP-MS has facilitated a considerable enlargement of the series of elements that can be determined directly and, thus, is of great importance for speciation and bioavailability studies. Normal levels of a number of trace constituents in clinical samples have been determined by ICP-MS [550]. In urine good agreement of ICP-MS results with those of other techniques was obtained for elements with masses beyond 81 (Pb, Cd, Hg and Tl). Deviations were found for As, Fe and Se, which could be partially eliminated by precipitation of the chlorides from the measurement solution. With ICP-MS Pb can be determined in blood [551] and the bioavailability of Zn has been studied [552]. For the analysis of small samples, as shown for blood analysis [553], or for the analysis of samples with high salt contents, flow injection can be applied. By coupling ICP-MS with chromatographic techniques, metals bound to different protein fractions can also be determined separately [554]. In addition, metabolic studies can be performed by isotopic dilution work, which is very promising for medical applications.

For applications in the life sciences, where limited sample volumes or fairly complex mixtures are the norm, chromatographic techniques enabling high chromatographic resolution, such as capillary zone electrophoresis, coupled on-line to ICP-MS are very powerful. Applications such as the determination of organoselenium compounds and of metalloproteins in serum [556] or the separation of the six relevant arsenic compounds in water at their 1–2 µg/L level [557] should be mentioned. The use of low-consumption high-efficiency nebulizers, such as the direct injection high efficiency nebulizer [558] are of great value in these applications.

Environmental work

ICP-MS is very promising in the area of environmental studies. Many elements can be determined directly in drinking water. In waste water analysis sample decomposition by treatment with HNO_3–H_2O_2 is often required and the most frequent isobaric interferences have been described [559]. For seawater analysis, the salt contents makes sample pretreatment necessary, which can be done by chelate extraction. Beauchemin et al. [560] obtained a preconcentration of a factor of 50 by sorption of the trace elements onto an SiO_2 column treated with 8-hydroxyquinoline and determined Ni, Cu, Zn, Mo, Cd, Pb and U in seawater. In river water Na, Mg, K, Ca, Al, V, Cr, Mn, Cu, Zn, Sr, Mo, Sb, Ba and U could be determined directly and Co, Ni, Cd and Pb after the above mentioned preconcentration procedure. For As, preconcentration by evaporation of the sample was sufficient. Isotope dilution delivers the highest accuracy [561] and the procedure has been applied to the characterization of a standard reference sample [560]. ICP-MS, subsequent to sample decomposition with HNO_3–H_2O_2, has also been used for trace determinations in marine sediments [562] and for the trace characterization of marine biological samples [563, 564]. Owing to the extremely high power of detection ICP-MS can be used to determine very low background concentrations in unpolluted areas. With high-resolution ICP-MS and a microconcentric high-efficiency nebulizer Rh, Pd and Pt can be detemined in ice samples with detection limits down to 0.02, 0.08 and 0.008 pg/g, respectively [565] as can the actinides in environmental samples [566].

ICP-MS coupled with chromatography has become very important for speciation of environmentally relevant elements.

For the speciation of chromium in waste water samples from the galvanic industry, Andrle et al. [567] fixed the Cr^{III} and Cr^{VI} species present through a reaction with dithiocarbamates. In this way, stable complexes were formed, which could then be separated by chromatography. The preparation of these complexes can be performed in a flow system which includes a thermostated reactor and fixation of the complexes on a solid phase. The separation can be performed by on-line coupling with reversed-phase chromatography and detection by ICP-MS using hydraulic high-pressure nebulization, as introduced for atomic absorption spectrometry by Berndt [143]. Calibration of the procedure is performed by standard additions so as to correct for any shifts in the Cr^{III}–Cr^{VI} equilibrium during the complexation reaction. The chromatograms, in the case of a waste water sample, showed that multielement speciation is certainly possible, as elements other than Cr may be present in different valence states and accordingly also form different complexes with dithiocarbamates.

Other methods of speciation of chromium in water samples lie in the use of anion exchange resins, which were shown by Barnowski et al. [568] to retain both the Cr^{III} and the Cr^{VI} species. This approach has the advantage that chloride ions which could possibly be present are retained and accordingly do not cause spectral interferences, e.g. when determining preconcentrated iron species.

Gas chromatography coupled with ICP-MS enables the determination of volatile organometal compounds, such as the organolead compounds, to be performed, with a high power of detection. In other cases a derivatization has to be performed, e.g. ethylation with tetraethylborate or a Grignard reaction (for a discussion see Ref. [452]). Suitable coupling systems have been described combining GC and quadrupole or sector field ICP-MS (see e.g. Prange and Jantzen [569] and Heisterkamp et al. [570]), where the transfer line is heated. A drawback is the risk of skewing the gas chromatography peaks when using these mass spectrometers in the multielement mode. As with gas chromatography separations can be accelerated still further using multi-capillary gas chromatography [571]; skewing of peaks then becomes an even greater risk. The related systematic errors can be eliminated by using TOF-ICP-MS, as shown by Leach et al. [572]. For GC use of a large and bulky ICP is also no longer required and savings in investment and operating costs can be made by switching to microwave plasma sources.

6.1.4
Outlook

Further trends in mass spectrometry with discharges at atmospheric pressure lie in the use of alternative plasmas, in the progress of developments to mass spectrometric equipment and in the improvement of sample introduction.

Without loss in analytical performance ICPs can be operated at lower gas flows and power consumption especially at higher frequencies (up to 100 MHz) (see e.g. Ref. [573]). The addition of molecular gases such as N_2 has been thoroughly investigated by Lam and Horlick [574], who found a decrease in the formation of cluster ions in a number of cases. Helium ICPs have been investigated by Chan and Montaser [575] and these could be very useful for the determination of the halogens, as shown in Ref. [576].

MIPs have also been used as sources for mass spectrometry, as e.g. described by Satzger et al. [577]. Whereas sampling capabilities are practically limited to coupling with gas chromatography or electrothermal atomization, MIPs at higher power may be useful alternative sources, as they can also be operated in nitrogen [578]. This has been shown for a high-power nitrogen discharge by Okamoto [579], where a number of interferences known from argon plasmas could be avoided. Moreover, the operating costs of such a system are lower and for specific applications, such as process analysis, it might be of interest.

Other plasmas at atmospheric pressure, such as the FAPES (furnace atomic plasma emission source) developed by Blades [580] have been used as ion sources for mass spectrometry. With FAPES detection limits in the fg range can be obtained, as microsamples can be analyzed with virtually no transport losses (Fig. 117) [581]. However, further investigations on interferences certainly still need to be made.

By using high-resolution systems instead of quadrupole based instruments spectral interferences can be eliminated in a number of cases. To this aim, sector field instruments, provided they become cheaper, and time-of-flight instruments, especially for the case of transient signals, begin to find uses [75]. As on-line coupling of high resolving separation systems with ICP-MS becomes of more and more use and multielement speciation is requested not only in volatile but also in liquid samples, plasma-TOF-MS becomes of greater interest. This is shown by the first results obtained with capillary zone electrophoresis coupled with TOF-ICP-MS by Bings et al. [582]. Advanced types of mass spectrometers, well-known from work in organic mass spectrometry, such as ion cyclotron resonance mass spectrometry [583] and ion traps may become of use. Both constitute approaches for achieving extremely high spectral resolution and, in the case of soft plasmas, also work with clusters. With ICP Fourier transform ion cyclotron resonance mass spectrometry a resolving power of up to 88 000 has been realized and detection limits at the sub-mg/L level found.

A microwave plasma torch can be operated very stably with argon as well as with helium and can be used as an ion source for time-of-flight mass spectrometry. Such a system, as described by Pack et al. [76], is very useful for element-specific detec-

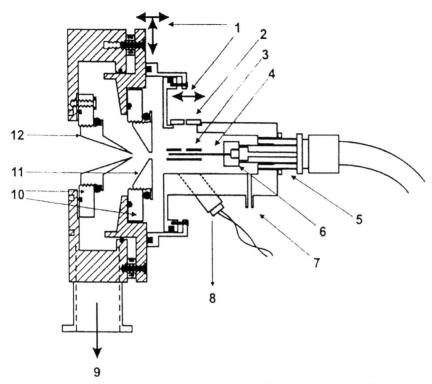

Fig. 117. FAPIMS (furnace atomization plasma ionization MS) workhead and sampling interface, (1): x, y, z translation; (2): viewing/ sampling port; (3): ICC furnace; (4): rf center electrode; (5): type N rf connector; (6): brass collet; (7): plasma gas (He) inlet; (8): photodiode temperature sensor; (9): to roughing pump; (10): MACOR support rings; (11): modified sampler cone; (12): standard skimmer cone. (Reprinted with permission from Ref. [581].)

tion in gas chromatography, as peak skewing is absent. It is easy to bring the GC column up immediately behind the plasma and to heat the complete transfer line by resistance heating of a copper wire wound around the capillary (Fig. 118). In this way there is no chance of deterioration of the chromatographic resolution and, as helium can be used as the working gas, the detection limits for halogens are very low. When recording the chromatograms of the halogenated hydrocarbons, carbon and chlorine can be monitored simultaneously (Fig. 119). The stability of the plasma is obvious from the flat signal for chlorine at the elution time for methanol. The detection limits for chlorinated compounds were shown to be in the fg range and are considerably lower than in a low-pressure ICP and a quadrupole mass spectrometer. This may partly be due to the small size of the microwave discharge as compared with the ICP, resulting in a lower analyte dilution. The coupling of time-of-flight mass spectrometry with a helium microwave plasma torch and gas chromatography may accordingly become a real alternative to the microwave in-

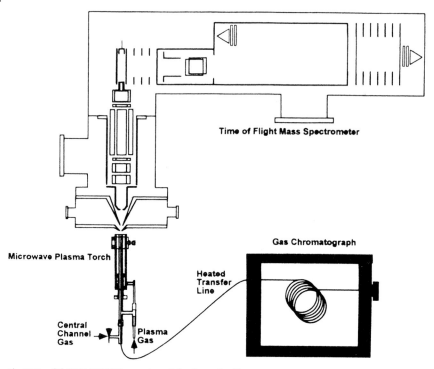

Fig. 118. GC-MPT-TOFMS experimental set-up. Capillary
column extends from the GC oven to the tip of the MPT, where
a plasma is formed. The plasma is sampled through a 0.5 mm
orifice into the TOFMS for mass analysis. (Reprinted with
permission from Ref. [76].)

duced plasma atomic emission detector (MIP-AED) and also to ICP-MS coupled
with gas chromatography for the determination of organometallic or halogenated
compounds.

In the field of sample introduction, hyphenated systems and devices allowing on-
line preconcentration, automated sample introduction or speciation will become
more and more important.

Summarizing, the potential of ICP-MS lies in the fact that analyses can now be
performed with the flexible sampling of an ICP, with true multielement capacity
and a high power of detection.

Manufacturers

- Agilent Technologies, Palo Alto, CA, USA
- Finnigan MAT, Bremen, Germany
- GBC, Australia
- Hitachi, Tokyo, Japan

Fig. 119. Isotope-specific chromatograms of halogenated hydrocarbons (chlorobutane to chlorohexane) in methanol obtained by GC-MPT-TOFMS. Twin boxcar averagers are used for data collection. (a): Signal from ^{12}C, (b): signal from ^{35}Cl. (Reprinted with permission from Ref. [76].)

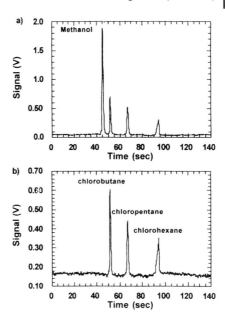

- Leco, St Joseph, MI, USA
- Micromass Ltd., UK
- PerkinElmer Sciex, Norwalk, CT, USA
- Spectro Analytical Instruments, Kleve, Germany
- Thermo Jarrell Ash, Waltham, USA
- Varian Ass., Mulgrave, Australia
- VG Elemental, Sussex, UK

6.2
Glow discharge mass spectrometry

Glow discharges [584], known from their use as radiation sources for atomic emission spectrometry have also became recognized as powerful ion sources for mass spectrometry. This development started with spark source mass spectrometry, where continuous efforts were made to arrive at more stable sources with the added advantage that the matrix dependency of the analyte signals would be lower than in spark sources [69].

In glow discharge mass spectrometry the analyte is volatilized by sputtering and the ions are produced in the negative glow of the discharge as a result mainly of collisions of the first kind with electrons and Penning ionization. Subsequently, an ion beam is extracted which in its composition is representative of the sample. Between the glow discharge and the spectrometer a reduction in gas pressure is required. As the glow discharge is operated at a pressure of around 1 mbar, a two-stage vacuum system is required. A cathodic extraction can be done at the cathode plume, by taking the sample as the cathode and drilling a hole in it. The aperture should reach the hot plasma center. In most cases the extraction of the ions, however, is done anodically.

As a result of impurities in the filler gas and the complexity of the processes taking place, not only do analyte and filler gas ions occur in the mass spectra but also many other ions. Spectral interferences therefore can occur and for various reasons:

- isobaric interferences of isotopes of different elements, e.g. $^{40}Ar^+$ with $^{40}Ca^+$ or $^{92}Zr^+$ with $^{92}Mo^+$;
- interferences of analyte ion signals with doubly charged ions, e.g. $^{56}Fe^{2+}$ with $^{28}Si^+$ (this type of interference, however, is much rarer than in earlier work with high vacuum sparks);
- interferences of analyte ions with cluster ions formed from analyte and gas species, e.g. $^{40}Ar^{16}O^+$ with $^{56}Fe^+$;
- interferences by signals from residual gas impurities, e.g. $^{14}N^{16}O^1H^+$ with $^{31}P^+$ forming, for instance, metal argides.

Mass spectrometric measurements contributed to clarifying the mechanisms of reactions with reactive species such as air and and water vapor. Indeed, the presence of water vapor may seriously influence ionization and excitation in the glow discharge plasma, thus leading to a decrease in analytical signals such that they are barely observable [585]. The major consequence would be an oxidation of the sample surface, quenching of argon metastables, low sputtering, and loss of analyte atoms due to enhanced gas-phase reactions. Therefore, the presence of water vapor should be limited by using high-purity gases, optimized vacuum systems and getters as appropriate. Cryogenic cooling may also be helpful. In a subsequent study, water was pulse injected into the discharge, to clarify the behavior of various cathodic materials (Cu, Fe, and Ti) [586]). Depending on the reactivity of these metals, oxides were formed during the pulsed injection and shortly afterwards. The effects for Ti were found to be larger than for Cu and Fe.

General models of GD-MS, as presented by Vieth and Huneke [587], include electron ionization three-body recombination and compare fairly well with experimental measurements of singly- and multiply-charged Ar ions. Penning ionization was found to predominate when the ion species had ionization energies below the energy of the Ar metastables. It can be shown with an electrostatic probe how GD-MS parameters are influenced by the voltage of an auxiliary electrode inserted into the plasma [588]. The plasma potential and the ion energies seem to follow the bias potentials well whereas the electron temperatures behave in a more complex fashion. The formation and use of signals for doubly charged analyte ions in GD-MS, when the singly charged ions suffer from interferences, has been treated by Goodner et al. [589]. A study of Ar, He, Kr, Ne and N_2, from the point of view of spectral interferences and sputtering, showed that Ar gave the best sputtering while with respect to signals, memory effects and cost, Kr was the worst [590]. The dependence of the ion current, ion intensity and energy distribution on power, carrier gas pressure and sampling distance in the case of a magnetron rf-GD for MS was investigated for a borosilicate glass cathode, and ion intensities were found to depend strongly on pressure and distance [591].

The possibility of separating the interfering signals depends on the instrument resolution. In the case of quadrupoles the mass resolution at best is unity and these interferences can only be corrected for by mathematical procedures. However, even with high-resolution instruments having a resolution of 5000 many inteferences remain, especially those with hydrides in the mass range 80–120 Da.

6.2.1
Instrumentation

Glow discharges operated at pressures in the 0.1–1 mbar range can be coupled to mass spectrometers by using an aperture between both chambers with a size of about 1 mm.

Much work has been done with dc discharges. The Grimm-type source with flat cathodes is usually used. The glow discharge may, however, also use pin-shaped samples which can be introduced into the source without even admitting any air. Harrison and Bentz [584] accordingly coupled discharges with a pin cathode as well as hollow cathode plumes to a high-resolution sector field mass spectrometer. This geometry was found in 1984 in the early glow discharge mass spectrometers (VG 9000) available through VG Instruments (Fig. 120A). A double-focussing spectrometer in so-called reversed Nier–Johnson geometry is used, with which a spectral resolution of about 10^4 can be obtained. The mass range up to 280 Da can be covered by using an acceleration voltage of 8 kV. A Faraday-cup and a so-called Daly–Multiplier are used as detectors with the possibility of continuous switching. The vacuum of 10^{-6} mbar is maintained with the aid of oil diffusion or turbomolecular pumps. To lower the background, the source housing can be cooled down to act as a cryo-pump.

However, with further developments quadrupole mass spectrometers have also been used to detect the ions produced in a glow discharge. No high voltage is required to realize a sufficiently high ion extraction yield and rapid scans can easily be performed, which enables quasi-simultaneous multielement measurements to be made. An electrostatic analyzer is often used in front of the spectrometer as an energy filter. Jakubowski et al. [592] described a combination of a Grimm-type glow discharge with a quadrupole mass spectrometer (Fig. 120B) and studied the basic influence of the working and construction parameters. Glow discharge mass spectrometers using quadrupoles are now commercially available.

A special dc GD for combination with a sector-field instrument has been designed for the analysis of high-purity Si [593]. The Si^+ ion yield was found to increase with gas pressure, probably as a consequence of the enhancement in the Ar metastable population. The signals for the Si_2^+ decrease perhaps as a result of an increase in dissociative collisions. Detection limits for elements such as Al, As, B, Cl, Fe, P and U are at a level of $6 \times 10^{10} – 6 \times 10^{13}$ atoms/cm^3.

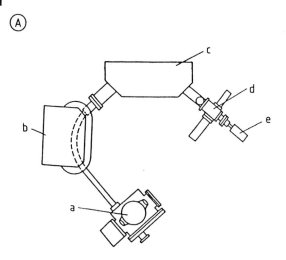

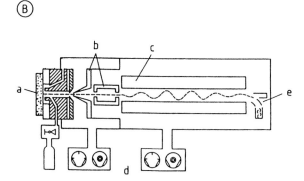

Fig. 120. Glow discharge mass spectrometry. A: Sector field based instrument (a): glow discharge source; (b): magnetic sector field; (c): electrostatic energy analyzer; (d): Daly detector; (e): Faraday cup (VG 9000, courtesy of VG Instru-ments) and B: a quadrupole based low resolution glow discharge mass spectrometer (a): source; (b): ion optics; (c): quadrupole mass filter; (d): differential pumping system; (e): detector. Similar to described in Ref. [611].

A modified version of the Grimm-type GDL has been described by Shao and Horlich [594]. The source has a floating restrictor and is designed so as to replace an ICP torch in an ICP-MS. Therefore, the anode is slightly positive with respect to the earthed skimmer interface of the MS system. The simultaneous analysis of an unknown sample and a reference material was carried out by means of a system based on two pulsed GD sources housed within the same tube [595]. Optimization of the relative position of the two cathodes was achieved by evaluating the signals produced in GD-MS when using the same specimen for each of them.

The introduction of rf-powered glow discharges in MS has led to a large broadening of the applicability of the method [596]. In initial studies it was shown that the bias potential is directly proportional to rf power and inversely proportional to the discharge gas pressure. When coupled to a high-resolution double-focussing mass spectrometer [597], problems of rf shielding, grounding and coupling to the accelerating potential were faced. To this end a source capable of handling both pin and flat samples was built. It operated stably at 13.56 MHz, 10–50 W and 0.1–1 mbar. Ion currents up to 10^{-9} A and a mass resolution of 8000 could be obtained.

Depth profiling is also possible as a result of the radial flow of support gas onto the sample surface. A successful extraction and focussing of ions necessitated parametric studies with respect to ion kinetic energies.

The analytical figures of merit of rf-GD-MS also with respect to depth profiling have been evaluated by Marcus et al. [598]. Two current versions of the lamp were considered, namely one for flat and one for pin samples.

An rf-GD-TOF mass spectrometer has been described by Myers et al. [599]. The ion optics which focus ions toward the entrance of the TOF instrument are the same as those used in the original ICP-TOF-MS. By means of pin-shaped brass samples of varying lengths, the sample–skimmer distance in the GD-TOF-MS instruments was optimized at 4 mm, while discharge gas pressure and power provided the best results in the 50–60 W and 0.3 mbar range, respectively. The application of small negative potentials to the skimmer cone (extraction orifice) was found to improve signals only slightly. However, higher negative potentials reduced both signals and resolving power. The skimmer potential was also found to affect the final kinetic energies of the ions before their extraction in the TOF. At 0.3 mbar all ions extracted for mass analysis were found to have approximately the same kinetic energy and detection limits were stated to be at a level of 1 µg/g.

Work has also been carried out with TOF-MS using pulsed GD sources. The effect of the distribution of the kinetic energy was lowered by resorting to a two-stage acceleration field by using a dc microsecond-pulsed GD as described by Hang et al. [600]. Here the sample was placed on the tip of a direct insertion probe. The configuration permitted the use of current intensities as high as 500 mA with 200 ns pulses being applied. During the active cycle of pulsed operation a larger net signal (by up to 2 orders of magnitude) can be produced than could be generated through a comparable power level in the dc mode. Furthermore, the fast pulsed discharge operation may permit a diagnostic evaluation of the plasma processes and give improved analytical performance when used with TOF-MS.

An rf-planar-magnetron GD has also been coupled with TOF-MS. For rapid sample changing without venting the mass spectrometer, a sliding PTFE seal was placed at the interface. The seal in turn holds a Macor ring, shields it from the plasma and supports the sample [601]. Detection limits for conducting and insulating materials were of the order of 0.1 and 10 µg/g for B and Mg in Macor, and Bi, Cr, Mn, Ni, and Pb in aluminum, respectively. The source–spectrometer combination still needs improvements at the interface with respect to the extraction location for analyte ions, the scattered ion noise and the extraction repetition rate.

When applying a hollow cathode (HC) discharge as the ion source, two groups of ions (respectively with high and low kinetic energy) were detected in the plasma of a reversed HC source combined with an MS [602]. The former group of ions are produced in the negative glow region, while the latter are formed in the extraction orifice of the cathode base. Discrimination between these two groups by setting the acceleration voltage so as to separate the high-energy group (analyte atoms) from the low-energy group (Ar carrier gas and cluster ions) could lead to improved analytical performance.

Hollow cathodes have also been used for the direct analysis of liquids. As reported by You et al. [603] a particle beam liquid chromatography (LC)-MS device can be interfaced with a heated HC unit. The high efficiency of the thermal nebulization system ensures that analyte particles from the aqueous solution can be transported into the heated HC cell for subsequent vaporization and excitation as well as ionization. The discharge characteristics were investigated in the case of Ar and He as support gas, the latter giving less line-rich emission spectra. To explore in detail the behavior of the system, signals for Cs and Na were studied as a function of the discharge pressure, current, solvent flow rate, tip temperature, etc. Optimum results were obtained at low flow rates (even down to 0.2 mL/min) and temperatures of 200–220 °C, respectively. The system can be used successfully as an ion source for mass spectrometry. The transport of analytes was found to be enhanced by the addition of HCl to the solutions due to an enhancement in analyte size favored by this reagent, as revealed by scanning electron microscopy. Typically, the particle size was in the 2–8 μm range with transport efficiencies of 4–18%, as found for Cu, Fe and Na test solutions [604].

Gas sampling GD was first described by McLuckey et al. as a soft ionization source [490]. It was shown to facilitate the ionization of gaseous compounds while giving both elemental as well as molecular information, as shown for the case of AsH_3 produced by hydride generation (Fig. 121) [488] and can be used successfully for the ionization of any atomic as well as molecular vapor. By alternating from soft to harsher conditions, it is possible to change from atomic to molecular information. This can be illustrated by the spectra obtained from ferrocene vapor [605].

McLuckey proposed the use of glow discharges coupled with quadrupole ion-trap MS for the determination of high-explosive substances in the vapor phase. The gas sampling GD system was found to be very effective at forming negative ions, e.g. from mono-, di- and trinitrophenols, mono-, di- and trinitrotoluenes, S, and others.

Also low pressure ICPs and MIPs, often used as ion sources (see e.g. Creed et al. [606]) are very similar in their performance to glow discharges, except for them being electrodeless discharges. They have shown particular merits for element-specific detection in chromatography, where detection limits down to the pg/s level

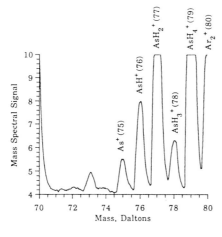

Fig. 121. Mass spectrum obtained from argon gas sampling glow discharge coupled with hydride generation for the determination of As. Solution: 10 μg/mL As; argon pressure: 0.2 Torr; voltage: 760 V; current: 20 mA; quadrupole mass spectrometer (Balzers), vacuum in 3rd stage: 2×10^{-5} Torr, signal detection: secondary electron multiplier. (Reprinted with permission from Ref. [488].)

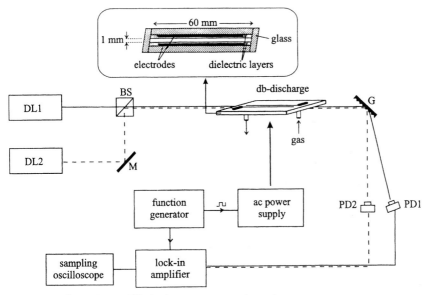

Fig. 122. Dielectric barrier (db) discharge microstrip plasma in set-up for diode laser AAS. (DL1, DL2): diode lasers, (BS): beam splitter, (M): mirror, (G): grating, (PD1, PD2): photodiodes. (Reprinted with permission from Ref. [608].)

can be obtained. These values, however, were found to be surpassed for halogens by microwave discharges at atmospheric pressure coupled with TOF-MS.

Glow discharges have also now been miniaturized. In the discharge in a micro-structured system, described by Eijkel et al. [607], molecular emission is obtained and the system can be used successfully to detect down to 10^{-14} g/s methane with a linear response over two decades. A barrier-layer discharge for use in diode laser atomic spectrometry has also been described recently (Fig. 122) [608].

6.2.2
Analytical performance

With flat as well as pin form cathodes the discharge is usually operated at about 1 mbar, a gas flow rate of a few mL/min and the current is about 2–5 mA with a burning voltage of 500–1500 V.

Normally the power is stabilized and the pressure, the current, the voltage or the power are taken as optimization parameters. All electrically-conductive samples and even semi-conductors can be analyzed directly. After a preburn phase the discharge can be stable for many hours. Detection limits down to the ng/g range are obtained for most of the elements.

Mass spectrometric methods are relative methods and need to be calibrated with known reference samples. To this aim so-called relative sensitivity factors are used. They are defined as:

$$RSF = x_{el}/x_{ref} \cdot c_{ref}/c_{el} \tag{300}$$

where x_{el} and x_{ref} are the isotope signals for the element to be determined and a reference element, respectively, and c is their concentration. When matrix effects are absent the factors are 1. In GD-MS the RSFs are found to be much more closer to 1 as compared for instance with spark source mass spectrometry [609] and secondary ion mass spectrometry.

Particularly in quadrupole systems spectral interferences may limit the power of detection as well as the analytical accuracy. This is shown by a comparison of a high- and a low-resolution scan of a mass spectrum (Fig. 123) [609, 610]. In the first scan (A), the signals of $^{56}Fe^{2+}$, $^{28}Si^+$, $^{12}C^{16}O^+$ and $^2N_2^+$ are clearly separated, which is not, however the case in the quadrupole mass spectra. In the latter scan, interferences can be recognized from comparisons of the isotopic abundances. Also physical means such as the use of neon as an alternative discharge gas [611] can be used to overcome the spectral interference problem.

Similar approaches can be followed with all types of glow discharges discussed under Section 6.2.1.

6.2.3
Analytical applications

The use and topical applications of GD-MS as sources for MS has been well covered in a review by Caroli et al. [612].

Bulk concentrations

GD-MS is of use for the direct determination of major, minor and trace bulk concentrations in electrically-conductive and semi-conductor solids down to concentrations of 1 ng/g. Because of the limited ablation (10–1000 nm/s), the analysis times especially when samples are inhomogeneous are long. It has been applied specifically to the characterization of materials such as Al, Cd, Ga, In, Si, Te, GaAs and CdTe.

A comparative study of spark-ablation ICP-MS and GD-MS in the case of steel has been reported by Jakubowski and coworkers [536, 613]. The RSFs for a number of trace elements and the measurement precision are very similar in both cases. Steel analysis by GD-MS benefits from the addition of 1% of H_2 to the Ar discharge gas [614], the explanation for which is certainly complex. For certified reference steels, including superalloys, reliable analysis results can be obtained. The determination of Mo, Nb and Zr in steels by GD-MS was found to be affected by the formation of multiply-charged cluster ions (metal argides) [615]. A correction based on the assumption that the rate of formation of the singly-charged argide is the same for all analytes and coincident with that of $FeAr^+$ was used. The capabilities of low resolution GD-MS were shown by the example of steel analysis [616], where detection limits were down to 1 ng/g and up to 30 elements could be determined.

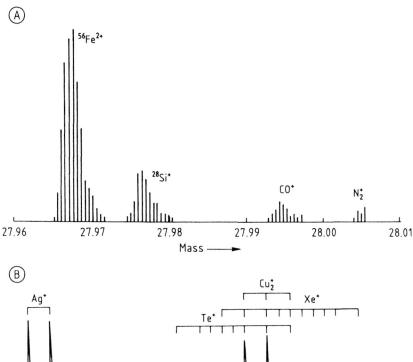

Fig. 123. Glow discharge mass spectrum at high resolution
(A) (reprinted with permission from Ref. [610]) and at low
resolution (B) (reprinted with permission from Ref. [592]).

Rare earth elements have been determined in metallic Gd, La, Nd, Pt and Tb by
GD-MS. Surface contamination can be removed by a 10 min predischarge and
careful assessment of polyatomics interferences performed. At the 1 μg/g level
RSDs are up to 40%. Also rf-GD-MS has been applied to the analysis of metals and
its figures of merit discussed [599].

Both for high-purity substances [617] and semiconductor-grade materials [618]
up to 16 elements can be determined in 5N and 6N type samples. At high resolu-

tion the determination of Ca and K is also possible [611]. Mykytiuk [619] showed that GD-MS can now effectively replace spark source mass spectrometry for a number of applications with excellent detection limits (often at the ng/g level) and with greater independence from the matrix composition. Interferences are still limiting in a number of cases, but Oksenoid et al. [620] showed that a substantial improvement by several orders of magnitude in analyte-to-interfering Ar gas intensity could be obtained by optimizing the lamp geometry, the working voltages and the gas flow. This allowed elements such as Br, K and Rb to be determined down to 1 µg/g, 1 µg/g and 10 ng/g, respectively. Routine bulk trace determinations in high-purity metals often revealed differences of one order of magnitude [621], which was attributed to sample inhomogeneity. Vieth and Huneke [609] determined 56 elements in six alloys and found good agreement between the RSFs and those calculated on the basis of a semi-empirical model accounting for diffusion, ionization recombination and ion-extraction phenomena.

High-purity and alloyed Al can be analyzed with a presputtering time of 30 min at a voltage higher than 1000 V and a current of 5.5 mA. For the actual analysis 1000 V and 2 mA were found to be suitable [622]. An investigation of background spectral interferences revealed little influence of gaseous, aqueous and solution species, although the formation of metal argides and multiply charged Ar species could not be disregarded in the evaluation of mass spectra [623]. For Al the detection limits of the rf-GD were found to be higher than those stated for commercially available high-resolution GD mass spectrometers, namely, in µg/g: 0.61 for Mn, 0.53 for Ni, 0.19 for Pb, 0.17 for Cr and 0.15 for Bi [601]. In Al samples Th and U concentrations of 20 ng/g can still be determined [617]. Limits of detection in the ng/g range and below with precisions of between 7 and 30% were obtained in the analysis of Al- and Co-based alloys by GD-MS [624], where the importance of ultraclean working conditions was stressed. Saito et al. [625] determined B in high-purity Mo down to 0.1 ng/g. For the elements C, N and O in steel at concentrations of 20 µg/g and below the lamp must be cooled to reduce the background species intensities, evacuated over a period of 20 min to allow the determination of N, presputtering should be used for the determination of C and the Ar purified with a ZrO catalyst to improve the quantification of O. When using cooling in the case of high-purity Ga, the detection limits were below the blank values of chemical analysis [626]. This also seemed to be useful in the analysis of $YBa_2Cu_3O_x$ and other high-temperature superconductors with a pin GD [627].

For metal alloys isotope ratios in solid samples can be determined for U and Os by GD-MS [628]. To avoid interferences Kr was selected as the discharge gas, but the results were still seriously affected by multiply-charged ions of Ca, Kr and Si. Isotopic measurements of Pd charged electrolytically with protium and deuterium were performed with a precision of better than 0.07%, however, only with a high-resolution instrument could interferences be prevented [629]. Also cooling of the discharge cell was required so as to eliminate hydride interferences from adsorbed H_2O. Trace elements were determined in Pt powder by GD-MS after ascertaining their RSFs [630]. The interference of PtH could not be disregarded in the case of

Au and Ir, requiring cryocooling to eliminate it. Accuracy and precision were in the range 10–15 and 5–10%, respectively, while the power of detection turned out to be better by 1–2 orders of magnitude than in ICP atomic emission spectrometry.

High-purity Ti has been analyzed by GD-MS to ascertain the Sc content [631]. The interference of $^{50}Ti^{40}Ar^{2+}$ limited the power of detection to 25 ng/g. Magnetron GD plasmas have been used for high-precision analysis of pure Cu, Mg, Al, Zn-based and Mn alloys and showed potential for high-precision analyses [632]. Copper-based alloys have been analyzed by quadrupole GD-MS using a system with optimized orifice-to-cathode distance, energy analyzer setting and bias voltage and trace elements were determined down to the µg/g level with a precision of 2–5%. Hutton and Raith [633] reported accuracies and precision similar to those of conventional arc and spark analysis. In 5N and 6N gallium Ca and K can be determined at high resolution with detection limits below the blanks of chemical analysis [634].

For compact SiC samples the analysis results were in good agreement with those of neutron activation analysis [618].

Powder samples

Electrically non-conducting powder samples can be analyzed after mixing with a metal powder such as Cu, Ag or Au and briquetting, as is known from experience with GD-OES. However, degassing and oxygen release may necessitate long preburn times and introduce instabilities. Also blanks arising from the metal powder may be considerable, e.g. for Pb in the case of copper powder.

The compacted pellet approach for powders was used by De Gendt et al. [635], who investigated the influence of the host matrix type (Ag, Cu), cooling and cathode shape (pins, disks) for some Fe ore certified reference materials. While binder and sample matrix had little effect on experimental RSFs, cooling and sample geometry played a significant role. Careful optimizing and controlling the working parameters allows accuracy and precision of the measurements of better than 10% to be obtained. The use of a secondary cathode for GD-MS of non-conducting samples such as solid glasses or sintered iron ores has been reported [636]. It was found that different sample types require ad hoc optimization of the measurement parameters, e.g. by changing the ratio between the sample and the secondary cathode signals. Furthermore, the purity of the secondary cathode may influence blanks and detection limits and the electrical resistivity of the sample is important too.

Iron meteorites have been analyzed for major (Co, Fe, Ni) as well as for minor and trace elements (up to 53 elements with detection limits down to the ng/g level) by GD-MS [637]. The isotopic composition was also shown to enable some information on the origin of the outer shell to be obtained and anomalies for C and N concentrations could be seen.

Teng et al. [638] examined some factors affecting trace determinations in soils, in particular red clay and forest soil. Satisfactory results were obtained by re-sorting the RSFs from soils CRMs. In particular the influence of sample oxygen content and host binder identity were examined and GD-MS was found to be less prone to matrix effects than laser ablation ICP-MS.

Betti et al. [639] used GD-MS for the determination of trace isotopes in soil, sediments and vegetation by blending with a conductive host matrix, namely Ag and using the secondary cathode technique so as to achieve a stable discharge. In this way detection limits in the pg/g range could be obtained for the radioisotopes ^{137}Cs, ^{239}Pu, ^{241}Pu, ^{90}Sr and ^{232}Th using an integration time of 1 h and a mass resolution of 100. Both total elemental concentrations and organometallic species could be determined in soil by means of GD-MS and GC-MS [640].

Suspended particulate matter can be analyzed after depositing it on high-purity In [641]. With detection limits at the sub-μg/g level data for 53 elements in a 10 mg sample were obtained, which proved to agree by better than a factor of 2 for 34 elements with results from other techniques. Atmospheric particulate matter was subjected to analysis by GD-MS by Schelles and Van Grieken [642]. Air was pumped through a single-orifice impactor stage in which the aerosol was collected on a metal support, which was then used as the cathode in the GD unit.

The maturity of GD-MS as a technique capable of routinely providing complete chemical analyses at the ultratrace level for insulating solid materials has clearly been demonstrated by quantification of a full range of elements (from Li to U) in coal and coal fly ash [643]. The samples were mixed with high-purity Ag powder as the binder and pressed into a pin shaped pellet by means of a polypropylene mold. Critical steps in the determination process were the inhomogeneous distribution of elements within and among the fly ash particles and the purity of the binder. The presence of highly volatile compounds hindered the application to bituminous coal.

The analysis of Al_2O_3 powders was possible after mixing with copper powder, but the presputtering was found to be critical. Also both argon and neon can be used as working gases, allowing detection limits down to the μg/g level to be obtained and optimization towards freedom from interferences [644].

Compact non-conductive samples

The ability of rf-GD sources to sputter or atomize compact non-conductive samples directly has been investigated intensively. The effects of cryogenic cooling, power, pressure and distance between sample and exit orifice have been investigated to improve the performance of rf-GD-MS in the analysis of oxides [645]. The first parameter was found to be crucial to removing gaseous interfering species. After careful optimization a precision of 5% could be obtained and the RSFs were 0.5–3 depending on the matrix, allowing semiquantitative analyses only. The rf-GD-MS analysis of glass and ceramic samples has also been reported [646]. By adjusting the ion transfer optics of the double-focussing mass spectrometer the ratio of analyte-to-background contaminant ion intensities can be optimized. Macor, a fairly insoluble non-conducting glass ceramic used as an insulating material capable of withstanding high voltages, can be analyzed directly [647] and O, Si, Mg, Al and B determined.

Pigmented polymer coatings on steel and PTFE as well as PTFE-based copolymers can be analyzed by rf-GD-MS in order to fingerprint them [648]. Advantages compared with SIMS and x-ray photoelectron spectroscopy are that the method is fast and does not require dissolution of the sample, and thermal volatilization processes do not appear to take place.

A tandem system consisting of pulsed dye laser ablation and ionization in a GD source for MS have been described by Barshick and Harrison [649]. The role played by the working gas (Ar, He, Ne) on redeposition of sputtered material has also been clarified. Removal of interfering species in GD-MS is possible through the use of getters such as Ag, C, Ta, Ti and W [650]. This approach has been applied successfully in the determination of rare earths so as to avoid oxide ion formation.

Depth profiles

These can be recorded just as in GD-OES, however, with a much higher resolution so that here thin multilayers can still be characterized [613] (Fig. 124]. Also the conversion of the intensity scale into a concentration scale is easier and the sensitivity higher. By the use of different working gases (see e.g. Ref. [651]) the scope of the method can be continuously adapted compared to other methods for depth-profiling.

In the depth-profiling analysis of steel by GD-MS, initial degassing caused serious interference problems independent of whether fast or slow erosion rates were adopted. Thermal degassing under vacuum conditions in the ion source before igniting the discharge has been shown to be helpful.

In frictional brass-coated steels quantification of continuously varying concentrations of Cu, Fe and Zn can be performed based on a linear combination of the RSFs and/or sputter rates for both Fe and brass [652]. Technical layers in a Cr–Ni system with a thickness of 30 nm to 10 μm can be characterized with respect to composition and thickness with GD-MS using a penetration rate of up to 0.1 μm/s and a depth resolution of 10 nm [613].

Dry solution residues

Analysis of dry solution residues can also be performed successfully with GD-MS. This has been shown with the aid of hollow cathodes, with which Sr, Ba and Pb can be determined [653] and where in 10–20 μL up to 70 elements can be determined with detection limits in the pg range [654]. With a Grimm-type discharge solution residues can be formed on the cathode by evaporating microvolumes of analyte solutions having low total salt contents. With noble metals such as Pt and Ir [655], use can be made of cementation to fix the analyte onto a copper target, so that it is preconcentrated and can be sputtered reproducibly. This has been found to be a way of fixing pg amounts of Ir on a copper plate prior to analysis with GD-MS.

A GD-MS method has been developed for the analysis of microliter volumes of aqueous solutions which permitted the long-term acquisition of data [656]. Samples were either adsorbed on pin-shaped electrodes prepared by pressing high-

(I)

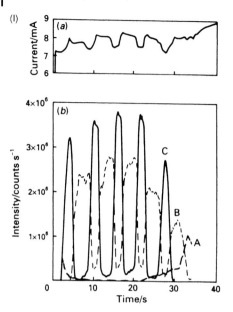

(II)

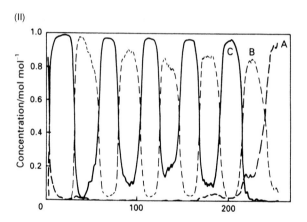

Fig. 124. Intensity versus time profile obtained in GD-MS analysis of a Cr–Ni multilayer system on Si substrate: (a): discharge current and (b): single-ion monitoring profiles of A: Si, B: Cr and C: Ni (I), and concentration versus depth profile calculated from measurements (II). (Reprinted with permission from Ref. [613]).

purity Ag powder or by preparing a slurry of Ag powder and the solution subsequently dried and pressed into a pin. In both instances homogeneity and particle size of the individual materials are critical. Tests were carried out with the NIST SRMs 3171 and 3172 multielement solutions of Al, Ba, Be, Cd, Cr, Cu, Fe, K, Mg, Mn, Na, Ni, Pb, Se, Sr and Zn. The average relative error was about 14% (2–30%) indicating the need for further work. For the determination of Pt in urine of patients treated with Cisplatin, the residues were dissolved in water and aliquots dried on the tips of carbon electrodes allowing the determination of both Pb and Pt down to the ng/g level. It was thus possible to conclude that Pb was displaced from body compartments and mobilized by administering Pt [657].

Acid digestion is recommended for the analysis of waste oils from of vehicles. A few µL of the resulting aqueous leachates can then be pipetted onto Ag powder and the slurries dried and pressed into polyethylene slugs to produce pins that can eventually be submitted to GD-MS analysis for the assay of their Pb contents [658]. Determinations can be performed by isotope dilution and concentrations as low as 3 µg/g Pb determined with a precision of better than 5%. GD-MS has also been used for the analysis of crude oils [659] and Cr, Cu, Fe. Mg, Na, Ni, Pb, Si, Sn and Ti can be determined in NIST SRMs, SPEX organometallic standard oils and refined oil composites. The method performs very well for limited amounts of sample, but the polyatomic interferences are a drawback.

Gases and vapors

Glow discharges are also very useful to excite and ionize dry vapors as generated by electrothermal vaporization or hydride generation. In the first case, Guzowski et al. [605] showed that by electrothermal vaporization of dry solution aliquots in a graphite cup atomizer, vapor sampling was possible and particularly in the case of TOF-MS allowed reliable determinations to be made. The system is very useful for studying vaporization effects, as signals of different elements can actually be monitored simultaneously and this also applies for isotope signals. It also was found that by alternating hard and soft conditions, even molecular information from bromoform or ferrocene can be obtained.

Hydride generation can also be used for sample introduction into a helium, argon or neon discharge and depending on the pressure and current more molecular ions or atomic species are found. The technique has excellent sensitivity and a fairly good dynamic range and precision [660].

In speciation, glow discharges are excellent detectors for GC work as shown earlier. In addition to the low power and pressure ICPs they can be used success- fully for element-specific detection for gas chromatography. An rf-GD-MS system has been used as a detector for GC by Olson et al. [661]. The set-up should consist of a temperature-controlled transfer line of stainless steel from the exit of the GC to the inlet of the GD source. The system has been tested with tetraethyl-Pb, tetraethyl-Sn and tetrabutyl-Sn and provided useful structural information for the identification of these compounds through the observation of fragment peaks; the detection limits were down to 1 pg.

Manufacturers

Glow discharge (GD) mass spectrometers with high (h) and low (l) resolution are available from several manufacturers.

- Extrel, Pittsburgh, PA, USA: GD(l)
- Turner Scientific, Finnigan, Bremen, Germany: ICP + GD(l)
- VG Instruments, Fisons, Cheshire, UK: GD(h), GD(l)

7

Atomic Fluorescence Spectrometry

In atomic fluorescence spectrometry the analyte is brought into an atom reservoir (flame, plasma, glow discharge, furnace) and excited by absorbing monochromatic radiation emitted by a primary source. This can be a continuous source (xenon lamp) or a line source (hollow cathode lamp, electrodeless discharge lamp or a tuned laser). Subsequently, the fluorescence radiation emitted, which may be of the same wavelength (resonance fluorescence) or of a longer wavelength (non-resonance fluorescence) is measured with or without being spectrally resolved.

7.1
Principles

A simplified model for the fluorescence process can be drawn up for two levels [662]. When a two-level system is considered (Fig. 125) and excitation is expected to occur only as a result of absorption of radiation with radiant density ρ_ν, without any contributions from collision processes to the excitation, the population of the excited level (n_2) can be given by:

$$\mathrm{d}n_2/\mathrm{d}t = n_1 \cdot B_{12} \cdot \rho_\nu - n_2[A_{21} + k_{21} + B_{21} \cdot \rho_\nu] \tag{301}$$

where B_{21} is the transition probability, ν the frequency and n_1 is the population of the lower level. A_{21} is the transition probability for spontaneous emission, B_{21} the transition coefficient for stimulated emission and k_{21} is the coefficient for decay by collisions. The exciting radiation is supposed to be "white", this means a constant radiation density over the absorption profile. When the statistical weights of n_1 and n_2 are equal, $B_{21} = B_{12}$ and when in addition $n_T = n_2 + n_1$, this being the condition for a closed two-level system in thermal equilibrium, Eq. (301) can be solved as:

$$n_2 = n_T \cdot B \cdot \rho_\nu \cdot t_r[1 - \exp(-t/t_r)] \tag{302}$$

because $n_2 = 0$ when $t = 0$.

Fig. 125. Transitions in a two-level system for atomic fluorescence spectrometry. (Reprinted with permission from Ref. [662].)

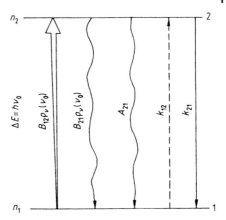

$$t_r = (2 \cdot B \cdot \rho_v + A_{21} + k_{21})^{-1} \tag{303}$$

and is referred to as the response time of the pumping process.

In the case of primary sources with low intensity and atom reservoirs at low pressure, e.g. glow discharges in which quenching is low:

$$n_2(t) = n_T \cdot B \cdot \rho_v \cdot \tau_{sp} \cdot [1 - \exp(-t/\tau_{sp})] \tag{304}$$

or in equilibrium:

$$n_2(t)_{eq} = n_T \cdot B \cdot \rho_v \cdot \tau_{sp} \tag{305}$$

When the primary source is of a low intensity and quenching occurs in the atom reservoir, as is the case in an atmospheric pressure furnace, flame or plasma:

$$n_2(t) = n_T \cdot B \cdot \rho_v \cdot \tau \cdot [1 - \exp(-t/\tau)] \tag{306}$$

with

$$\tau = (A_{21} + k_{21})^{-1} \tag{307}$$

being a longer response time.

When the absorption of radiation increases up to a certain value A_{21} and k_{21} become negligible. Then $n_2 = n_T/2$ and becomes independent of the radiant density of the exciting radiation and a state of saturation is reached. This situation can be realized when lasers are used as primary sources.

The radiaton density required to obtain saturation can be calculated as:

$$\rho_v^s = (A_{21} + k_{21})/2B \tag{308}$$

this being the radiant density with which an equilibrium population at the point of half saturation is obtained.

Apart from fluorescence with two levels, cases where more than two levels are involved in the process are also known.

The radiating level may often be somewhat below the level to which pumping occurs and the transition between the two levels can be radiationless and take place through heat losses. Also the fluorescence transition sometimes does not end at the ground level. With two levels resonance fluorescence occurs, where stray radiation from the exciting radiation limits the power of detection. In the case of three-level systems there is non-resonant fluorescence, where this limitation does not apply. However, here the radiant densities are much lower. Therefore, populating the excited level will only be sufficiently successful when using very intensive primary sources such as tunable lasers.

In atomic fluorescence work, the radiant density of the primary source can vary with:

- time: pulsed or continuous sources;
- frequency: when the source has a certain spectral profile, i.e. white or line radiation with a certain line width;
- space: the source can be spatially inhomogeneous, through which suitable optics for the sampling of the fluorescence radiation is required.

The absorption profile of the atoms in the atom reservoir is a function of the different line broadening mechanisms and contibutions to the physical line width.

When self-absorption is negligible, the intensity of the fluorescence radiation is given by:

$$B_F = n_2 \cdot h\upsilon_0 \cdot (1/4\pi) \cdot A_{21} = n_2 \cdot h\upsilon_0 \cdot (1/4\pi) \cdot (1/\tau_{sp}) \tag{309}$$

Where l is the thickness of the fluorescence layer. By combination with Eq. (304), it becomes obvious that:

- at primary sources with low intensities the fluorescence increases linearly with the radiant density of the primary source;
- when the radiant density of the primary source is sufficient to obtain saturation (e.g. a tunable laser) the radiation density of the fluorescence radiation cannot be increased further by increasing the intensity of the primary source.

Equilibrium equations for three-level systems can be set up just as for two-level systems.

The models must be corrected because of the following.

- The absorption of the exciting radiation is only proportional to the density of analyte atoms and the depth of the atom reservoir in the first approximation. As the intensity decreases with the depth because of absorption, a correction factor must be used and the primary intensity is given by:

$$\rho_\nu(\nu)\{1 - \exp[-k(\nu) \cdot L]\}\,d\nu \tag{310}$$

- The fluorescence radiation itself is also only proportional to the atom number densities and the pathlength through the atom reservoir in a first approximation because of self-absorption.

Both of these conditions depend on the set-up and also on the fact that a continuous source or a line source is used.

As especially in the case of non-resonant AFS the fluorescence intensities are low, special precautions have to be taken for signal acquisition. Indeed, the signal-to-noise ratios can be improved considerably by using pulsed signals and phase-sensitive amplification, as is easily done e.g. in the case of laser sources (see Section 2.1).

7.2
Instrumentation

A system for AFS consisits of a primary source, an atom reservoir and a detection system for measuring the fluorescence radiation. In some cases the detection can actually be performed without the need for spectral dispersion, e.g. by using filters only.

Primary sources

As primary sources, continuous sources such as a tungsten halogenide or a deuterium lamp can be used. They have the advantage that multielement determinations are possible. However, because of the low radiant densities saturation is not obtained and the power of detection is not fully exploited. With line sources such as hollow cathode sources and electrodeless discharge lamps much higher radiances can be obtained. Even ICPs into which a concentrated solution is introduced can be used as a primary source, through which multielement determinations become possible.

With laser sources in the continuous as well as in the pulsed mode saturation can be obtained. In most cases dye lasers are used. They have a dye (mostly a substituted coumarine) as the active medium. This liquid is continuously pumped through a cuvette and has a broad wavelength band (ca. 100 nm), within which a wavelength can be selected with the aid of a grating or prism. By using several dyes the whole wavelenth range down to 440 nm can be covered. The laser is often pumped by a second laser (e.g. an excimer laser) and can be operated in a pulsed mode. As its energy output is very high, it is possible to apply frequency doubling and Raman shifting, with which the accessible wavelength range can be decreased down to the 200 nm range. Accordingly, with some flexibility saturation can be realized for almost any term. In the case of resonant transitions ($f = 0.1$–1) the energy required therefore is 10–100 mW, whereas for a non-resonant transition ($f = 10^{-3}$) it is up to 1 W. Tunable dye lasers allow frequencies of up to around 10 Hz to be achieved and the pulse length can be down to 2–10 ns.

Fluorescence volumes

Chemical flames are very useful as fluorescence volumes, because of their good sample uptake capabilities and temporal stability. As the temperature is limited (3000 K for the C_2H_2–N_2O flame) problems arise for elements with thermally stable oxides, as is known from experience with flame AAS. These problems do not occur when an ICP is used as the atom reservoir [663]. In the ICP the measurements should be made in the higher part of the tail flame as otherwise the ionized and excited analyte fraction is too high. To keep the formation of oxides and the sideways diffusion of analyte low, a burner with an extended outer tube should be used for AFS work. In a graphite furnace the atomization efficiency is high and due to the limited expansion of the vapor cloud, the residence time of the analyte atoms is high as well. Accordingly, graphite furnace atomization laser AFS has extremely low absolute detection limits. Discharges under reduced pressure can also be used as the fluorescence volume. They have the advantage that quenching of the fluorescent levels, which may limit the power of detection but can also cause matrix effects, is lower [664].

Elements with low intensity fluorescence lines (e.g. Eu, Tm and Y) have been determined in aqueous solutions by depositing and drying nanoliter amounts of sample on the Ni cathode of a miniature GD source used as the atom reservoir [665]. The atomic cloud thus formed was excited by a Cu-vapor laser-pumped dye laser to detect fluorescence directly. Absolute detection limits of 2 fg for Eu, 0.08 fg for Tm and 1.2 pg for Y were achieved and the total time for analysis from sample probing to data acquisition did not exceed 5 min.

Furthermore, laser-produced plasmas may also be used as the fluorescence volume [226], which has been shown to be particularly useful for direct trace determinations in electrically non-conducting solids.

Detection

The detection of the fluorescence radiation differs in resonant and in non-resonant AFS. In the first case, the radiation is measured in a direction perpendicular to that of the incident exciting radiation. However the system will suffer from stray radiation and emission of the flame. The latter can be eliminated by using pulsed primary sources and phase-sensitive detection. In the case of non-resonant fluorescence, stray radiation problems are not encountered, although the fluorescence intensities are lower, which necessitates the use of lasers as primary sources and spectral apparatus that will isolate the fluorescence radiation. A set-up for laser excited AFS (Fig. 126) may make use of a pulsed dye laser pumped by an excimer laser. The selection of the excitation line is then done by the choice of the dye and

Fig. 126. Experimental set-up for laser induced atomic fluorescence work. (a): Flame, ICP, etc., (b): dye laser; (c): pumping laser, (d): photomultiplier, (e): monochromator, (f): boxcar integrator, (g): data treatment and display.

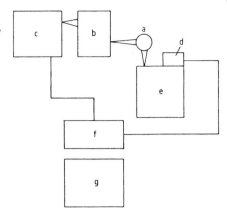

the tuning of the laser. Pulses with a length of 2–10 ns occur at a frequency of up to 25 Hz. A mirror can be placed behind the source so that the exciting radiation passes through the fluorescence volume twice and to lead the largest possible fraction of the fluorescence radiation into the monochromator. The photomultiplier signal is then amplified and measured with a box-car integrator of which a ns window is synchronized with the pumping laser.

7.3
Analytical performance

Detection limits

The detection limits in AFS are low particularly for elements with high excitation energies, such as Cd, Zn, As, Pb, Se and Tl.

In flame AFS, elements which form thermally stable oxides such as Al, Mg, Nb, Ta, Zr and the rare earths are hampered by insufficient atomization. This is not the case when an ICP is used as the fluorescence volume. Here the detection limits for laser excitation and non-resonant fluorescence are lower than in ICP-AES (Table 18) [663]. ICP-AFS can be performed for both atomic and ionic states [664].

Using a graphite furnace as the fluorescence volume and laser excitation, the detection limits are very low; e.g. Ag: 3×10^{-3}, Co: 2×10^{-3}, Pb: 2×10^{-3} ng/mL [665, 666].

Very sensitive determinations of Mo can be performed by dry solution residue analysis with laser-excited AFS in a hollow cathode discharge as the atomizer, as shown by Grazhulene et al. [667]. Bolshov et al. [668] showed that very low levels of lead in Antartic ice samples could be determined by laser-excited AFS using dry solution residue analysis with graphite furnace atomization.

When performing laser-excited AFS at a laser plume, it would appear to be useful to produce the laser plasma at pressures below atmospheric pressure, as then the ablation depends only slightly on the matrix [226, 669].

Tab. 18. Detection limits in laser atomic fluorescence spectrometry.

Element	Detection limit (ng/mL)	
	Laser AFS ICP [663]	*Laser AFS furnace [665]*
Ag		0.003
Al	0.4	
B	4	
Ba	0.7	
Co		0.002
Cu		0.002
Eu		10
Fe		0.003
Ga	1	
Ir		0.2
Mn		0.006
Mo	5	
Na		0.02
Pb		0.000025
Pt		4
Si	1	
Sn	3	
Ti	1	
Tl	7	
V	3	
Y	0.6	
Zr	3	

Spectral interferences

In atomic fluorescence spectrometry spectral interferences are low as the fluorescence spectra are not line rich.

Linear dynamic range

The linear dynamic range of AFS is similar to that of OES or even larger. Indeed there is no background limitation nor is there a serious limitation because of self-reversal, both of which limit the linear dynamic range in OES at the lower end and at the higher end, respectively. It may well be that the exciting radiation becomes less intensive when penetrating into the fluorescence volume and also that fluorescence radiation is absorbed by ground-state atoms before leaving the atom reservoir. These limitations, however, are less stringent than self-reversal in OES.

The application of laser-excited AFS in the analytical laboratory up to now has been seriously hampered by the complexity and cost of tunable lasers, which may change with the availability of less expensive but powerful and also tunable diode lasers, which can be operated in the complete analytically important wavelength range.

8
Laser Enhanced Ionization Spectrometry

Laser enhanced ionization spectrometry (LEI) [670] is based on the optogalvanic effect, which has been known since the early work on gas discharges. It was found that the current in a discharge under reduced pressure increases when irradiating with monochromatic radiation of a wavelength which corresponds to a transition in the term diagram of the discharge gas. As an analytical method LEI is based on a measurement of the increase of the analyte ionization in an atom reservoir resulting from a selective excitation of atoms or molecules with the aid of laser radiation [669, 670].

8.1
Principles

Ionization of atoms X in an atom reservoir can result from collisions with energy-rich atoms of another species M (Fig. 48) and the rate constant for the reaction

$$M + X \overset{k_{1i}}{\rightleftharpoons} M + e + X^+$$

can be written as:

$$k_{1i} = n_X \cdot [8kT/(\pi \cdot m_r)]^{1/2} \cdot Q_{MX} \cdot \exp(-E_i/kT) \tag{311}$$

where n_X is the concentration of analyte atoms, Q_{MX} is the cross section for collisions between X and M, which are atoms of another element, k is Boltzmann's constant, E_i is the ionization energy and T is the temperature in the atom reservoir. When the atoms X are in an excited state E_2 above the ground level, it can easily be seen that k_{1i} will change into

$$k_{2i} = \exp(E_2/kT) \cdot k_{1i} \tag{312}$$

and the increase in ionization for a level that is 1 eV above the ground level in the case of a flame with $T = 2500$ K is $e^5 = 150$. The ionization equilibrium will change although the recombination rate remains the same. Moreover, states with high energies (so-called Rydberg states) have a longer lifetime ($\tau \approx n^3$ with n being the main quantum number) and therefore ionization through collisions or through field ionization increases. However, the probability of ionization does not increase in an unlimited way with E_2 in one-step processes as the transition probability for the transition $1 \rightarrow 2$ decreases with $E_2 - E_1$. The increase in ionization also depends on the population of the level 2. In order to get a sufficiently high population in the excited level, to which the transition probability should be sufficiently high, a laser must be used. Indeed, for a closed system and two terms n_1 and n_2, the equilibrium is characterized by the equation:

$$B_{12} \cdot \rho_v \cdot n_1 = B_{21} \cdot \rho_v \cdot n_2 + A_{21} \cdot n_2 + k_{21} \cdot n_2 \tag{313}$$

where $B_{21} = (g_1/g_2) B_{12}$ and $n_T = n_1 + n_2 + n_i$.

B_{12} and B_{21} are the transition probabilities for absorption and stimulated emission, respectively, and n_1, n_2, n_i and n_T the analyte atom densities for the lowest state, the excited state, the ionized state and the total number densities. g_1 and g_2 are the statistical weights, A_{21} is the transition probability for spontaneous emission and k_{21} the coefficient for collisional decay. Accordingly,

$$n_2 = B_{12} \cdot \rho_v (n_T - n_i) / [(1 + g_1/g_2) B_{12} \cdot \rho_v + A_{21} + k_{21}] \tag{314}$$

n_2 will increase until $A_{21} + k_{21}$ becomes negligible and then $n_2/n_1 = g_2/g_1$. The required radiation densities are therefore about 10–100 mW for a resonant transition ($f = 0.1$–1) and 1 W/cm^3 for a non-resonant transition ($f = 0.001$). Furthermore, the use of a continuous wave laser allows n_i to be maximum and the use of multistep excitation may be helpful especially when high-energy terms (Rydberg states) have to be populated.

Indeed, when F is defined as:

$$F = n_2/(n_i + n_2) = n_2/(n_T - n_i) = n_2/n_t \tag{315}$$

$$-dn_2/dt = k_{2i} \cdot n_2 = k_{2i} \cdot F(n_t - n_i) = k_{2i} \cdot F \cdot n_t \tag{316}$$

Integration leads to:

$$\ln n_t = -k_{2i} \cdot F \cdot t + C \tag{317}$$

and as $n_t = n_T - n_i$ and at $t = 0$ $n_t = n_T$:

$$n_i = n_T [1 - \exp(-k_{2i} \cdot F \cdot t)] \tag{318}$$

For an irradiation over the time $\Delta t_1 n_i \rightarrow n_T$ when:

$$\Delta t_1 \gg 1/(k_{2i} \cdot F) \tag{319}$$

Accordingly, the laser irradiation should last long enough therefore cw (continous wave) lasers might be more appropriate than pulsed lasers, provided the radiant densities are sufficiently high.

It could also be that it is more beneficial to populate the Rydberg level in two steps, because it is easier to reach a level that is near to the ionization energy when very short wavelengths are not used. For an ionization potential of 7 eV, radiation with $\lambda < 220$ nm is required, as for 1 eV $= 1.6 \times 10^{-16}$ erg, λ is 1240 nm (because of $E = h \cdot c / \lambda$). In addition, the oscillator strength decreases with $E_2 - E_1$ and in the case of two-step procedures the selectivity also increases.

LEI as an analytical technique thus requires a tunable laser, an atom reservoir, a galvanic detection system and read-out electronics including a boxcar integrator. It was first realized experimentally with a flame as the atom reservoir using pneumatic nebulization for sample uptake [670].

A pulsed dye laser pumped with flash lamps is used in most cases, as here the selection of different dyes, frequency doubling and Raman shifting, which is possible as a result of the high energies available, allows the whole spectral range down to 200 nm to be covered. However, diode lasers, particularly for multistep excitation, and dye lasers pumped with an excimer or a nitrogen laser can also be applied.

The atom reservoir can be a flame, furnace or also a plasma discharge at atmospheric and at reduced pressure.

In the case of a flame (Fig. 127), the sample is usually introduced by pneumatic nebulization. The laser enters the flame a few mm above the burner head and a large collector electrode is situated a few mm above the laser beam location. This allows the ions formed to be extracted with a high efficiency. The collector electrode should be large so as to minimize changes in the field as a result of changes in electron number densities due to different concomitant concentrations. It should also be positioned inside the space containing the ions produced by LEI so as to minimize analyte signal losses. The measurement unit includes phase-sensitive amplification to ensure high S/N ratios, but does not require any spectrometric equipment as only an ion current is being measured.

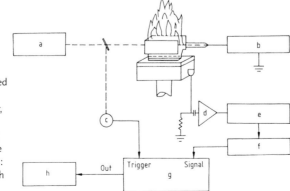

Fig. 127. Flame laser enhanced ionization spectrometry [670, 671]. (a): Flashlamp/dye laser, (b): high voltage, (c): trigger photodiode, (d): preamplifier, (e): pulse amplifier, (f): active filter, (g): boxcar averager, (h): chart recorder. (Reprinted with permission from Ref. [671]).

8.2
Figures of merit

The sensitivity in LEI depends on the properties of the laser system, the terms populated, the atom reservoir and the detection system. It can be written as:

$$S = c \cdot X_B \cdot \beta(A)[\exp(-E_i^*/kT) \cdot B_{12}(A) \cdot \rho_v^P(A)]$$ (320)

where c is a constant, X_B is the number density of analyte atoms in the lower level according to Boltzmann, β is the atomization efficiency in the flame, $\rho_v^P(A)$ is the spectral radiation density per laser pulse, E_i^* is the ionization energy of the excited level and S is the sensitivity in nA/(ng/mL). The noise includes contributions from flicker noise, nebulizer noise and also detector noise, and when applying phase-sensitive amplification detection limits are in the sub-ng/mL range (Table 19) [670, 671]. They are lowest when using a furnace as an atom reservoir because here the highest atom number densities and the longest residence times are encountered. In the case of pulsed lasers the radiation densities enhance the sensitivities, whereas in the case of cw lasers the n_i values increase as a result of improved excitation efficiencies.

In the so-called thermionic diode, a heated collector electrode is simply brought into the vicinity of the laser beam and isolated towards the vessel walls. No extraction voltage is applied and the ion current flowing as a result of the decreasing negative space charge around the collector, as a result of the ionization by the laser,

Tab. 19. Comparison of limits of detection for cw and pulsed laser LEI. (Reprinted with permission from Ref. [671].)

Cw		Pulsed	
element/wavelength (nm)	limit of detection (ng/mL)	dynamic range (orders of magnitude)	limit of detection (ng/mL)
Ca 422.7	1	5	
Ca 300.7			0.1
Ga 417.2	60	>3	
Ga 287.4			0.07
In 451.1	0.1	>4	
In 410.1	16	4.5	
In 303.9			0.006
K 766.5	0.1	>4	
K 294.3			1
Li 460.3	20	>4	
Li 670.8			0.001
Na 589.0	0.025	5	
Na 285.3			0.05
Rb 420.2	0.7	3	
Rb 780.0	0.09	>5	
Rb 629.8			2
Sr 460.7	0.4	>5	

is taken as the analytical signal. The whole system is usually operated at reduced pressure and allows detection of 10^2 atoms to be made (for Refs., see [672]).

In LEI the matrix effects are low. This applies in particular to samples with complicated atomic spectra, as no spectral interferences are encounterd. Elements with low ionization energies may cause matrix effects. The latter could be due to a change in the ion current as a result of changes in the electron number density in the flame. This can be compensated for by measuring alternately with and without laser irradiation. Ionization of the matrix by the laser may also lead to errors in the ion currents measured. These can be eliminated by modulation of the frequency of the exciting laser radiation around v, which corresponds to $E_2 - E_1 = hv$, where E_2 is the intermediate level (Rydberg state) and E_1 the ground level. Furthermore, the ion production as a result of the laser irradiation may influence the collection effi-cieny at the electrode as the electric field around the electrode changes. This effect can be minimized by using a large electrode and by collecting all of the ions pro-duced. This sets limits on the pulsing frequency. Indeed, for a distance of 5 mm between the collector and the burner, a field of 100 V/cm and a mobility of 20 cm^2/ V s, this being typical for a flame in which $T = 2500$ K, the highest allowed pulsing rate is 4000 Hz. Even so, for Na concentrations of 100 µg/mL matrix effects are still negligible.

The linear dynamic range in flame LEI is of the order of 3 decades. However, the multielement capacity is limited as monochromatic laser radiation is required.

When using two lasers and applying two-photon spectroscopy, only those atoms that do not have a velocity component in the observation direction will undergo LEI. Then the absorption signals become very narrow (Doppler-free spectroscopy). This enhances the selectivity and the power of detection, however, it also makes iso-tope detection possible. Uranium isotopic ratios can thus be detected, similarly to with atomic fluorescence [673] or diode laser AAS. Thus for dedicated applications a real alternative to isotope ratio measurements with mass spectrometry is available.

8.3
Analytical applications

LEI has been applied successfully to the trace determination of Tl [674] for cer-tification purposes, and for combinations with laser evaporation and all other atomization techniques represents a powerful approach to detection. Laser photo-ionization and galvanic detection have been applied to hollow cathode dark space diagnostics [675]. Photoionization is produced to measure the dark space widths of linear field distributions directly. A theoretical model has been developed and its predictions verified with experimental findings for a uranium hollow cathode dis-charge operated in neon or xenon. Variations in the ground-state densities of sputtered neutrals have also been measured.

Apart from the galvanic detection of the ion currents, direct mass spectrometric detection of the ions can also be applied, as is the case with resonance ionization mass spectrometry (RIMS) [676]. In addition, ionization can be performed by multi-photon absorption, which requires very intense primary sources.

9
Sample Preparation for Atomic Spectrometry

In a large number of atomic spectrometric methods sample preparation is required but depending on the method or sample introduction technique used, it can vary widely from a simple polishing step to a complete sample dissolution and subsequent analyte–matrix separation.

9.1
Sample preparation in direct compact sample analysis

In the direct analysis of metal samples, as in spark or x-ray fluorescence spectrometric analysis for process or product control, as treated in the pertinent literature (see e.g. Ref. [677]) treatment of the sample surface by turning-off or polishing with suitable emery paper may be adequate. However, in order to obtain the highest precision and accuracy, special procedures such as turning-off with a diamond in the case of glow discharge atomic spectrometry for bulk analysis [476] may be very helpful. New procedures such as laser ablation with a moving laser beam (LINA) no longer require any treatment of either metal or of ceramic or glass surfaces.

9.2
Grinding, sieving and compaction of powders

Grinding is of paramount importance since it allows homogenization in the case of direct solids analysis, and when applying sample dissolution it is often indispensable so as to increase the effectiveness of treatment with the aggressive reagents required for decomposition and dissolution. Sample grinding can be done in rotation mills using vessels and pestles made of different materials. Stainless steel or very hard materials such as tungsten carbide or agate are very useful. When grinding hard materials the difference in hardness between the sample materials and the grinding tools should at be least 2 on the Mohs hardness scale. In the case of SiC for example, which is produced as a solid material, grinding with WC introduces considerable amounts of W and with steel mills, Fe is found in the sample materials. When grinding SiC granulates in a high-purity mill made of SiC the abrasion of the mill could be up to a few % of the sample [546]. Grinding allows the mean particle size to be brought down to around 10 μm and in the case of wet grinding to around 5 μm. After grinding, particle sizes of powders can easily be determined with laser diffraction of a slurry of the material, or by automated electron probe microanalysis after depositing a layer on Nuclepore filters and coating with C or Au to render the sample electrically conductive (see Fig. 62 for comparative measurements) [203].

In order to sieve samples, sets of sieves with increasing mesh number and made of different materials, including very inert ones, can be used.

For the compaction of powders it is often necessary to mix the powder with a binder, to homogenize the mixture and to apply high pressure. Some powders such as Cu, Au, Ag and graphite powder may be mixed in a ratio of up to of 5:1 with the sample powder and electrically- as well as heat-conductive and even vacuum-tight pellets can be obtained. To reduce the amount of sample material required, it is often sufficient just to place the sample binder mixture in the middle of

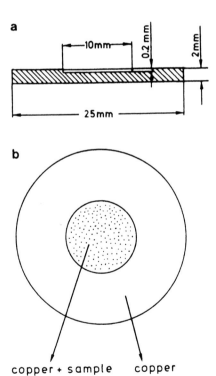

Fig. 128. Preparation of electrically- and heat-conductive as well as vacuum-tight pellets for the analysis of powders by direct solids sampling atomic spectrometric methods. (a) Side-on view, (b): top view (according to Ref. [210]).

the pellet (Fig. 128) [210]. Pressures required to obtain stable pellets can be up to 80 T/cm^2.

9.3
Sample dissolution

Sample dissolution can be performed with liquid or solid reagents at elevated temperatures causing a chemical reaction or melting of the mixture. The dissolution aids may introduce considerable contamination, as they are present in large excess. In addition, the vessels and also the laboratory atmosphere may constitute a source of contamination, whereas analyte losses particularly due to volatilization or adsorption onto the vessel walls may occur. Dissolution procedures for many types of materials can be found in the literature. They are used together with methodology for trace matrix separations based on classical chemical principles.

9.3.1
Wet chemical methods

Wet chemical dissolution is often performed with acids and the effectiveness increases with the oxidizing capacity. The latter depends on the acid (HCl < HNO$_3$ < HClO$_4$ < HCl–HNO$_3$) and the maximum temperature, depending on the boiling point of the acid (HCl < HNO$_3$ < H$_2$SO$_4$ < H$_3$PO$_4$). As the maximum temperature increases with the pressure work is often carried out in PTFE vessels in steel autoclaves at high pressure (up to 200 atm). The acids that give very soluble salts, such as HNO$_3$ or HClO$_4$ (not for K) are advantageous. Acids such as HNO$_3$ and HCl can be very efficiently purified by isothermic distillation. Quartz is preferred over glass as a vessel material. However, when HF is required for the treatment of silicon containing materials, vessels made of Pt or PTFE should be used. Whereas quartz and glass can be cleaned successfully by purging them with steam or HNO$_3$ vapors, Pt is rinsed most efficiently with dilute chlorine free HNO$_3$. In order to keep contamination risks low during chemical treatment in open systems, work in cleanrooms including the use of laminar-flow clean benches is advisable [678].

9.3.2
Fusion procedures

Fusion with alkali fluxes (Na$_2$CO$_3$, K$_2$CO$_3$, Na$_2$B$_4$O$_7$, etc.) as well as acid fluxes (KHSO$_4$) or more volatile salts such as NH$_4$HSO$_4$ in an excess of often 10:1 is most effective for the dissolution of many refractory powders such as Al$_2$O$_3$ or ZrO$_2$. However, these methods are hampered by the purity of the fluxes or abrasion of the vessels when it comes to applications for trace analysis. Recrystallization is a way forward, except for chemicals such as NH$_4$HSO$_4$, which can be prepared from pure chemicals (ClHSO$_3$ and pure NH$_3$) [312]. In the case of ZrO$_2$ powders such fluxes are very useful, as the excess can be volatilized easily and the fusion can be made in quartz vessels at no more than 500 °C.

Special equipment for the parallel fusion of several samples including automatic pouring of the molten mixtures into water for dissolution or leaching are commercially available.

9.3.3
Microwave-assisted methods

Microwave-assisted treatment with acids, sometimes at pressures of up to 80 atm in vessels made of fiber enforced plastic is very efficient for increasing the speed and effectiveness of dissolution procedures. This is related to the fact that the liquid mixture is heated directly and not by convection through the wall. Many procedures have been described in a monograph [679] and in the appropriate literature. Also procedures for slurries with flow-through systems have been proposed [680] and used very efficiently, e.g. for powdered biological materials. Systems enabling the dissolution of several samples to be carried out simultaneously and under rigid pressure and temperature control [681] are now commercially available. Resistance-heated systems have also been proposed for on-line dissolution as an alternative to microwave-heated systems, mainly for the treatment of slurries of biological samples [130].

9.3.4
Combustion techniques

Combustion in an oxygen atmosphere of samples that have an organic matrix has long been used for sample decomposition. In the so-called Trace-O-mat [682] cellulose powder is often added to the sample so that a number of inorganic samples such as soils can even be decomposed. The decomposition products are then collected in a liquid nitrogen cooled trap. As a further development for inorganic refractory samples, fluorination has been proposed. This can be done with minute (about 10 mg) amounts of elemental fluorine in nickel reactors [683, 684]. The volatile reaction products can be determined on-line by IR spectrometry [685] or gas mass spectrometry [686]. In order to eliminate spectral interferences a controlled distillation of the condensed reaction products from a liquid nitrogen cooled system can be applied and this allows, for example, the main contaminants of Si_3N_4 to be determined with reasonable precision.

9.4
Flow injection analysis

Trace matrix separations and concentration procedures are required in combined procedures, which are a prerequisite to characterizing reference materials, as then calibration with solutions containing only the elements to be determined can be applied. Such procedures are now used in flow injection based systems.

9.5
Leaching sample preparation methods

The isolation of analytes by leaching, as is often done in soil analysis, is often dictated by the difficulties encountered when dissolving samples completely. However, also for speciation work, where the analytes are labile such procedures are often the only way to obtain the analyte solutions (see e.g. Ref. [687] for the isolation of organomercury compounds from fish samples).

10
Comparison with Other Methods

Atomic absorption, optical emission and atomic fluorescence as well as plasma mass spectrometry and new approaches such as laser enhanced ionization now represent strong tools for elemental analysis including speciation and are found in many analytical laboratories. Their power of detection, reliability in terms of systematic errors and their costs reflecting the economic aspects should be compared with those of other methods of analysis, when it comes to the development of strategies for solving analytical problems (Table 20).

10.1
Power of detection

AAS using furnaces remains a most sensitive method, certainly for easily volatilized elements with high excitation energies and this also applies to AFS using laser excitation. AAS will definitely benefit from the development of diode lasers, as it is then possible to eliminate the use of spectrometers for spectral isolation. Moreover, the dynamic range problems could be decreased by measuring in the wings of the absorption lines and the background measurement also could be performed by tuning the laser. This necessitates, however, obtaining lasers with which the analytical wavelengths down to the VUV can be covered. LEI is another most powerful approach with respect to power of detection, especially when applying Doppler-free spectroscopy. However, these methods have limited multielement capacity. From this point of view, plasma atomic spectrometric methods are more powerful. ICP-OES will remain an important tool for multielement determinations in liquid samples and its power of detection is between that of flame and furnace AAS. Thus it is the appropriate method for tasks such as waste water analysis. As a result of the availability of new detectors enabling multiline calibration and actual simultaneous acquisition of line and spectral background intensities, ICP-OES is now capable of high-precision analyses, which previously could only be done with laborious wet chemical and spectrophotometric methods.

ICP-MS has provided further progress particularly with respect to power of detection, provided the matrix loading of the sample solutions remains low. There may also be limitations in the low mass range (below mass 80), due to interfer-

Tab. 20. Figures of merit of analytical methods.

Method	Detection limit	Matrix effect	Cost
Atomic absorption			
Flame	++	+	+
Furnace	+++	+++	++
Atomic emission			
Dc arc	++	+++	+++
Spark	+	++	++
DE plasma	++	++	+
ICP	++	+	++
MIP	++	+	++
Glow discharge	++	+	++
Laser	+	+	++
Atomic fluorescence (laser)	+++	++	+++
LEI	+++	+	+++
X-ray spectrometry			
XRF	++(+[a])	+	++
Electron microprobe	++	++	+++
PIXE	+++	++	+++
Auger electron spectroscopy	++	++	+++
Mass spectrometry			
Spark	+++	+++	+++
ICP	+++	++	+++
Glow discharge	+++	+	+++
SIMS	+++	++	+++
Laser	+++	++	+++
Activation analysis	+++	+	+++
Electrochemistry	+++	+++	+
Spectrophotometry	++	+++	+
Spectrofluorimetry	+++	+++	+
Chromatography			
Gas	+++	+++	++
Liquid	++	+++	++

+ Low; ++ medium; +++ very high.
[a]   TRXRF = total reflection x-ray fluorescence.

ences by cluster ions. The use of sample introduction techniques generating dry aerosols has certainly improved matters, e.g. for Fe, As and Se. Further improvements are possible by the use of high-resolution mass spectrometry using sector field instruments. However, lower cost instruments are required so as to make the technique affordable to a broader circle of users. Time-of-flight mass spectrometry will definitely help significantly wherever transient signals are obtained and multi-element determinations are required, such as in electrothermal evaporation, single shot laser ablation and gas chromatographic detection. Therefore, multicollector mass spectrometers eventually with multichannel detection are the ultimate solution to precise isotope ratio determinations as are required in tracer or isotope dilution work.

Whereas ICP atomic spectrometry is certainly a most useful method for multi-element determinations in liquid samples at low concentration levels, glow discharges are very useful for the analysis of solids. Glow discharge atomic emission gained its merits specifically for depth profiling analysis and in addition is useful for bulk analysis together with standard methods such as arc and spark emission spectrometry. Glow discharge mass spectrometry was first found to be of interest for the analysis of ultrapure metals, as required in microelectronics. However, quantification is difficult and certified samples are required for calibration. When applying dry solution residue analysis, GD-MS has extremely low absolute detection limits. This approach is very useful for many applications in the life sciences, as it combines a high power of detection with the possibility of isotope dilution. In mass spectrometry this approach offers new possibilities, as more classical thermionic techniques are only applicable to elements of which the species can be evaporated. The use of soft discharges is interesting, as molecular information can be obtained. With solids, soft laser ablation combined with time-of-flight mass spectrometric detection is also useful.

Optical atomic and mass spectrometric methods can be used for the determination of the light elements, which is an advantage over x-ray spectrometric methods. Apart from this restriction, total reflection x-ray fluorescence provides a high power of detection, high accuracy and high multielement capacity, all with a minimum of sample preparation [688]. It should be considered together with developments such as work with polarized x-rays enabling much lower detection limits to be achieved, also with energy-dispersive systems and local analysis with focussed x-ray radiation, as well as with the use of soft x-rays facilitating speciation in a number of cases [689].

10.2
Analytical accuracy

The analytical accuracy of methods can only be discussed with regard to the complete analytical procedure applied. Therefore, it is necessary to optimize sample preparation procedures and trace–matrix separations specifically to the requirements of the analytical results in terms of accuracy, power of detection, precision, cost and the number of elements and increasingly of the species to be determined. However, the intrinsic sensitivity to matrix interferences of the different methods of determination remains important.

In optical emission and in mass spectrometry, spectral interferences remain an important limitation to the analytical accuracy achievable. In atomic emission this applies particularly to the heavier elements as they have the more line rich atomic spectra. When these heavy metals are present as the matrix, as is often the case in metal analysis, the necessity of matrix separations is obvious when trace analyses have to be performed. In order to overcome limitations by spectral interferences, high resolution Echelle spectrometers are finding more and more uses. They are compact and thus combine high resolution together with excellent stability and further enable multielement determinations to be made by using advanced detector technology, e.g. with CCDs. The latter are also very useful for classical Paschen–Runge spectrometers with many CCDs covering the whole analytical range.

In mass spectrometry spectral interferences limit the accuracy in the low mass range in particular. Progress could be expected from the use of sector field and time-of-flight mass spectrometry. However, in the first case transmission and in the second case dynamic range problems must be given attention. Signal depressions and enhancements are a further main cause of interferences in ICP-MS. They can be succesfully taken care of by using standard additions, as in ICP-MS the spectral background is low. Furthermore, internal standardization may well allow compensatation to be made for easily ionized elements effects. This is more difficult in ICP-AES, where the spectral background especially in trace analysis is considerable and may be influenced strongly by changes in the concentrations of easily ionized elements.

In AAS and AFS, limitations to the analytical accuracy are mostly related to physical and chemical interferences and are due less to spectral interferences. In furnace AAS thermochemical processes limit the achievable accuracy and necessitate temperature programs to be carefully worked out in order to cope with errors arising from thermochemical effects. In AFS and also in LEI, it is necessary to control matrix influences relating to quenching when analyzing real samples.

Because of the necessity to characterize reference materials traceability in the measurements is very important. In atomic spectrometry background acquisition methods have improved so much that although it is not an absolute methodology, every step in the calibration and in the measurement processes can be extremely well characterized.

10.3
Economic aspects

The power of detection and the accuracy of analytical methods cannot be discussed without considering the expense arising from instruments and operating costs as well as from the laboratory personnel. Methods allowing multielement analyses to be performed and achieving a high throughput of samples are certainly advantageous for routine laboratories. In this respect, ICP spectrometric methods in particular offer good possibilities despite the high instrument costs and the high consumption of power and gases. Miniaturized spectrometers using new detector technology are both very powerful and at the same time much less expensive. Microplasmas are cheap to construct, as are the plasma generation and operation, and for well defined purposes such as element-specific detection in chromatography are already proving to be useful.

In many cases, however, the costs arising from sample preparation will become decisive, which favors x-ray spectrometric methods, provided the earlier mentioned limitations are not encountered. Future progress will certainly depend on the availabilty of on-line sample treatment using, for example, flow injection and eventually on-line sample dissolution as is possible in some cases with microwave-assisted heating. Also the realization of separations in miniaturized systems and with minute amounts of reagents is very promising. In each instance the question of which method should be selected will have to be discussed for each type of analytical task to be solved.

Literature

1 KIRCHHOFF G. R. and BUNSEN R. (1860) Chemical analysis by spectrum-observation, *Philos Mag* 20: 89–98.

2 WALSH A. (1955) The application of atomic absorption spectra to chemical analysis, *Spectrochim Acta* 7: 108–117.

3 GROTRIAN W. (1928) *Graphische Darstellung der Spektren von Atomen mit ein, zwei und drei Valenzelektronen*, Springer, Berlin.

4 GRIEM H. R. (1964) *Plasma Spectroscopy*, McGraw-Hill, New York.

5 DE GALAN L., SMITH R. and WINEFORDNER J. D. (1968) The electronic partition functions of atoms and ions between 1500 K and 7000 K, *Spectrochim Acta, Part B* 23: 521–525.

6 PEARCE W. J. and DICKERMAN P. J. (eds) (1961) *Symposium on Optical Spectrometric Measurements of High Temperatures*, University of Chicago Press, Chicago.

7 BOUMANS P. W. J. M. (1966) *Theory of Spectrochemical Excitation*, Hilger & Watts, London.

8 ROBINSON D. and LENN P. D. (1967) Plasma diagnostics by spectroscopic methods, *Appl Opt* 6: 983–1000.

9 KALNICKI D. J., KNISELEY R. N. and FASSEL V. A. (1975) Inductively coupled plasma optical emission spectroscopy. Excitation temperatures experienced by analyte species, *Spectrochim Acta, Part B* 30: 511–525.

10 CORLISS C. H. and BOZMAN W. R. (1962) *Experimental transition probabilities for spectral lines of 70 elements derived from the NBS tables of spectral line intensities. The wavelengths, energy levels transition probability and oscillator strength of 25000 lines between 200 and 900 nm for 112 spectra of 70 elements*, NBS monograph 53, Washington DC.

11 UNSÖLD A. (1955) *Physik der Sternatmosphären*, Springer, Berlin.

12 JUNKES J. and SALPETER E. W. (1961) Linienbreiten in der photographischen Spektralphotometrie II. Das optische Linienprofil, *Ric Spettrosc* 2: 255–483.

13 BRUCE C. F. and HANNAFORD P. (1971) On the width of atomic resonance lines from hollow cathode lamps, *Spectrochim Acta, Part B* 26: 207–235.

14 BROEKAERT J. A. C., LEIS F. and LAQUA K. (1979) Application of an inductively coupled plasma to the emission spectroscopic determination of rare earths in mineralogical samples, *Spectrochim Acta, Part B* 34: 73–84.

15 LOCHTE-HOLTGREVEN W. (ed) (1968) *Plasma Diagnostics*, North-Holland Publ. Company, Amsterdam.

16 ELWERT G. (1952) Verallgemeinerte Ionisationsformel eines Plasmas, *Z Naturforsch* 7a: 703–711.

17 MERMET J. M., Spectroscopic diagnostics: basic concepts, in Boumans P. W. J. M. (ed) (1987) *Inductively coupled plasma emission spectroscopy, Part II*, Wiley-Interscience, New York, 353–386.

18 DIERMEIER R. and KREMPL H. (1967) Thermische Anregungsfunktionen und Normtemperaturen von Atom- und Ionenlinien in Zweikomponentenplasmen, *Z Phys* 200: 239–248.

19 HERZBERG G. (**1950**) *Molecular spectra and molecular structure, 2nd edition,* Van Nostrand Reinhold Ltd., New York.

20 PEARCE R. W. B. and GAYDON A. G. (**1953**) *The identification of molecular spectra,* Chapman and Hall Ltd., London.

21 BROEKAERT J. A. C. (**1977**) Determination of rotational temperatures in a transitional type hollow cathode, *Bull Soc Chim Belges* 86: 895–906.

22 RAEYMAEKERS B., BROEKAERT J. A. C. and LEIS F. (**1988**) Radially resolved rotational temperatures in nitrogen-argon, oxygen-argon, air-argon and argon ICPs, *Spectrochim Acta, Part B* 43: 941–949.

23 ISHII I. and MONTASER A. (**1991**) Radial excitation temperatures in argon–oxygen and argon–sair ICP, *Spectrochim Acta, Part B* 46: 1197–1206.

24 FERREIRA N. P., HUMAN H. G. C. and BUTLER L. R. P. (**1980**) Kinetic temperatures and electron densities in the plasma of a side view Grimm-type glow discharge, *Spectrochim Acta, Part B* 35: 287–95.

25 BOCKASTEN K. (**1961**) Transformation of observed radiances into radial distribution of a plasma, *JOSA* 51: 943–947.

26 HIEFTJE G. M. (**1992**) Plasma diagnostic techniques for understanding and control, *Spectrochim Acta, Part B* 47: 3–25.

27 ELWERT G. (**1952**) Über die Ionisations- und Rekombinationsprozesse in einem Plasma und die Ionisationsformel der Sonnenkorona, *Z Naturforsch* 7a: 432–439.

28 BROEKAERT J. A. C. (**1987**) Trends in optical spectrochemical trace analysis with plasma sources, *Anal Chim Acta* 196: 1–21.

29 BORER M. and HIEFTJE G. M. (**1991**) Tandem sources for atomic emission spectrometry, *Spectrochim Acta Rev* 14: 463–486.

30 DOERFFEL K. (**1984**) *Statistik in der analytischen Chemie,* Verlag Chemie, Weinheim.

31 SACHS L. (**1969**) *Statistische Auswertungsmethoden,* Springer Verlag, Berlin.

32 NALIMOV V. V. (**1963**) *The application of mathematical statistics to chemical analysis,* Pergamon Press, Oxford.

33 MILLER J. C. and MILLER J. N. (**1989**) *Statistics for analytical chemistry,* Ellis Horword Ltd. Publishers, Chichester.

34 WENDE M. C., unpublished work.

35 INGLE Jr. J. D. and CROUCH S. R. (**1988**) *Spectrochemical analysis,* Prentice-Hall International, Inc., Englewood Cliffs.

36 HIEFTJE G. M. (**1972**) Signal-to-noise enhancement through instrumental techniques, *Anal Chem* 44: 81A–88A.

37 VAN BORM W. A. and BROEKAERT J. A. C. (**1990**) Noise characteristics in inductively coupled plasma optical emission spectrometry using slurry nebulization and direct powder introduction techniques, *Anal Chem* 62: 2527–2532.

38 WINGE R. K., ECKELS D. E., DeKALB E. L. and FASSEL V. A. (**1988**) Spatiotemporal characteristics of the inductively coupled plasma, *J Anal At Spectrom* 3: 849–855.

39 PEREIRO R., MIN WU., BROEKAERT J. A. C. and HIEFTJE G. M. (**1994**) Direct coupling of continuous hydride generation with microwave plasma torch atomic emission spectrometry for the determination of arsenic, antimony and tin, *Spectrochim Acta, Part B* 49: 59–73.

40 BROEKAERT J. A. C., BRUSHWYLER K. R., MONNIG C. A. and HIEFTJE G. M. (**1990**) Fourier-transform-atomic emission spectrometry with a Grimm-type glow-discharge source, *Spectrochim Acta, Part B* 45: 769–778.

41 POLLMANN D., PILGER C., HERGENRÖDER R., LEIS F., TSCHÖPEL P. and BROEKAERT J. A. C. (**1994**) Noise power spectra of inductively coupled plasma mass spectrometry using a cooled spray chamber, *Spectrochim Acta, Part B* 49: 683–690.

42 OMENETTO N. and WINEFORDNER J. D. (**1979**) Atomic fluorescence spectrometry basic principles and applications, *Prog Anal Spectrosc* 2: 1–183.

43 KAISER H. **(1947)** Die Berechnung der Nachweisempfindlichkeit, *Spectrochim Acta* 3: 40–67.

44 KAISER H. and SPECKER H. **(1956)** Bewertung und Vergleich von Analysenverfahren, *Z Anal Chem* 149: 46–66.

45 HUBAUX A. and VOS G. **(1970)** Decision and detection limits for linear calibration curves, *Anal Chem* 42: 849–855.

46 BEHRENDS K. **(1967)** Folgen analytische Fehlerkurven eine Gaußverteilung? *Fresenius' Z Anal Chem* 235: 391–401.

47 HIMMELBLAU D. **(1969)** *Process analysis by statistical methods*, Wiley, New York.

48 GROVE E. L. **(1971)** *Analytical Emission Spectrometry*, Vol. 1, Parts I and II. Decker Inc., New York.

49 WILLIAMS C. S. and BECKLUND O. A. **(1972)** *Optics*, Wiley, New York.

50 KELIHER P. N. and WOHLERS C. C. **(1976)** Echelle grating spectrometers in analytical spectrometry, *Anal Chem* 48: 333A–340A.

51 KAISER H. **(1948)** Über Schwärzungstransformationen, *Spectrochim Acta* 3: 159–190.

52 BOUMANS P. W. J. M. **(1971)** *Off-line data processing in emission spectrography*, XVI Coll Spectrosc Intern, Heidelberg, Preprints, Vol. II, 247–253.

53 SIEDENTOPF H. **(1934)**, *Z Phys* 35: 454.

54 ZWORYKIN V. K. and RAMBERG E. G. **(1949)** *Photoelectricity and its application*, Wiley, New York.

55 TALMI Y. **(1979)** *Multichannel image detectors*, American Chemical Society Symposium Series 102.

56 SWEEDLER J. V., JALKIAN R. F. and DENTON M. B. **(1989)** A linear charge coupled device detector system for spectroscopy, *Appl Spectrosc* 43: 953–961.

57 HIEFTJE G. M. and BRUSHWYLER K. R. **(1991)** Glow-discharge atomic emission spectrometry with a spectrally segmented photodiode-array spectrometer, *Appl Spectrosc* 45: 682–691.

58 BROEKAERT J. A. C., BRUSHWYLER K. R. and HIEFTJE G. M., unpublished work.

59 DANIELSON A., LINDBLOM P. and SÖDERMANN E. **(1974)** The IDES spectrometer, *Chem Scr* 6: 5–9.

60 SWEEDLER J. V., BILHORN R. B., EPPERSON P. M., SIMS G. R. and DENTON M. B. **(1988)** High-performance charge transfer device detectors, *Anal Chem* 60: 282A–291A.

61 EPPERSON P. M. and DENTON M. B. **(1989)** Binning spectral images in a charge-coupled device, *Anal Chem* 61: 1513–1519.

62 BILHORN R. B. **(1987)** *PhD Dissertation*, University of Arizona.

63 BILHORN R. B., EPPERSON P. M., SWEEDLER J. V. and DENTON M. B. **(1987)** Charge transfer device detectors for analytical optical spectroscopy – operation and characteristics, *Appl Spectrosc* 41: 1114–1124.

64 FURUTA N., BRUSHWYLER K. R. and HIEFTJE G. M. **(1989)** Flow-injection analysis utilizing a spectrally segmented photodiode array ICP emission spectrometer I. Micro-column preconcentration for the determination of molybdenum, *Spectrochim Acta, Part B* 44 (1989) 349–358.

65 ROSENKRANZ B., BREER C. B., BUSHER W., BETTMER J. and CAMMANN K. **(1997)** The plasma emission detector – a suitable detector for speciation and sum parameter analysis, *J Anal At Spectrom* 12: 993–996.

66 TREADO P. J. and MORRIS M. D. **(1989)** A thousand points of light: the Hadamard transform in chemical analysis and instrumentation, *Anal Chem* 61: 723A–734A.

67 FAIRES L. M. **(1986)** Fourier transforms for analytical atomic spectroscopy, *Anal Chem* 58: 1023A–1043A.

68 ADAMS F., GIJBELS R. and VAN GRIEKEN R. (eds) **(1988)** *Inorganic mass spectrometry*, Wiley, New York.

69 BACON J. R. and URE A. **(1984)** Spark source mass spectrometry: recent developments and applications, *Analyst* 109: 1229–1254.

70 HEUMANN K. G., BEER F. and WEISS H. **(1983)** Chlorid Spurenbestimmung in Reinstchemikalien durch negative Thermionen-Massenspektrometrie, *Mikrochim Acta* I: 95–108.

71 BRUNNEE C. (1987) The ideal mass analyzer: fact or fiction, *Int J Mass Spectrom Ion Processes* 76: 125–237.

72 BENNINGHOVEN A., RÜDENAUER F. G. and WERNER H. W. (1986) *Secondary ion mass spectrometry*, Wiley, New York.

73 DENOYER E., VAN GRIEKEN R., ADAMS F. and NATUSCH D. F. S. (1982) Laser microprobe mass spectrometry I: basic principles and performance characteristics, *Anal Chem* 54: 26A–41A.

74 MAHONEY P. P., RAY S. J. and HIEFTJE G. M. (1997) Time-of-flight mass spectrometry for elemental analysis, *Appl Spectrosc* 51: 16A–28A.

75 MYERS D. P., LI G., YANG P. and HIEFTJE G. M. (1994) An inductively coupled plasma time-of-flight mass spectrometer for elemental analysis. Part I: optimization and character-ization, *J Am Soc Mass Spectrom* 5: 1008–1016.

76 PACK B. W., BROEKAERT J. A. C., GUZOWSKI J. P., POEHLMANN J. and HIEFTJE G. M. (1998) Determination of halogenated hydrocarbons by helium microwave plasma torch time-of-flight mass spectrometry coupled to gas chromatography, *Anal Chem* 70: 3957–3963.

77 HARRISON W. W. and HANG W. (1996) Pulsed glow discharge time-of-flight mass spectrometry, *J Anal At Spectrom* 11: 835–840.

78 MAJIDI V., MOSER M., LEWIS C., HANG W. and KING F. L. (2000) Explicit speciation by microsecond pulsed glow discharge time-of-flight mass spectrometry: concurrent acquisition of structural, molecular and elemental information, *J Anal At Spectrom* 15: 19–25.

79 HANG W., YANG P., WANG X., YANG C., SU Y. and HUANG B. (1994) Microsecond pulsed glow discharge time-of-flight mass spectrometer, *Rapid Commun Mass Spectrom* 8: 590–594.

80 MAHONEY P. P., RAY S. J., LI G. and HIEFTJE G. M. (1999) Preliminary investigation of electrothermal vaporization sample introduction for inductively coupled plasma time-of-flight mass spectrometry, *Anal Chem* 71: 1378–1383.

81 TURNER P. J., MILLIS D. J., SCHRÖDER E., LAPITAJS G., JUNG G., IACONE L. A., HAYDAR D. A. and MONTASER A. (1998) *Instrumentation for low- and high-resolution ICPMS*, in: Montaser A. (ed) Inductively coupled plasma mass spectrometry. Wiley-VCH, New York, 421–501.

82 STALFORD G. C., KELLEY P. E., SYKA J. E. P., REYNOLDS W. E. and TODD J. F. J. (1984) Recent improvements in and applications of advanced ion trap technology, *Int J Mass Spectrom Ion Processes* 60: 85.

83 KOPPENAAL D. W., BARINAGA C. J. and SMITH M. R. (1994) Perfor-mance of an inductively coupled plasma source ion trap mass spectrometer, *J Anal At Spectrom* 9: 1053–1058.

84 EIDEN G. C., BARINAGA C. J. and KOPPENAAL D. W. (1997) Beneficial ion–molecule reactions in elemental mass spectrometry, *Rapid Commun Mass Spectrom* 11: 37–42.

85 WILKINS C. L. and GROSS M. L. (1981) Fourier transform mass spectrometry for analysis, *Anal Chem* 53: 1661A–1676A.

86 MILGRAM K. E., WHITE F. M., GOODNER K. L., WATSON C. H., CLIFFORD H., KOPPENAAL D. W., BARINAGA C. J., SMITH B. H., WINEFORDNER J. D., MARSHALL A. G., HOUK R. S. and EYLER J. R. (1997) High-resolution inductively coupled plasma Fourier transform ion cyclo-tron resonance mass spectrometer, *Anal Chem* 69: 3714–3721.

87 WARREN A. R., ALLEN L. A., PANG H. M., HOUK R. S. and JANGHORBANI M. (1994) Simultaneous measurement of ion ratios by inductively coupled plasma mass spectrometry with a twin-quadrupole instrument, *Appl Spectrosc* 48: 1360–1366.

88 GRAY A. L. and DATE A. R. **(1983)** ICP source mass spectrometry using continuous flow ion extraction, *Analyst* 108: 1033–1050.

89 LICHTE F. E., MEIER A. L. and CROCK J. G. **(1987)** Determination of the rare-earth elements in geological materials by ICP-MS, *Anal Chem* 59: 1150–1157.

90 JEFFREYS H. and JEFFREYS B. **(1962)** *Methods of mathematical physics, statistics,* Cambridge University Press, Cambridge, 488.

91 HORLICK G. and SHAO Y. **(1992)** *Inductively coupled plasma mass spectrometry for elemental analysis,* in: Montaser A. and Golightly D. W. (eds) Inductively coupled plasmas in analytical atomic spectrometry, 2nd edition. VCH, New York.

92 VAUGHAN M. A. and HORLICK G. **(1990)** Ion trajectories through the input optics of an inductively coupled plasma mass spectrometer, *Spectrochim Acta, Part B* 45: 1301–1311.

93 MERTEN D., BROEKAERT J. A. C. and LeMARCHAND A. **(1999)** Spectrum scanning in rapid sequential atomic emission spectrometry with the inductively coupled plasma, *Spectrochim Acta, Part B* 54: 1377–1382.

94 SNEDDON J. (ed) **(1990)** *Sample introduction in atomic spectroscopy,* Elsevier, Amsterdam.

95 BROEKAERT J. A. C. and BOUMANS P. W. J. M. **(1987)** *Sample introduction techniques in ICP-AES,* in: Boumans P. W. J. M. (ed) Inductively coupled plasma emission spectroscopy, Vol. I. Wiley, New York, 296–357.

96 MONTASER A. and GOLIGHTLY D. W. (eds) **(1992)** *Inductively coupled plasmas in analytical atomic spectrometry, 2nd edition,* VCH Publishers, Inc., Weinheim.

97 BROWNER R. F. and BOORN A. W. **(1984)** Sample introduction, the Achilles heel of atomic spectrometry? *Anal Chem* 56: 786A–798A.

98 BROWNER R. F. and BOORN A. W. **(1984)** Sample introduction techniques for atomic spectrometry, *Anal Chem* 56: 875A–888A.

99 BORER M. W. and HIEFTJE G. M. **(1991)** Tandem sources for atomic emission spectrometry, *Spectrochim Acta Rev* 14: 463–486.

100 AZIZ A., BROEKAERT J. A. C., LAQUA K. and LEIS F. **(1984)** A study of direct analysis of solid samples using spark ablation combined with excitation in an inductively coupled plasma, *Spectrochim Acta, Part B* 39: 1091–1103.

101 CIOCAN A., ÜBBING J. and NIEMAX K. **(1992)** Analytical applications of the nmicrowave induced plasma used with laser ablation of solid samples, *Spectrochim Acta, Part B* 47: 611–617.

102 BROEKAERT J. A. C. and TÖLG G. **(1987)** Recent developments in atomic spectrometry methods for elemental trace determinations, *Fresenius' J Anal Chem* 326: 495–509.

103 BROEKAERT J. A. C. **(1987)** Trends in optical spectrochemical trace analysis with plasma sources, *Anal Chim Acta* 196: 1–21.

104 SCOTT R. H., FASSEL V. A. and KNISELEY R. N. **(1974)** Inductively coupled plasmas for optical emission spectroscopy, *Anal Chem* 46: 75–80.

105 SCHUTYSER P. and JANSSENS E. **(1979)** Evaluation of spray chambers for use in inductively coupled plasma atomic emission spectrometry, *Spectrochim Acta, Part B* 34: 443–449.

106 WU M. and HIEFTJE G. M. **(1992)** A new spray chamber for ICP spectrometry, *Appl Spectrosc* 46: 1912–1918.

106a SCHALDACH G., BERGER L. and BERNDT H. **(2000)** *Optimization of a spray chamber design using computer simulation and evolution strategies,* 2000 Winter Conf on Plasma Spectrometry, Fort Lauderdale, Abstracts, p. 43.

107 GOUY G. L. **(1879)** Photometric research on colored flames, *Ann Chim Phys* 18: 5–101.

108 MEINHARD J. E. **(1979)** *Pneumatic nebulizers, present and future,* in: Barnes R. M. (ed): Applications of plasma emission spectroscopy. Heyden, London, 1–14.

109 WOHLERS C. C. **(1977)** Comparison of nebulizers for the inductively coupled plasma, *ICP Inf Newsl* 3: 37–51.

110 LIU H. and MONTASER A. **(1994)** Phase-Doppler diagnostic studies of primary and tertiary aerosols produced by a high-efficiency nebulizer, *Anal Chem* 66: 3233–3242.

111 VANHAECKE F., VAN HOLDERBEKE M., MOENS L. and DAMS R. **(1996)** Evaluation of a commercially available microconcentric nebulizer for inductively coupled plasma mass spectrometry, *J Anal At Spectrom* 11: 543–548.

112 KRANZ E. **(1972)** Untersuchungen über die optimale Erzeugung und Förderung von Aerosolen für spektrochemische Zwecken, *Spectrochim Acta, Part B* 27: 327.

113 VALENTE S. E. and SCHRENK W. G. **(1970)** The design and some emission characteristics of an economical dc arc plasmajet excitation source for solution analysis, *Appl Spectrosc* 24: 197.

114 FUJISHIRO, KUBOTA M. and ISHIDA R. **(1984)** A study of designs of a cross flow nebulizer for ICP atomic emission spectrometry, *Spectrochim Acta, Part B* 39: 617–620.

115 BABINGTON R. S. **(1969)** *Method of atomizing liquid in a mono-dispersed spray*, U.S. Patent 3,421,692.

116 EBDON L. and CAVE M. R. **(1982)** A study of pneumatic nebulisation systems for ICP emission spectrometry, *Analyst* 107: 172–178.

117 RAEYMAEKERS B., GRAULE T., BROEKAERT J. A. C., ADAMS F. and TSCHÖPEL P. **(1988)** Characteristics of nebulized suspensions of refractory oxide powders used for the production of ceramics and their evaporation behaviour in an inductively coupled plasma, *Spectrochim Acta, Part B* 47: 923–940.

118 LATHEN C., unpublished work.

119 LAYMAN L. R. and LICHTE F. E. **(1982)** Glass frit nebulizer for atomic spectrometry, *Anal Chem* 54: 638–642.

120 LIU H., CLIFFORD R. H., DOLAN S. P. and MONTASER A. **(1996)** Investigation of a high-efficiency nebulizer and a thimble glass frit nebulizer for elemental analysis of biological materials by inductively coupled plasma atomic emission spectrometry, *Spectrochim Acta, Part B* 51: 27–40.

121 KIMBERLEY E. L., RICE G. W. and FASSEL V. A. **(1984)** Direct liquid injection sample introduction for flow injection analysis and liquid chromatography with inductively coupled argon plasma spectrometric detection, *Anal Chem* 56: 289–292.

122 MCLEAN J. A., MINNICH M. G., LACONE L. A., LIU H. and MONTASER A. **(1998)** Nebulizer diagnostics: fundamental parameters, challenges, and techniques on the horizon, *J Anal At Spectrom* 13: 829–842.

123 WANG L., MAY S. W., BROWNER R. F. and POLLOCK S. H. **(1996)** Low-flow interface for liquid chromatography-inductively coupled plasma mass spectrometry speciation using an oscillating capillary nebulizer, *J Anal At Spectrom* 11: 1137–1146.

124 B'HYMER C., SUTTON K. and CARUSO J. A. **(1998)** Comparison of four nebulizer–spray chamber interfaces for the high-performance liquid chromatographic separation of arsenic compounds using inductively coupled plasma mass spectrometric detection, *J Anal At Spectrom* 13: 855–858.

125 SEBASTIANI E., OHLS K. and RIEMER G. **(1973)** Ergebnisse zur Zerstäubung dosierter Lösungsvolumina bei der AAS, *Fresenius' Z Anal Chem* 264: 105–109.

126 BERNDT H. and SLAVIN W. **(1978)** Automated trace analysis of small samples using the "injection method" of flame atomic absorption spectrometry, *At Absorpt Newsl* 17: 109.

127 GREENFIELD S. and SMITH P. B. **(1972)** The determination of trace metals in microlitre samples by plasma torch emission, *Anal Chim Acta* 59: 341–348.

128 AZIZ A., BROEKAERT J. A. C. and LEIS F. **(1982)** The optimization of an ICP injection technique and the application to the direct analysis of small-volume serum samples, *Spectrochim Acta, Part B* 37: 381–389.

129 RUZICKA J. and HANSEN E. H. **(1988)** *Flow injection analysis*, Wiley, New York.

130 HAIBER S. and BERNDT H. **(2000)** A novel temperature (360 °C)/high-pressure (30 Mpa) flow system for online digestion applied to ICP spectrometry, *Fresenius' J Anal Chem* 368: 52–58.

131 POLLMANN D., LEIS F., TÖLG G., TSCHÖPEL P. and BROEKAERT J. A. C. **(1994)** Multielement trace determinations in Al₂O₃ ceramic powders by inductively coupled plasma mass spectrometry with special reference to on-line preconcentration, *Spectrochim Acta, Part B* 49: 1251–1258.

132 LOBINSKI R., BROEKAERT J. A. C., TSCHÖPEL P. and TÖLG G. **(1992)** Inductively-coupled plasma atomic emission spectroscopic determination of trace impurities in ZrO₂-powder, *Fresenius' J Anal Chem* 342: 569–580.

133 VOGT T., unpublished work.

134 HARTENSTEIN S. D., RUZICKA J. and CHRISTIAN G. D. **(1985)** Sensitivity enhancements for flow injection analysis ICP-AES using an on-line preconcentration ion-exchange column, *Anal Chem* 57: 21–25.

135 FANG Z. **(1993)** *Flow injection separation and preconcentration*, VCH, Weinheim.

136 SCHRAMEL P., XU L., KNAPP G. and MICHAELIS M. **(1993)** Multi-elemental analysis in biological samples by on-line preconcentration on 8-hydroxy-quinoline–cellulose microcolumn coupled to simultaneous ICP-AES, *Fresenius' J Anal Chem* 345: 600–606.

137 NUKUYAMA S. and TANASAWA Y. **(1938, 1939)** Experiments on the atomization of liquids in air stream, *Trans Soc Mech Eng* (Japan) 4: 86 and 5: 68.

137 ALLEN T. **(1981)** *Particle size measurement*, Chapman & Hall, London.

138 VAN BORM W., BROEKAERT J. A. C., KLOCKENKÄMPER R., TSCHÖPEL P. and ADAMS F. C. **(1991)** Aerosol sizing and transport studies with slurry nebulization in inductively coupled plasma spectrometry, *Spectrochim Acta, Part B* 46: 1033–1049.

139 OLSEN S. D. and STRASHEIM A. **(1983)** Correlation of the analytical signal to the characterized nebulizer spray, *Spectrochim Acta* 38: 973–975.

140 BOORN A. W. and BROWNER R. **(1982)** Effects of organic solvents in ICP-AES, *Anal Chem* 54: 1402–1410.

141 KREUNING G. and MAESSEN F. J. M. J. **(1987)** Effects of the solvent plasma load of various solvents on the excitation conditions in medium power ICP, *Spectrochim Acta, Part B* 42: 677–688.

142 DOHERTY M. P. and HIEFTJE G. M. **(1984)** Jet-impact nebulization for sample introduction in ICP spectrometry, *Appl Spectrosc* 38: 405–411.

143 BERNDT H. **(1988)** High-pressure nebulization: a new way of sample introduction for atomic spectroscopy, *Fresenius' J Anal Chem* 331: 321–323.

144 POSTA J. and BERNDT H. **(1992)** A high-performance flow/hydraulic high pressure nebulization system (HPF/HHPN) in flame AAS for improved elemental trace determinations in highly concentrated salt solutions, *Spectrochim Acta, Part B* 47: 993–999.

145 BERNDT H. and YANEZ J. **(1996)** High-temperature hydraulic high-pressure nebulization: a recent nebulization principle for sample introduction, *J Anal At Spectrom* 11: 703–712.

146 POSTA J., BERNDT H., LUO S. K. and SCHALDACH G. **(1993)** High-performance flow flame atomic absorption spectrometry for automated on-line separation and determination of Cr(III)/Cr(VI) and preconcentration of Cr VI, *Anal Chem* 65: 2590–2595.

147 SYTI A. **(1974)** Developments in methods of sample injection and atomization in atomic spectrometry, *CRC Crit Rev Anal Chem* 4: 155–225.

148 YANG J., CONVER T. S., KOROPCHAK J. A. and LEIGHTY D. A. **(1996)** Use of a multi-tube Nafion membrane dryer for desolvation with thermospray sample introduction to inductively coupled plasma-atomic emission spectrometry, *Spectrochim Acta, Part B* 51: 1491–1504.

149 WENDT R. and FASSEL V. E. **(1965)** Induction coupled plasma spectrometric excitation source, *Anal Chem* 37: 920–922.

150 FASSEL V. A. and BEAR B. R. **(1986)** Ultrasonic nebulization of liquid samples for analytical ICP atomic spectroscopy, *Spectrochim Acta, Part B* 41: 1089–1113.

151 BOUMANS P. W. J. M. and DE BOER F. J. **(1975)** Studies of an ICP for OES. II Compromise conditions for simultaneous multielement analyses, *Spectrochim Acta, Part B* 30: 309–334.

152 BARNETT N. W., CHEN L. S. and KIRKBRIGHT G. F. **(1984)** The rapid determination of arsenic by OES using a microwave induced plasma source and a miniature hydride generation device, *Spectrochim Acta, Part B* 39: 1141–1147.

153 MANNING D. C. **(1971)** A high sensitivity arsenic–selenium sampling system for atomic absorption spectroscopy, *At Absorpt Newsl* 10: 123.

154 THOMPSON M., PAHLAVANPOUR B., WALTON S. J. and KIRKBRIGHT G. F. **(1978)** Simultaneous determination of trace concentrations of arsenic, antimony, bismuth, selenium and tellurium in aqueous solution by introduction of the gaseous hydrides into an ICP source for emission spectrometry, *Analyst* 103: 568–579.

155 BROEKAERT J. A. C. and LEIS F. **(1980)** Application of two different ICP-hydride techniques to the determination of arsenic, *Fresenius' Z Anal Chem* 300: 22–27.

156 BULSKA E., TSCHÖPEL P., BROEKAERT J. A. C. and TÖLG G. **(1993)** Different sample introduction systems for the simultaneous determination of As, Sb and Se by microwave-induced plasma atomic emission spectrometry, *Anal Chim Acta* 271: 171–181.

157 TAO H. and MIYAZAKI A. **(1991)** Determination of germanium, arsenic, antimony, tin and mercury at trace levels by continuous hydride generation-helium microwave-induced plasma atomic emission spectrometry, *Anal Sci* 7: 55–59.

158 SCHICKLING C., YANG J. and BROEKAERT J. A. C. **(1996)** Optimization of electrochemical hydride generation coupled to microwave-induced plasma atomic emission

spectrometry for the determination of arsenic and its use for the analysis of biological tissue samples, *J Anal At Spectrom* 11: 739–746.

159 STRINGER C. S. and ATTREP Jr. M. **(1979)** Comparison of digestion methods for determination of organoarsenicals in wastewater, *Anal Chem* 51: 731–734.

160 WELZ B. and MELCHER M. **(1983)** Investigations on atomisation mechanisms of volatile hydride-forming elements in a heated quartz cell, *Analyst* 108: 213–224.

161 VIEN S. H. and FRY R. C. **(1988)** Ultrasensitive simultaneous determination of arsenic, selenium, tin and antimony in aqueous solution by hydride generation gas chromatography with photoionisation detection, *Anal Chem* 60: 465–472.

162 ZHANG Li., ZHE-MING N. and XIAO-QUAN S. **(1989)** In situ preconcentration of metallic hydrides in a graphite furnace coated with palladium, *Spectrochim Acta, Part B* 44: 339–346.

163 ZHANG Li., ZHE-MING N. and XIAO-QUAN S. **(1989)** In situ concentration of metallic hydride in a graphite furnace coated with palladium – determination of bismuth, germanium and tellurium, *Spectrochim Acta, Part B* 44: 751–758.

164 STURGEON R. E., WILLIE S. N., SPROULE G. I., ROBINSON P. T. and BERMAN S. S., Sequestration of volatile element hydrides by platinum group elements for graphite furnace atomic absorption, *Spectrochim Acta, Part B* 44: 667–682.

165 HAASE O., KLARE M., KRENGEL-ROTHENSEE K. and BROEKAERT J. A. C. **(1998)** Evaluation of the determination of mercury at the trace and ultra-trace levels in the presence of high concentrations of NaCl by flow injection-cold vapour atomic absorption spectrometry using $SnCl_2$ and $NaBH_4$, *Analyst* 123: 1219–1222.

166 NAKAHARA T., YAMADA S. and WASA T. **(1990)** Continuous-flow determination of trace iodine by atmospheric-pressure helium microwave-induced plasma atomic emission spectrometry using generation of volatile iodine from iodide, *Appl Spectrosc* 44: 1673–1678.

167 NAKAHARA T., MORI T., MORIMOTO S. and ISHIKAWA H. (1995) Continuous-flow determination of aqueous sulfur by atmospheric-pressure helium microwave-induced plasma atomic emission spectrometry with gas-phase sample introduction, *Spectrochim Acta, Part B* 50: 393–403.

168 RODRIGUEZ J., PEREIRO R. and SANZ-MEDEL A. (1998) Glow discharge atomic emission spectrometry for the determination of chlorides and total organochlorine in water samples via on-line continuous generation of chlorine, *J Anal At Spectrom* 13: 911–915.

169 SANZ-MEDEL A., VALDES-HEVIA Y TEMPRANO M. C., BORDEL GARCIA N. and FERNANDEZ DE LA CAMPA M. R. (1995) Generation of cadmium atoms at room temperature using vesicles and its application to cadmium determination by cold vapor atomic spectrometry, *Anal Chem* 67: 2216–2223.

170 PREUß E. (1940) Beiträge zur spektralanalytischen Methodik. II Bestimmung von Zn, Cd, Hg, In, Tl, Ge, Sn, Pb, Sb und Bi durch fraktionnierte Distillation, *Z Angew Mineral* 3: 1.

171 L'VOV B. V. (1961) The analytical use of atomic absorption spectra, *Spectrochim Acta* 17: 761–770.

172 MAßMANN H. (1968) Vergleich von Atomabsorption und Atomfluo-reszenz in der Graphitküvette, *Spectrochim Acta* 23: 215–226.

173 L'VOV B. V., PELIEVA L. A. and SHARNOPOL'SKII A. I. (1977) Decrease in the effect of the base during the atomic-absorption analysis of solutions in tube furnaces by vaporization of samples from a graphite substrate, *Zh Prikl Spektrosk* 27: 395–399.

174 McNALLY J. and HOLCOMBE J. A. (1987) Existence of microdroplets and dispersed atoms on the graphite surface in electrothermal atomizers, *Anal Chem* 59: 1105–1112.

175 WEST T. S. and WILLIAMS X. K. (1969) Atomic absorption and fluorescence spectroscopy with a carbon filament atom reservoir. Part I. Construction and operation of atom reservoir, *Anal Chim Acta* 45: 27–41.

175a SYCHRA V., KOLIHOVA D., VYSKOCILOVA O., HLAVAC R. and PÜSCHEL P. (1979) Electrothermal atomization from metallic surfaces. Part 1. Design and performance of a tungsten-tube atomizer, *Anal Chim Acta* 105:263–270.

176 DELVES H. T. (1970) A micro-sampling method for the rapid determination of lead in blood by atomic-absorption spectrophotometry, *Analyst* 95: 431–438.

177 BERNDT H. and MESSERSCHMIDT J. (1979) Eine Mikromethode der Flammen-Atomabsorptions- und Emissionsspektrometrie (Platinschl-aufen-Methode) I. Die Grundlagen der Methode – Anwendung auf die Bestimmung von Blei und Cadmium in Trinkwasser, *Spectrochim Acta, Part B* 34: 241–256.

178 BROOKS E. I. and TIMMINS K. J. (1985) Sample introduction device for use with a microwave-induced plasma, *Analyst* 110: 557–558.

179 L'VOV B. V. (1978) Electrothermal atomization – the way toward absolute methods of atomic absorption analysis, *Spectrochim Acta, Part B* (1978) 153–193.

180 AZIZ A., BROEKAERT J. A. C. and LEIS F. (1982) A contribution to the analy-sis of microamounts of biological samples using a combination of graphite furnace and microwave induced plasma atomic emission spectroscopy, *Spectrochim Acta* 37: 381–389.

181 MILLER-IHLI N. J. (1989) Automated ultrasonic mixing accessory for slurry sampling into a graphite furnace atomic absorption spectrometer, *J Anal At Spectrom* 4: 295–297.

182 GROBENSKI Z., LEHMANN R., TAMM R. and WELZ B. (1982) Improvements

in graphite furnace atomic absorption microanalysis with solid sampling, *Mikrochim Acta [Wien]* I: 115–125.

183 VOLYNSKY A. B. **(1998)** Applications of graphite tubes modified with high-melting carbides in electrothermal atomic absorption spectrometry I. General approach, *Spectrochim Acta, Part B* 53: 509–535.

184 VOLYNSKY A. B. **(1998)** Applications of graphite tubes modified with high-melting carbides in electrothermal atomic absorption spectrometry II. Practical aspects, *Spectrochim Acta, Part B* 53: 1607–1645.

185 MILLARD D. L., SHAN H. C. and KIRKBRIGHT G. F. **(1980)** Optical emission spectrometry with an inductively coupled radiofrequency argon plasma source and sample introduction with a graphite rod electrothermal vaporization device. Part II. Matrix, inter-element and sample transport effcts, *Analyst* 105: 502–508.

186 AZIZ A., BROEKAERT J. A. C. and LEIS F. **(1982)** Analysis of microamounts of biological samples by evaporation in a graphite furnace and inductively coupled plasma atomic emission spectroscopy, *Spectrochim Acta, Part B* 37: 369–379.

187 BROEKAERT J. A. C. and LEIS F. **(1985)** An application of electrothermal evaporation using direct solids sampling coupled with microwave induced plasma optical emission spectroscopy to elemental determinations in biological matrices, *Mikrochim Acta* II: 261–272.

188 KANTOR T. **(2000)** Sample introduction with graphite furnace electrothermal vaporization into an inductively coupled plasma: effects of streaming conditions and gaseous phase additives, *Spectrochim Acta, Part B* 55: 431–448.

189 SOMMER D. and OHLS K. **(1980)** Direkte Probeneinführung in ein stabil brennendes ICP, *Fresenius' Z Anal Chem* 304: 97–103.

190 SALIN E. D. and HORLICK G. **(1979)** Direct sample insertion device for ICP emission spectrometry, *Anal Chem* 51: 2284–2286.

191 ZARAY Gy., BROEKAERT J. A. C. and LEIS F. **(1988)** The use of direct sample insertion into a nitrogen–argon inductively coupled plasma for emission spectrometry I. Technique optimization and application to the analysis of aluminium oxide, *Spectrochim Acta, Part B* 43: 241–253.

192 KARNASSIOS V. and HORLICK G. **(1989)** A computer-controlled direct sample insertion device for ICP-MS, *Spectrochim Acta, Part B* 44: 1345–1360.

193 BLAIN L. and SALIN E. D. **(1992)** Methodological solutions for the analysis of sediment samples by direct sample insertion ICP atomic emission, *Spectrochim Acta, Part B* 47: 205–217.

194 HAUPTKORN S. and KRIVAN V. **(1994)** Determination of silicon in boron nitride by slurry sampling electrothermal atomic absorption spectrometry, *Spectrochim Acta, Part B* 49: 221–228.

195 GRÉGOIRE D. C., MAAWALI S. A. I. and CHAKRABARTI C. L. **(1992)** Use of Mg/Pd chemical modifiers for the determination of volatile elements by electrothermal vaporization ICP-MS: effect on mass transport efficiency, *Spectrochim Acta, Part B* 47: 1123–1133.

196 NICKEL H. and BROEKAERT J. A. C. **(1999)** Some topical applications of plasma atomic spectrochemical methods for the analysis of ceramic powders, *Fresenius' J Anal Chem* 363: 145–155.

197 LOBINSKI R., VAN BORM W., BROEKAERT J. A. C., TSCHÖPEL P. and TÖLG G. **(1992)** Optimization of slurry nebulization inductively-coupled plasma atomic emission spectrometry for the analysis of ZrO$_2$-powder, *Fresenius' J Anal Chem* 342: 563–568.

198 MCKINNON P. W., GIESS K. C. and KNIGHT T. V. **(1981)** *A clog-free nebulizer for use in inductively coupled plasma-atomic emission spectroscopy*, in: Barnes R. M. (ed) Developments in atomic plasma spectrochemical analysis. Heyden, London, 287–301.

199 BARNES R. M. and NIKDEL S. (1976) Temperature and velocity profiles and energy balances for an inductively coupled plasma discharge in nitrogen, *J Appl Phys* 47: 3929–3934.

200 HIEFTJE G. M., MILLER R. M., PAK Y. and WITTIG E. P. (1987) Theoretical examination of solute particle vaporization in analytical atomic spectrometry, *Anal Chem* 59: 2861–2872.

201 MERTEN D., HEITLAND P. and BROEKAERT J. A. C. (1997) Modelling of the evaporation behaviour of particulate material for slurry nebulization inductively coupled plasma atomic emission spectrometry, *Spectrochim Acta, Part B* 52: 1905–1922.

202 RAEYMAEKERS B., VAN ESPEN P. and ADAMS F. (1984) The morphological characterization of particles by auto-mated scanning electron microscopy, *Mikrochim Acta* II: 437–454.

203 BROEKAERT J. A. C., BRANDT R., LEIS F., PILGER C., POLLMANN D., TSCHÖPEL P. and TÖLG G. (1994) Analysis of aluminium oxide and silicon carbide ceramic materials by inductively coupled plasma mass spectrometry, *J Anal At Spectrom* 9: 1063–1070.

204 DOCEKAL B. and KRIVAN V. (1992) Direct determination of impurities in powered silicon carbide by electro-thermal atomic absorption spectrometry using the slurry sampling technique, *J Anal At Spectrom* 7: 521–528.

205 KRIVAN V., BARTH P. and SCHNÜRER-PATSCHAN C. (1988) An electro-thermal atomic absorption spectrom-eter using semiconductor diode lasers and a tungsten coil atomizer: design and first applications, *Anal Chem* 70: 3625–3632.

206 LUCIC M. and KRIVAN V. (1999) Analysis of aluminium-based ceramic powders by electrothermal vaporization inductively coupled plasma atomic emission spectrometry using a tungsten coil and slurry sampling, *Fresenius' J Anal Chem* 363: 64–72.

207 MIN H. and XI-EN S. (1989) Analysis of ultrafine ZrO_2 for minor and trace elements by a slurry sampling inductively coupled plasma atomic emission spectrometric method, *Spectrochim Acta, Part B* 44: 957–964.

208 OBENAUF R. H., BOSTWICK R., McCANN M., McCORMACK J. D., MERRIMAN R. and SALEM D. (1994) *Spex handbook of sample preparation and handling*, Spex, Metuchen.

209 JOHNES J. L., DAHLQUIST R. L. and HOYT H. E. (1971) A spectroscopic source with improved analytical properties and remote sampling capacity, *Appl Spectrosc* 25: 629–635.

210 EL ALFY S., LAQUA K. and MAßMANN H. (1973) Spektrochemische Analysen mit einer Glimmentladungslampe III. Entwicklung und Beschreibung einer Universalmethode zur Bestimmung von Hauptbestandteilen in elektrisch nichtleitenden pulverförmigen Substanzen, *Fresenius' Z Anal Chem* 263: 1–14.

211 HUMAN H. G. C., SCOTT R. H., OAKES A. R. and WEST C. D. (1976) The use of a spark as a sampling–nebulising device for solid samples in atomic absorption spectrometry, atomic fluorescence and ICP emission spectroscopy, *Analyst* 101: 265–271.

212 ENGEL U., KEHDEN A., VOGES E. and BROEKAERT J. A. C. (1999) Direct solid atomic emission spectrometric analysis of metal samples by an argon microwave plasma torch coupled to spark ablation, *Spectrochim Acta, Part B* 54: 1279–1289.

213 LEMARCHAND A., LABARRAQUE G., MASSON P. and BROEKAERT J. A. C. (1987) Analysis of ferrous alloys by spark ablation coupled to inductively coupled plasma atomic emission spectrometry, *J Anal At Spectrom* 2: 481–484.

215 BROEKAERT J. A. C., LEIS F. and LAQUA K. (1985) *A contribution to the direct analysis of solid samples by spark erosion combined to ICP-OES*, in: Sansoni B. (ed) Instrumentelle Multielementanalyse. Verlag Chemie, Weinheim.

216 RAEYMAEKERS B., VAN ESPEN P., ADAMS F. and BROEKAERT J. A. C. **(1988)** A characterization of spark-produced aerosols by automated electron probe micro-analysis, *Appl Spectrosc* 42: 142–150.

217 KEHDEN A., FLOCK J., VOGEL W. and BROEKAERT J. A. C. **(2000)** Direct solids atomic spectrometric analysis of metal samples by laser induced argon spark ablation coupled to ICP-OES, *Appl Spectrosc*, in press.

218 GOLLOCH A. and SIEGMUND D. **(1997)** Gliding spark spectroscopy – rapid survey analysis of flame retardants and other additives in polymers, *Fresenius' J Anal Chem* 358: 804–811.

219 BARNES R. M. and MALMSTADT H. V. **(1974)** Liquid-layer-on-solid-sample spark technique for emission spectrochemical analysis, *Anal Chem* 46: 66–72.

220 BINGS N., ALEXI H. and BROEKAERT J. A. C., *Universität Dortmund*, unpublished work.

221 OMENETTO N. **(1979)** *Analytical laser spectroscopy*, Wiley, New York.

222 MOENKE-BLANKENBURG L. **(1989)** *Laser micro analysis*, Wiley, New York.

223 LAQUA K., Analytical spectroscopy using laser atomizers. **(1985)** in: Martellucci S. and Chester A. N. (eds) *Analytical laser spectroscopy*, Plenum Publishing Corporation, New York, 159–182.

224 LEIS F., SDORRA W., KO J. B. and NIEMAX K. **(1989)** Basic investigations for laser microanalysis: OES of laser-produced sample plumes, *Mikrochim Acta /Wien/* II: 185–199.

225 GAGEAN M. and MERMET J. M. **(1998)** Study of laser ablation of brass materials using inductively coupled plasma atomic emission spectrometric detection, *Spectrochim Acta, Part B* 53: 581–591.

226 SDORRA W., QUENTMEIER A., NIEMAX K. **(1989)** Basic investigations for laser microanalysis. II. Laser-induced fluorescence in laser-produced sample plumes, *Mikrochim Acta /Wien/* II: 201–218.

227 UEBBING J., BRUST J., SDORRA W.,

LEIS F. and NIEMAX K. **(1991)** Reheating of a laser-produced plasma by a second laser, *Appl Spectrosc* 45: 1419–1423.

228 HEMMERLIN M. and MERMET J. M. **(1996)** Determination of elements in polymers by laser ablation ICP AES: effect of the laser beam wavelength, energy and masking on the ablation threshold and efficiency, *Spectrochim Acta, Part B* 51: 579–589.

229 MARGETIC V., PAKULEV A., STOCKHAUS, BOLSHOV M., NIEMAX K. and HERGENRÖDER R. **(2000)** A comparison of nanosecond and femtosecond laser-induced plasma spectroscopy of brass samples, *Spectrochim Acta, Part B* 55: 1771–1785.

230 KAMINSKY M. **(1965)** *Atomic and ionic impact phenomena on metal surfaces*, Springer Verlag, Berlin.

231 PENNING F. M. **(1957)** *Electrical discharges in gases*, Philips Technical Library, Eindhoven, p. 41.

232 BOUMANS P. W. J. M. **(1973)** Sputtering in a glow discharge for spectrochemical analysis, *Anal Chem* 44: 1219–1228.

233 BROEKAERT J. A. C. **(1987)** State-of-the-art of glow discharge lamp spectrometry, *J Anal At Spectrom* 2: 537–542.

234 MARCUS R. K. **(1996)** Radiofrequency powered glow discharges: opportunities and challenges, *J Anal At Spectrom* 11: 821–828.

235 ANDERSON G. S., MEYER W. N. and WEHNER G. K. **(1962)** Sputtering of dielectrics by high-frequency fields, *J Appl Phys* 33: 2991–2992.

236 CHRISTOPHER S. J., YE Y. and MARCUS R. K. **(1997)** Ion kinetic energy distributions and their relationship to fundamental plasma parameters in a radio frequency glow discharge, *Spectrochim Acta, Part B* 52: 1627–1644.

237 LAQUA K. **(1991)** *Spektralanalyse mit Glimmentladungen–Gegenwärtiger Stand und Zukunftsaussichten*, in: Borsolorf R; Fresenius W., Günzler H., Huber W., Lüdenwald L., Tölg G. and Wisser H. (eds) Analytiker

Taschenbuch, Band 10, Springer, Berlin, 297–349.

238 Broekaert J. A. C., Bricker T., Brushwyler K. R. and Hieftje G. M. **(1992)** Investigations of a jet-assisted glow discharge lamp for optical emission spectrometry, *Spectrochim Acta, Part B* 47: 131–142.

239 Bogaerts A. and Gijbels R. **(1997)** Calculation of crater profiles on a flat cathode in a direct current glow discharge, *Spectrochim Acta, Part B* 52: 765–777.

240 Dogan M., Laqua K. and Maßmann H. **(1972)** Spectrochemische Analysen mit einer Glimmentladungslampe als Lichtquelle- I Elektrische Eigensch-aften, Probenabbau und spektraler Charakter, *Spectrochim Acta, Part B* 26: 631–649.

241 Marcus R. (ed) **(1993)** *Glow discharge spectrometries*, Plenum Press, New York.

242 Van Straaten M., Vertes A. and Gijbels R. **(1991)** Sample erosion studies and modeling in a glow discharge ionization cell, *Spectrochim Acta, Part B* 46: 283–280.

243 Bogaerts A. and Gijbels R. **(1998)** Modeling of direct current glow discharges and comparison with experiments: how good is the agreement? *J Anal At Spectrom* 13: 945–953.

244 Christopher S. J., Hartenstein M. L., Marcus R. K., Belkin M. and Caruso J. A. **(1998)** Characterization of helium/argon working gas systems in a radiofrequency glow discharge atomic emission source. Part I: Optical emission, sputtering and electrical characteristics, *Spectrochim Acta, Part B* 53: 1181–1196.

245 Heintz M. J., Mifflin K., Broekaert J. A. C. and Hieftje G. M. **(1995)** Investigations of a magnetically enhanced Grimm-type glow discharge source, *Appl Spectrosc* 49: 241–246.

246 Heintz M. J., Broekaert J. A. C. and Hieftje G. M. **(1997)** Analytical characterization of a planar magne-tron radio frequency glow-discharge source, *Spectrochim Acta, Part B* 52: 579–591.

247 Walsh A. **(1955)** The application of atomic absorption spectra to chemical analysis, *Spectrochim Acta* 7: 108–117.

248 Welz B. and Sperling M. **(1999)** *Atomic absorption spectrometry*, Wiley-VCH, Weinheim.

249 Price W. J. **(1979)** *Spectrochemical analysis by atomic absorption*, Heyden, London.

250 Ebdon L. **(1982)** *An introduction to atomic absorption spectroscopy*, Heyden, London.

251 Hergenröder R. and Niemax K. **(1988)** Laser atomic absorption spectroscopy applying semiconductor lasers, *Spectrochim Acta, Part B* 43: 1443–1449.

252 Ng K. C., Ali A. H., Barber T. E. and Winefordner J. D. **(1988)** Multiple mode semiconductor diode laser as a spectral line source for graphite furnace atomic absorption spectros-copy, *Anal Chem* 62: 1893–1895.

253 Koirthyohann R. S. and Pickett E. E. **(1965)** Background correction in long path atomic absorption spec-trometry, *Anal Chem* 37: 601–603.

254 Moulton G. P., O'Haver T. C. and Harnly J. M. **(1990)** Signal to noise ratios for continuous source atomic absorption spectrometry using a linear photodiode array to monitor sub-nanometre wavelength intervals, *J Anal At Spectrom* 5: 145–150.

255 Heitmann U., Schütz M., Becker-Roß H. and Florek S. **(1996)** Measurements on the Zeeman-splitting of analytical lines by means of a continuum source graphite furnace atomic absorption spectrom-eter with a linear charge coupled device array, *Spectrochim Acta, Part B* 51: 1095–1105.

256 Heitmann U., Schütz M., Becker-Roß H. and Florek S. **(1998)** *Spektroskopische Untersuchungen in der Graphitrohr-AAS mit Hilfe eines hochauflösenden Kontinuumstrahler-Spektrometers*, in: Vogt C., Wennrich R. and Werner G. (eds) CANAS 97, Colloquium Analytische Atomspek-troskopie, Universität Leipzig und Umweltforschungszentrum Leipzig-Halle GmbH, Leipzig.

257 NIEMAX K., ZYBIN A., SCHNÜRER-PATSCHAN C. and GROLL H. (1996) Semiconductor diode lasers in atomic spectrometry, *Anal Chem* 68: 351A–356A.

258 NIEMAX K., GROLL H. and SCHNÜRER-PATSCHAN C. (1993) Element analysis by diode laser spectroscopy, *Spectrochim Acta Rev* 15: 349–377.

259 ZYBIN A., SCHÜRER-PATSCHAN C. and NIEMAX K. (1992) Simultaneous multi-element analysis in a commercial graphite furnace by diode laser induced fluorescence, *Spectrochim Acta, Part B* 47: 1519–1524.

260 SILVER J. A. (1992) Frequency-modulation spectroscopy for trace species detection: theory and comparison among experimental methods, *Appl Opt* 31: 707–717.

261 GROLL H., SCHNÜRER-PATSCHAN C., KURITSYN Yu. and NIEMAX K. (1994) Wavelength modulation diode laser atomic absorption spectrometry in analytical flames, *Spectrochim Acta, Part B* 49: 1463–1472.

262 JAKUBOWSKI N., JEPKENS B., STÜWER D. and BERNDT H. (1994) Speciation analysis of chromium by inductively coupled plasma mass spectrometry with hydraulic high pressure nebulization, *J Anal At Spectrom* 9: 193–198.

263 BARSHICK C. M., SHAW R. W., YOUNG J. P. and RAMSEY J. M. (1994) Istopic analysis of uranium using glow discharge optogalvanic spectroscopy and diode lasers, *Anal Chem* 66: 4154–4158.

264 AMOS M. D. and WILLIS J. B. (1966) Use of high-temperature pre-mixed flames in atomic absorption spectroscopy, *Spectrochim Acta* 22: 1325–1343.

265 HERMANN R. and ALKEMADE C. T. J. (1983) *Chemical analysis by flame photometry, 2nd edition*, Interscience Publishers, New York.

266 ALKEMADE C. Th. J., HOLLANDER Tj., SNELLEMAN W. and ZEEGERS P. J. Th. (1982) *Metal vapours in flames*, Pergamon Press, Oxford.

267 DEAN J. A. and RAINS T. C. (eds) (1969) *Flame emission and atomi absorption spectrometry*, Vol. 1. Marcel Dekker, New York.

268 RUBESKA I. and MUSIL J. (1979) Interferences in flame spectroscopy: their elimination and control, *Prog Anal Spectrosc* 2: 309–353.

269 GREENWAY G. M., NELMS S. M., SKHOSANA I. and DOLMAN S. J. L. (1996) *A comparison of preconcentration reagents for flow injection analysis flame atomic spectrometry*, in: McIntosk S. and Brindle I. (eds) Special Issue: Flow injection analysis, state of the art applied to atomic spectrometry, *Spectrochim Acta, Part B* 51: 1909–1915.

270 FANG Z. I. and WELZ B. (1989) Optimisation of experimental parameters for flow injection flame atomic absorption spectrometry, *J Anal At Spectrom* 4: 83–89.

271 SLAVIN W. (1986) Flames, furnaces, plasmas, how do we choose? *Anal Chem* 58: 589A–597A.

272 FRECH W., BAXTER D. C. and HÜTSCH B. (1986) Spatially isothermal graphite furnace for atomic absorption spectrometry using side-heated cuvettes with integrated contacts, *Anal Chem* 58: 1973–1977.

273 BERNDT H. and SCHALDACH G. (1988) Simple low-cost tungsten-coil atomiser for electrothermal atomic absorption spectrometry, *J Anal At Spectrom* 3: 709–713.

274 L'VOV B. V., BAYONOV P. A. and RYABCHUK (1981) A macrokinetic theory of sample vaporization in electrothermal atomic absorption spectrometry, *Spectrochim Acta, Part B* 36: 397–425.

275 PAVERI-FONTANA S. L., TESSARI G. and TORSI G. C. (1974) Time-resolved distribution of atoms in flameless atomic absorption spectrometry. A theoretical calculation, *Anal Chem* 46: 1032–1038.

276 SLAVIN W., MANNING D. C. and CARNRICK G. R. (1981) The stabilized temperature platform furnace, *At Spectrosc* 2: 137–145.

277 VAN DEN BROEK W. M. G. T. and DE GALAN L. (1977) Supply and removal of sample vapor in graphite thermal atomizers, *Anal Chem* 49: 2176–2186.

278 SCHLEMMER G. and WELZ B. **(1986)** Palladium and magnesium nitrates, a more universal modifier for graphite furnace atomic absorption spectrometry, *Spectrochim Acta, Part B* 41: 1157–1166.

279 WELZ B., CURTIUS A. J., SCHLEMMER G., ORTNER H. M. and BIRZER W. **(1986)** Scanning electron microscopy studies on surfaces from electrothermal atomic absorption spectrometry-III. The lanthanum modifier and the determination of phosphorus, *Spectrochim Acta, Part B* 41: 1175–1201.

280 KRIVAN V. **(1992)** Application of radiotracers to methodological studies in atomic absorption spectrometry, *J Anal At Spectrom* 7: 155–164.

281 KANTOR T., BEZUR L., PUNGOR E. and WINEFORDNER J. D. **(1983)** Volatilization studies of magnesium compounds by a graphite furnace and flame combined atomic absorption method. The use of a halogenating atmosphere, *Spectrochim Acta, Part B* 38: 581–607.

282 L'VOV B. V. **(1990)** Recent advances in absolute analysis by graphite furnace atomic absorption spectrometry, *Spectrochim Acta, Part B* 45: 633–655.

283 BULSKA E. and ORTNER H. M. **(2000)** Intercalation compounds of graphite in atomic absorption spectrometry, *Spectrochim Acta, Part B* 55: 491–499.

284 DE JONGHE W. G. C. and ADAMS F. C. **(1979)** The determination of organic and inorganic lead compounds in urban air by atomic absorption spectrometry with electrothermal atomization, *Anal Chim Acta* 108: 21–30.

285 CHAKRABORTI D., DE JONGHE W. R. A., VAN MOL W. E., ADAMS F. C. and VAN CLEUVENBERGEN R. J. A. **(1984)** Determination of ionic alkyllead compounds in water by gas chromatography/atomic absorption spectrometry, *Anal Chem* 56: 2692–2697.

286 YAMAMOTO M., YASUDA M. and YAMAMOTO Y. **(1985)** Hydride generation AAS coupled with flow injection analysis, *Anal Chem* 57: 1382–1385.

287 JACKWERTH E., WILLMER P. G., HÖHN R. and BERNDT H. **(1979)** A simple accessory for the determination of mercury and the hydride-forming elements (As, Bi, Sb, Se and Te) using flameless atomic absorption spectroscopy, *At Absorpt Newsl* 18: 66–68.

288 KAISER G., GÖTZ D., SCHOCH P. and TÖLG G. **(1975)** Emissionsspektrometrische Elementbestimmung im Nano- und Picogramm-Bereich nach Verflüchtigung der Elemente in mit Mikrowellen induzierte Gasplasmen I. Extrem nachweisstarke Quecksilberbestimmung in wässrigen Lösungen, Luft, organischen und anorganischen Matrices, *Talanta* 22: 889–899.

289 PIWONKA J., KAISER G. and TÖLG G. **(1985)** Determination of selenium at ng/g- and pg/g-levels by hydride generation-AAS in biotic materials, *Fresenius' Z Anal Chem* 321: 225–234.

290 STURGEON R. E., WILLIE S. N. and BERMAN S. S. **(1985)** Hydride generation-atomic absorption determination of antimony in seawater with in-situ concentration in a graphite furnace, *Anal Chem* 57: 2311–2314.

291 WELZ B. and MELCHER M. **(1981)** Mutual interactions of elements in the hydride technique in atomic absorption spectrometry, *Anal Chim Acta* 131: 17–25.

292 SANZ-MEDEL A., DEL ROSARIO FERNANDEZ DE LA CAMPA M., GONZALEZ E. B. and FERNANDEZ-SANCHEZ M. L. **(1999)** Organised surfactant assemblies in analytical atomic spectrometry, *Spectrochim Acta, Part B* 54: 251–287.

293 BERNDT H. **(1984)** Probeneintragssystem mit Probenverbrennung oder Probenvorverdampfung für die direkte Feststoffanalyse und für die Lösungsspektralanalyse, *Spectrochim Acta, Part B* 39: 1121–1128.

294 DOCEKAL B. and KRIVAN V. **(1992)** Direct determination of impurities in powdered silicon carbide by electrothermal atomic absorption spectrometry using the slurry sampling technique, *J Anal At Spectrom* 7: 521–528.

295 HORNUNG D. and KRIVAN V. **(1999)** Solid sampling electrothermal atomic absorption spectrometry for analysis of high-purity tungsten trioxide and high-purity tungsten blue oxide, *Spectrochim Acta, Part B* 54: 1177–1191.

296 GOUGH D. S. **(1976)** Direct analysis of metals and alloys by atomic absorption spectrometry, *Anal Chem* 48: 1926–1930.

297 BERNHARD A. E. **(1987)** Atomic absorption spectrometry using sputtering atomization of solid samples, *Spectroscopy* 2: 118.

298 OHLS K. **(1987)** Ein neuer Weg zum Einsatz der Atomabsorptions-spektrometrie mit Hilfe der direkten Atomisierung fester Proben, *Fresenius' Z Anal Chem* 327: 111–118.

299 KANTOR T., POLOS L., FODOR P. and PUNGOR E. **(1976)** Atomic absorption spectrometry of laser-nebulized samples, *Talanta* 23: 585–586.

300 MIZUIKE A. **(1983)** *Enrichment techniques for inorganic trace analysis*, Springer, Berlin.

301 ZYBIN A. and NIEMAX K. **(1997)** GC analysis of chlorinated hydrocarbons in oil and chlorophenols in plant extracts applying element-selective diode laser plasma detection, *Anal Chem* 69: 755–757.

302 ZYBIN A., SCHNÜRER-PATSCHAN Ch. and NIEMAX K. **(1995)** Wavelength modulation diode laser atomic absorption spectrometry in modulated low-pressure helium plasmas for element-selective detection in gas chromatography, *J Anal At Spectrom*, 10: 563–567.

303 KOCH J. and NIEMAX K. **(1998)** Characterization of an element selective GC-detector based on diode laser atomic absorption spectrometry, *Spectrochim Acta, Part B* 53: 71–79.

304 KOCH J., MICLEA M. and NIEMAX K. **(1999)** Analysis of chlorine in polymers by laser sampling and diode-laser atomic absorption spectrometry, *Spectrochim Acta, Part B* 54: 1723–1735.

305 BERNDT H. and SOPCZAK D. **(1987)** Fehlererkennung und Programmoptimierung in der ET-AAS durch zeitaufgelöste Signale. Direkte Elementspurenbestimmung in Harn: Cd, Co, Cr, Ni, Tl and Pb, *Fresenius' Z Anal Chem* 329: 18–26.

306 DE LOOS-VOLLEBREGT M. T. C. and DE GALAN L. **(1985)** Zeeman atomic absorption spectrometry, *Prog Anal At Spectrom* 8: 47–81.

307 BROEKAERT J. A. C. **(1982)** Zeeman atomic absorption instrumentation, *Spectrochim Acta, Part B* 37: 65–69.

308 SMITH S. B. Jr. and HIEFTJE G. M. **(1983)** A new background-correction method for atomic absorption spectrometry, *Appl Spectrosc* 37: 419–414.

309 YAMAMOTO M., MURAYAMA S., ITO M. and YASUDA M. **(1980)** Theoretical basis for multielement analysis by coherent forward scattering atomic spectroscopy, *Spectrochim Acta, Part B* 35: 43–50.

310 WIRZ P., DEBUS H., HANLE W. and SCHARMANN A. **(1982)** Synchronous multielement detection by forward scattering in a transverse magnetic field (SYNFO), *Spectrochim Acta, Part B* 37: 1013–1020.

311 KINGSTON H. M. and JASSIE L. B. **(1988)** Introduction to microwave sample preparation: theory and practice, *American Chemical Society*, Washington DC.

312 KOHL F., JAKUBOWSKI N., BRANDT R., PILGER C. and BROEKAERT J. A. C. **(1997)** New strategies for trace analyses of ZrO_2, SiC and Al_2O_3 ceramic powders, *Fresenius' J Anal Chem* 359: 317–325.

313 TSCHÖPEL P. and TÖLG G. **(1982)** Comments on the accuracy of analytical results in nanogram and picogram trace analysis of the elements, *J Trace Microprobe Techn* 1: 1–77.

314 BERNDT H. and JACKWERTH E. **(1979)** Mechanisierte Mikromethode ("Injektionsmethode") der Flammen-Spektrometrie (Atomabsorption-Atomemission) für die Bestimmung der Serumelektrolyte und der Spurenelemente (Fe, Cu, Zn) – Teil 1, *J Clin Chem Clin Biochem* 19: 71–76.

315 ALT F. and MASSMANN **(1978)** Bestimmung von Blei in Blut mittels Atomabsorptionsspektrometrie, *Spectrochim Acta, Part B* 33: 337–342.

316 KONVAR U. K., LITTLEJOHN D. and HALLS D. J. **(1990)** Hot-injection procedures for the rapid analysis of biological samples by electrothermal atomic-absorption spectrometry, *Talanta* 37: 555–559.

317 BURBA P. and WILLMER P. G. **(1987)** Multielement-preconcentration for atomic spectroscopy by sorption of dithiocarbamate-metal complexes (e.g. HMDC) on cellulose collectors, *Fresenius' Z Anal Chem* 329: 539–545.

318 JACKWERTH E., MIZUIKE A., ZOLOTOV Y. A., BERNDT H., HÖHN R. and KUZMIN N. M. **(1979)** Separation and preconcentration of trace substances I. Preconcentration for ionorganic trace analysis, *Pure Appl Chem* 51: 1195–1211.

319 ALT F., JERONO U., MESSERSCHMIDT J. and TÖLG G. **(1988)** The determination of platinum in biotic and environmental materials, 1. μg/kg–g/kg-range, *Mikrochim Acta* 3: 299–304.

320 SUBRAMANIAN K. S. **(1988)** Determination of chromium(III) and chromium(VI) by ammonium pyrrolidinecarbodithioate-methyl isobutyl ketone furnace atomic absorption spectrometry, *Anal Chem* 60: 11–15.

321 DIRKX W. M. R., LOBINSKI R. and ADAMS F. C. **(1993)** A comparative study of gas chromatography with atomic absorption and atomic emission detection for the speciation analysis of organotin, *Anal Sci* 9: 273–278.

322 ODA C. E. and INGLE J. D. **(1981)** Speciation of mercury by cold vapor atomic absorption spectrometry with selective reduction, *Anal Chem* 53: 2305–2309.

323 SCHICKLING C. and BROEKAERT J. A. C. **(1995)** Determination of mercury species in gas condensates by on-line coupled high-performance liquid chromatography and cold-vapor atomic absorption spectrometry, *Appl Organomet Chem* 9: 29–36.

324 SCHERMER S., JURICKA L; PAUMARD J; BEIUROHR E., MOITYSIK F. M. and

BROEKAERT J. A. C. Optimization of electrochemical hydride generation in a miniaturized flow cell coupled to microwave-induced plasma, atomic emission spectrometry for the determination of selenium, Fresenius Journal of Analytical Chemistry, in press.

325 BUTCHER D. J., ZYBIN A., BOLSHOV M. A. and NIEMAX K. **(1999)** Speciation of methylcyclopentadienyl manganese tricarbonyl by high performance liquid chromatography–diode laser atomic absorption spectrometry, *Anal Chem* 71: 5379–5385.

326 GASPAR A. and BERNDT H. **(2000)** Beam injection flame furnace atomic absorption spectrometry: a new flame method, *Anal Chem* 72: 240–246.

327 GERLACH W. and SCHWEITZER E. **(1930)** *Die chemische Emissionsspektralanalyse*, Vol. 1. L. Voss, Leizig.

328 IDZIKOWSKI A., GADEK S. and MLECZKO W. **(1973)** *Atlas of the arc spectra for 70 elements for the grating spectrograph*, Monograph 3, Technical University of Wroclaw (Poland).

329 WINGE R. K., FASSEL V., PETERSON V. J. and FLOYD M. A. **(1985)** *ICP-AES – An atlas of spectral information*, Elsevier, Amsterdam.

330 HARRISON G. R. **(1991)** *Wavelength tables*, The MIT Press, Cambridge, MA.

331 SAIDEL A. N., PROKOFIEV V. K. and RAISKI S. M. **(1955)** *Spektraltabellen*, V E B Verlag Technik, Berlin.

332 GATTERER A. and JUNKES J. **(1945, 1947, 1959)** *Atlas der Restlinien, 3 volumes*, Specola Vaticana, Citta del Vaticano.

333 SALPETER E. W. **(1973)** *Spektren in der Glimmentladung von 150 bis 400 nm*, 1st to 5th volume. Specola Vaticana, Citta del Vaticano.

334 BOUMANS P. W. J. M. **(1984)** *Line coincidence tables for ICP-OES, Volumes 1 and 2*, Pergamon Press, Oxford.

335 BROEKAERT J. A. C. **(1983)** Evaluation of a flexible ICP-spectrometer using a SIT as radiation detector, Abstract 23 CSI, Amsterdam, *Spectrochim Acta, Part B* 38: Supplement,44.

336 SCHEIBE G. **(1933)** *Chemische Spektralanalyse*, in: Böttger W., *Physikalische Methoden der analytischen Chemie*, 1: 27.

337 LOMAKIN B. A. **(1930)** Quantitative Spektralbestimmung von Wismut in Kupfer, *Z Anorg Allg Chem* 187: 75–90.

338 LAQUA K. Optische Emissions-spektrometrie **(1980)** in: *Ullmann's Encyklopädie der technischen Chemie*, Vol. 5. Verlag Chemie, Weinheim, 441–500.

339 LAQUA K., HAGENAH W. D. and WÄCHTER H. **(1967)** Spektro-chemische Spurenanalyse mit spektralaufgelösten Analysenlinien, *Fresenius' Z Anal Chem* 221: 142–174.

340 AZIZ A., BROEKAERT J. A. C., LEIS F. **(1981)** The optimization of an ICP-injection technique and the application to the direct analysis of small-volume serum samples, *Spectrochim Acta, Part B* 36: 251–260.

341 WINGE R. K., DEKALB E. L. and FASSEL V. A. **(1985)** Comparative complexity of emission spectra from ICP, DC arc and spark excitation sources, *Appl Spectrosc* 39: 673–676.

342 MERTEN D., BROEKAERT J. A. C. and LEMARCHAND A. **(1997)** Application of a rapid sequential inductively coupled plasma optical emission spectrometric method for the analysis with linerich emission spectra by different means of sample introduction, *J Anal At Spectrom* 12: 1387–1391.

343 BROEKAERT J. A. C., WOPENKA B. and PUXBAUM H. **(1982)** Inductively coupled plasma optical emission spectrometry for the analysis of aerosol samples collected by cascade impactors, *Anal Chem* 54: 2174–2179.

344 BECKER-ROSS H. and FLOREK S. V. **(1997)** Echelle spectrometers and charge-coupled devices, *Spectrochim Acta, Part B* 52: 1367–1375.

345 BARNARD T. W., CROCKETT H., IVALDI J. C. and LUNDBERG P. L. **(1993)** Design and evaluation of an Echelle grating optical system for ICP-OES, *Anal Chem* 65: 1225–1230.

346 BUTLER-SOBEL C. **(1982)** Use of an extended torch with an ICP for the determination of nitrogen in aqueous solutions, *Appl Spectrosc* 36: 691–693.

347 BROEKAERT J. A. C. and ZEEMAN P. B. **(1984)** Some experiments with an extended torch for the determination of nitrogen in aqueous solution with the aid of ICP AES, *Spectrochim Acta, Part B* 39: 851–853.

348 HEITLAND P. **(2000)** Verwendung von Spektrallinien in der ICP-OES im Wellenlängenbereich 125–190 nm, *GIT Labor-Fachzeitschrift* 847–850.

349 CHRISTIAN G. D. and FELDMANN F. J. **(1971)** A comparison study of detection limits using flame-emission spectroscopy with the nitrous oxide–acetylene flame and atomic-absorption spectroscopy, *Appl Spectrosc* 25: 660–663.

350 ROSEN B. **(1970)** *International tables of selected constants 17*, Pergamon Press, Oxford.

351 ADDINK N. W. H. **(1971)** *DC arc analysis*, McMillan, London.

352 RIEMANN M. Z. **(1966)** Der Kaskadebogen als Anregungsquelle für die Spektralanalyse von Lösungen, *Z Anal Chem* 215: 407–424.

353 TODOROVIC M., VUKANOVIC V. and GEORGIJEVIC V. **(1968)** Optimal conditions for the use of a d.c. arc in an inhomogeneous magnetic field, *Spectrochim Acta, Part B* 24: 571–577.

354 BROEKAERT J. A. C. **(1973)** Calculation of detection limits in spectrographic analysis, *Bull Soc Chim Belg* 82: 561–567.

355 RADWAN Z., STRZYZEWSKA B. and MINCZEWSKI J. Spectrographic determination of trace amounts of rare earths I. Analysis of lanthanum and yttrium by the power sifting method. **(1963)**, *Appl Spectrosc* 17: 1–5.

356 RADWAN Z., STRZYZEWSKA B. and MINCZEWSKI J. **(1963)** Spectrographic determination of trace amounts of rare earths. II. Analysis of lanthanum, yttrium, europium, praesodymium neodymium and samarium by the carrier distillation method, *Appl Spectrosc* 17: 60–62.

357 RADWAN Z., STRZYZEWSKA B. and MINCZEWSKI J. **(1966)** Spectrographic determination of trace amounts of rare earths. III. Determination of all rare earth elements by the carrier distillation method, *Appl Spectrosc* 20: 236–240.

358 AVNI R. **(1978)** in Grove E. L. (ed): *Applied atomic spectroscopy*, Vol. 1. Plenum Press, New York.

359 Spitsberg A. T. and Otis J. S. **(2000)** Difficulties with direct-current/arc optical emission spectrometric analysis of molybdenum powder for chromium, iron and nickel, *Appl Spectrosc* 11: 1707–1711.

360 Slickers K. **(1992)** *Die automatische Atom-Emissions-Spektralanalyse*, Brühlsche Universitätsdrückerei, Gießen.

361 Koch K. H. **(1984)** Beiträge aus der Stahlindustrie zur angewandten Atomspektroskopie, *Spectrochim Acta, Part B* 39: 1067–1079.

362 Leis F., Broekaert J. A. C. and Laqua K. **(1987)** Design and properties of a microwave boosted glow discharge lamp, *Spectrochim Acta, Part B* 42: 1169–1176.

363 Jakubowski N. and Stuewer D. **(1988)** Comparison of Ar and Ne as working gases in analytical glow discharge mass spectrometry, *Fresenius' Z Anal Chem* 331: 145–149.

364 Tohyama K., Ono J., Onodera M. and Saeki M. **(1978)** Research and Development in Japan. Okochi Memorial Foundation, 31–35.

365 Coleman D. M., Sainz M. A. and Butler H. T. **(1980)** High frequency excitation of spark-sampled metal vapor, *Anal Chem* 52: 746–753.

366 Korolev V. V. and Vainshtein E. E. **(1959)** Plasma generator as excitation source in spectroscopic analysis, *Zh Anal Khim* 14: 658–662.

367 Margoshes M. and Scribner B. **(1959)** The plasma jet as a spectroscopic source, *Spectrochim Acta* 15: 138–145.

368 Kranz E. **(1964)** in: *Emissionsspektrokopie*, Akademie-Verlag, Berlin, p. 160.

369 Reednick J. **(1979)** A unique approach to atomic spectroscopy high energy plasma excitation and high resolution spectrometry, *Am Lab* May: 127–133.

370 Decker R. J. **(1980)** Some analytical characteristics of a three electrode DC argon plasma source for optical emission spectrometry, *Spectrochim Acta, Part B* 35: 19–35.

371 Leis F., Broekaert J. A. C. and Waechter H. **(1989)** A three-electrode direct current plasma as compared to an inductively coupled argon plasma, *Fresenius' Z Anal Chem* 333: 2–5.

372 Johnson G. W., Taylor H. E. and Skogerboe R. K. **(1979)** Determination of trace elements in natural waters by the DC argon plasma multi-element atomic emission spectrometer (DCP-MAES) Technique, *Spectrochim Acta, Part B* 34: 197–212.

373 Urasa I. T. and Ferede F. **(1987)** Use of direct current plasma as an element selective detector for simultaneous ion chromatographic determination of arsenic(III) and arsenic(V) in the presence of other common anions, *Anal Chem* 59: 1563–1568.

374 Reed T. B. **(1961)** Induction-coupled plasma torch, *J Appl Phys* 32: 821–824.

375 Greenfield S., Jones I. L. and Berry C. T. **(1964)** High-pressure plasmas as spectroscopic emission sources, *Analyst* 89: 713–720.

376 Wendt R. H. and Fassel V. A. **(1965)** Induction coupled plasma spectrometric excitation source, *Anal Chem* 37: 920–922.

377 Rezaaiyaan R., Hieftje G. M., Anderson H., Kaiser H. and Meddings B. **(1982)** Design and construction of a low-flow, low-power torch for ICP spectrometry, *Appl Spectrosc* 36: 626–631.

378 Van Der Plas P. S. C. and De Galan L. **(1984)** A radiatively cooled torch for ICP-AES using 1 L min^{-1} of argon, *Spectrochim Acta, Part B* 39: 1161–1169.

379 Van Der Plas P. S. C., De Waaij A. C. and De Galan L. **(1985)** Analytical evaluation of an air-cooled 1 L min^{-1} argon ICP, *Spectrochim Acta, Part B* 40: 1457–1466.

380 Meyer G. A. and Thompson M. D. **(1985)** Determination of trace element detection limits in air and oxygen inductively coupled plasmas, *Spectrochim Acta, Part B* 40: 195–207.

381 Bacri J., Gomes A. M., Fieni J. M., Thouzeau F. and Birolleau J. C. **(1989)** Transport coefficient calculation in air–argon mixtures: explanation of the improvement of detection limits of pollutants in air inductively coupled plasmas by argon addition, *Spectrochim Acta, Part B* 44: 887–895.

382 KALNICKI D. J., FASSEL V. A. and KALNICKY R. N. **(1977)** Electron temperatures and electron number densities experienced by analyte species in ICP with and without the presence of an easily ionized element, *Appl Spectrosc* 31: 137–150.

383 RAEYMAEKERS B., BROEKAERT J. A. C. and LEIS F. **(1988)** Radially resolved rotational temperatures in nitrogen–argon, oxygen–argon, air–argon and argon ICPs, *Spectrochim Acta* 43: 941–949.

384 ISHII I. and MONTASER A. **(1991)** A tutorial discussion on measurements of rotational temperature in ICP, *Spectrochim Acta, Part B* 46: 1197–1206.

385 BOUMANS P. W. J. M. and DE BOER F. J. **(1977)** An experimental study of a 1-kW, 50-MHz RF ICP with pneumatic nebulizer and a discussion of experimental evidence for a non-thermal mechanism, *Spectrochim Acta, Part B* 32: 365–395.

386 DE GALAN L. **(1984)** Some considerations on the excitation mechanism in the inductively coupled argon plasma, *Spectrochim Acta, Part B* 39: 537–550.

387 HUANG M., HANSELMANN D. S., YANG P. Y. and HIEFTJE G. M. **(1992)** Isocontour maps of electron temperature, electron number density and gas kinetic temperature in the Ar inductively coupled plasma obtained by laser-light Thomson and Rayleigh scattering, *Spectrochim Acta, Part B* 47: 765–786.

388 HUANG M., LEHN S. A., ANDREWS E. J. and HIEFTJE G. M. **(1997)** Comparison of electron concentrations, electron temperatures and excitation temperatures in argon ICPs operated at 27 and 40 MHz, *Spectrochim Acta, Part B* 52: 1173–1193.

389 HANSELMAN D. S., SESI N. N., HUANG M. and HIEFTJE G. M. **(1994)** The effect of sample matrix on electron density, electron temperature and gas

temperature in the argon inductively coupled plasma examined by Thomson and Rayleigh scattering, *Spectrochim Acta, Part B* 49: 495–526.

390 HANSELMAN D. S., WITHNELL R. and HIEFTJE G. M. **(1991)** Side-on photomultiplier gating system for Thomson scattering and laser-excited atomic fluorescence spectroscopy, *Appl Spectrosc* 45: 1553–1560.

391 MICHAUD-POUSSEL E. and MERMET J. M. **(1986)** Influence of the generator frequency and the plasma gas inlet area on torch design in ICP-AES, *Spectrochim Acta, Part B* 41: 125–132.

392 BRENNER J. B., ZANDER A., COLE M. and WISEMAN A. **(1997)** Comparison of axially and radially viewed inductively coupled plasmas for multielement analysis. Effect of sodium and calcium, *J Anal At Spectrom* 12: 897–906.

393 IVALDI J. C. and TYSON J. F. **(1995)** Performance evaluation of an axially viewed horizontal inductively coupled plasma for optical emission spectrometry, *Spectrochim Acta, Part B* 50: 1207–1226.

394 BINGS N. H., WANG C., SKINNER C. D., COLYER C. L., THIBAULT P. and HARRISON D. J. **(1999)** Microfluid devices connected to fused-silica capillaries with minimal dead volume, *Anal Chem* 71: 3292–3296.

395 BROEKAERT J. A. C. and LEIS F. **(1979)** An injection method for the sequential determination of boron and several metals in waste-water samples by inductively-coupled plasma atomic emission spectrometry, *Anal Chim Acta* 109: 73–83.

396 WALTERS F. H., PARKER L. R. Jr., MORGAN S. L. and DEMING S. N. **(1991)** *Sequential simplex optimization*, CRC Press, Inc., Boca Raton.

397 MOORE G. L., HUMPHRIES-CUFF P. J. and WATSON A. E. **(1984)** Simplex optimization of a nitrogen-cooled argon inductively coupled plasma for multielement analysis, *Spectrochim Acta, Part B* 39: 915–929.

398 PARSONS M. L., FORSTER A. R. and ANDERSON D. **(1980)** *An atlas of*

spectral interferences in ICP spectroscopy, Plenum Press, New York.

399 BROEKAERT J. A. C., LEIS F. and LAQUA K. **(1979)** Some aspects of matrix effects caused by sodium tetraborate in the analysis of rare earth minerals with the aid of inductively coupled plasma atomic emission spectrometry, *Spectrochim Acta, Part B* 34: 167–175.

400 BELCHAMBER R. M. and HORLICK G. **(1982)** Noise power spectra of optical and acoustic emission signals from an ICP, *Spectrochim Acta, Part B* 37: 17–27.

401 MANN S., GEILENBERG D., BROEKAERT J. A. C. and JANSEN M. **(1997)** Digestion methods for advanced ceramic materials and subsequent determination of silicon and boron by inductively coupled plasma atomic emission spectrometry, *J Anal At Spectrom* 12: 975–979.

402 GEILENBERG D., GERARDS M. and BROEKAERT J. A. C. **(2000)** Determination of the stoichiometric composition of high-temperature superconductors by ICP-OES for production control, *Mikrochim Acta* 133: 319–323.

403 KUCHARKOWSKI R., VOGT C. and MARQUARDT D. **(2000)** Accurate and precise spectrochemical analysis of Bi-Pb-Sr-Ca-Cu-O high-temperature superconductor materials, *Fresenius' J Anal Chem* 366: 146–151.

404 BARNETT W. B., FASSEL V. A. and KNISELEY R. N. **(1968)** Theoretical principles of internal standardization in analytical emission spectroscopy, *Spectrochim Acta* 23: 643–664.

405 MEYERS S. A. and TRACY D. H. **(1983)** Improved performance using internal standardisation in ICP emission spectroscopy, *Spectrochim Acta, Part B* 38: 1227–1253.

406 BROEKAERT J. A. C., KLOCKENKÄMPER R. and KO J. B. **(1983)** Emissionsspektrometrische Präzisionsbestimmung der Hauptbestandteile von Cu/Zn-Legierungen mittels Glimmlampe und ICP, *Fresenius' Z Anal Chem* 316: 256–260.

407 BROEKAERT J. A. C., LEIS F. and LAQUA K. **(1981)** The application of an argon/nitrogen inductively-coupled plasma to the analysis of organic solutions, *Talanta* 28: 745–752.

408 BOUMANS P. W. J. M. and LUX-STEINER M. Ch. **(1982)** Modification and optimization of a 50 MHz inductively coupled argon plasma with special reference to analyses using organic solvents, *Spectrochim Acta* 37: 97–126.

409 OLESIK J. W. and HOBBS S. E. **(1994)** Monodisperse dried microparticulate injector: a new tool for studying fundamental processes in inductively coupled plasmas, *Anal Chem* 66: 3371–3378.

410 TARR M. A., ZHU G. and BROWNER R. F. **(1993)** Microflow ultrasonic nebulizer for inductively coupled plasma atomic emission spectrometry, *Anal Chem* 65: 1689–1695.

411 VERMEIREN K., VANDECASTEELE C. and DAMS R. **(1990)** Determination of trace amounts of cadmium, lead, copper and zinc in natural waters by ICP-AES with thermospray nebulization after enrichment on Chelex-100, *Analyst* 115: 17–22.

412 DITTRICH K., BERNDT H., BROEKAERT J. A. C., SCHALDACH G. and TÖLG G. **(1988)** Comparative study of injection into a pneumatic nebuliser and tungsten coil electrothermal vaporisation for the determination of rare earth elements by inductively coupled plasma optical emission spectrometry, *J Anal At Spectrom* 3: 1105–1110.

413 LUCIC M. and KRIVAN V. **(1999)** Analysis of aluminium-based ceramic powders by electrothermal vaporization inductively coupled plasma atomic emission spectrometry using a tungsten coil and slurry sampling, *Fresenius' J Anal Chem* 363: 64–72.

414 KARANASSIOS V., ABDULLAH M. and HORLICK G. **(1990)** The application of chemical modification in direct sample insertion-inductively coupled plasma-atomic emission spectrometry, *Spectrochim Acta, Part B* 45: 119–130.

415 LIU X. R. and HORLICK G. **(1994)** In-situ laser ablation sampling for inductively coupled plasma atomic emission spectrometry, *Spectrochim Acta, Part B* 50: 537–548.

416 LIU Y., WANG X., YUAN D., YANG P., HUANG B. and ZHUANG Z. **(1992)** Flow-injection-electrochemical hydride generation technique for atomic absorption spectrometry, *J Anal At Spectrom* 7: 287–291.

417 HUEBER D. M. and WINEFORDNER J. D. **(1995)** A flowing electrolytic hydride generator for continuous sample introduction in atomic spectrometry, *Anal Chim Acta* 316: 129–44.

418 COX A. G., COOK L. G. and MCLEOD C. W. **(1985)** Rapid sequential determination of chromium III/chromium VI by flow injection analysis/ICP AES, *Analyst* 110: 331–333.

419 ISHIZUKA T. and UWAMINO Y. **(1983)** ICP emission spectrometry of solid samples by laser ablation, *Spectrochim Acta, Part B* 38: 519–527.

420 BRENNER I. B., WATSON A. E., STEELE T. W., JONES E. A. and GONCALVES M. **(1981)** Application of an argon–nitrogen inductively coupled radiofrequency plasma (ICP) to the analysis of geological and related materials for their rare earth contents, *Spectrochim Acta, Part B* 36: 785–797.

421 GARBARINO J. R. and TAYLOR M. E. **(1979)** An inductively-coupled plasma atomic-emission spectrometric method for routine water quality testing, *Appl Spectrosc* 33: 220–226.

422 U.S. Environmental Protection Agency, *Inductively coupled plasma-atomic emission spectrometric method for trace element analysis of water and wastes*, Washington, DC, 1979.

423 DIN 38 406 **(1987)** *Bestimmung der 24 Elemente Ag, Al, B, Ba, Ca, Cd, Co, Cr, Cu, Fe, K, Mg, Mn, Mo, Na, Ni, P, Pb, Sb, Sr, Ti, V, Zn und Zr durch Atom-emissionsspektrometrie mit induktiv gekoppeltem Plasma (ICP-AES) (E22)*, Beuthe Verlag, Berlin.

424 THOMPSON M., PAHLAVANPOUR B., WALTON S. J. and KIRKBRIGHT G. F. **(1978)** Simultaneous determination of trace concentrations of arsenic, antimony, bismuth, selenium and tellurium in aqueous solutions by introduction of the gaseous hydrides into an ICP source for emission spectrometry. Part II. Interference studies, *Analyst* 103: 705–713.

425 MIYAZAKI A., KIMURA A., BANSHO K. and UMEZAKI Y. **(1982)** Simultaneous determination of heavy metals in waters by ICP-AES after extraction into diisobutyl ketone, *Anal Chim Acta* 144: 213–221.

426 BERNDT H., HARMS U. and SONNEBORN M. **(1985)** Multielement-Spurenvoranreicherung aus Wässern an Aktivkohle zur Probenvorbereitung für die Atomspektroskopie (Flammen-AAS, ICP/OES), *Fresenius' Z Anal Chem* 322: 329–333.

427 HIRAIDE M., ITO T., BABA M., KAWAGUCHI H. and MIZUIKE A. **(1980)** Multielement preconcentration of trace heavy metals in water by preconcentration and flotation with indium hydroxide for ICP-AES, *Anal Chem* 52: 804–807.

428 BROEKAERT J. A. C., GUCER S. and ADAMS F. **(1990)** *Metal speciation in the environment*, Springer Verlag, Berlin.

429 GRAULE T., VON BOHLEN A., BROEKAERT J. A. C., GRALLATH E., KLOCKENKÄMPER R., TSCHÖPEL P. and TÖLG G. **(1989)** Atomic emission and atomic absorption spectrometric analysis of high-purity powders for the production of ceramics, *Fresenius' Z Anal Chem* 335: 637–642.

430 DOCEKAL B., BROEKAERT J. A. C., GRAULE T., TSCHÖPEL P. and TÖLG G. **(1992)** Determination of impurities in silicon carbide powders, *Fresenius' J Anal Chem* 342: 113–117.

431 MERTEN D., BROEKAERT J. A. C., BRANDT R. and JAKUBOWSKI N. **(1999)** Analysis of ZrO_2 powders by micro-wave assisted digestion at high pressure and ICP atomic spectrometry, *J Anal At Spectrom* 14: 1093–1098.

432 GROSS R., PLATZER B., LEITNER E., SCHALK A., SINABELL H., ZACH H. and KNAPP G. **(1992)** Atomic emission gas chromatographic detection – chemical and spectral

interferences in the stabilized capacitive plasma (SCP), *Spectrochim Acta, Part B* 47: 95–106.

433 LUGE S., BROEKAERT J. A. C., SCHALK A. and ZACH H. **(1995)** The use of different sample introduction techniques in combination with the low power stabilized capacitive plasma (SCP) as a radiation source for atomic emission spectrometry, *Spectrochim Acta, Part B* 50: 441–452.

434 LUGE S. and BROEKAERT J. A. C. **(1996)** Use of a low power, high frequency stabilized capacitive plasma combined with graphite furnace vaporization for the atomic emission spectrometric analysis of serum samples, *Anal Chim Acta* 332: 193–199.

435 HERWIG F. and BROEKAERT J. A. C. **(2000)** Optimization of a direct sample insertion into a stabilized capacitive plasma for optical emission spectrometry (SCP-OES), *Mikrochim Acta* 134: 51–56.

436 SKOGERBOE R. K. and COLEMAN G. N. **(1976)** Microwave plasma emission spectrometry, *Anal Chem* 48: 611A–622A.

437 MAVRODINEANU R. and HUGHES R. C. **(1963)** Excitation in radiofrequency discharges, *Spectrochim Acta* 19: 1309–1307.

438 BINGS N. H., OLSCHEWSKI M. and BROEKAERT J. A. C. **(1997)** Two-dimensional spatially resolved excitation and rotational temperatures as well as electron number density measurements in capacitively coupled microwave plasmas using argon, nitrogen and air as working gases by spectroscopic methods, *Spectrochim Acta, Part B* 52: 1965–1981.

439 DISAM A., TSCHÖPEL P. and TÖLG G. **(1982)** Emissionsspektrometrische Bestimmung von elementspuren in wässrigen Lösungen mit einem mantelgasstabilisierten kapazitiv gekoppelten Mikrowellenplasma (CMP), *Fresenius' Z Anal Chem* 310: 131–143.

440 HANAMURA S., SMITH B. W. and WINEFORDNER J. D. **(1983)** Speciation of inorganic and organometallic compounds in solid biological samples by thermal vaporization and plasma emission spectrometry, *Anal Chem* 55: 2026–2032.

441 SEELIG M., BINGS N. H. and BROEKAERT J. A. C. **(1998)** Use of a capacitively coupled microwave plasma (CMP) with Ar, N_2 and air as working gases for atomic spectrometric elemental determinations in aqueous solutions and oils, *Fresenius' J Anal Chem* 360: 161–166.

442 SEELIG M. and BROEKAERT J. A. C. **(2000)** Investigations on the on-line determination of metals in air flows by capacitively coupled microwave plasma atomic emission spectrometry (CMP-OES), *Spectrochim Acta, Part B* in press.

443 BEENAKKER C. I. M. **(1977)** Evaluation of a microwave-induced plasma in helium at atmospheric pressure as an element-specific detector for gas chromatography, *Spectrochim Acta, Part B* 32: 173–178.

444 KOLLOTZEK D., TSCHÖPEL P. and TÖLG G. **(1982)** Lösungsemissionsspektrometrie mit mikrowelleninduzierten Mehrfaden- und Hohlzylinderplasmen I.- Die radiale Verteilung von Signal- und Untergrundintensitäten, *Spectrochim Acta, Part B* 37: 91–96.

445 HELTAI Gy., BROEKAERT J. A. C., LEIS F. and TÖLG G. **(1990)** Study of a toroidal argon and a cylindrical helium microwave induced plasma for analytical atomic emission spectrometry – I. Configurations and spectroscopic properties, *Spectrochim Acta, Part B* 45: 301–311.

446 DEUTSCH R. D. and HIEFTJE G. M. **(1985)** Development of a microwave-induced nitrogen discharge at atmospheric pressure (MINDAP), *Appl Spectrosc* 39: 214–222.

447 WU M. and CARNAHAN J. W. **(1990)** Trace determination of cadmium, copper, bromine and chlorine into a helium microwave-induced plasma, *Appl Spectrosc* 44: 673–678.

448 MOUSSANDA P. S., RANSON P. and MERMET J. M. **(1985)** Spatially resolved spectroscopic diagnostics of an argon MIP produced by surface wave propagation (Surfatron), *Spectrochim Acta, Part B* 40: 641–651.

449 KOLLOTZEK D., OECHSLE D., KAISER G., TSCHÖPEL P. and TÖLG G. **(1984)** Application of a mixed-gas microwave induced plasma as an on-line element-specific detector in high-performance liquid chromatography, *Fresenius' Z Anal Chem* 318: 485–489.

450 KOLLOTZEK D., TSCHÖPEL P. and TÖLG G. **(1984)** Three-filament and toroidal microwave induced plasmas as radiation sources for emission spectrometric analysis of solutions and gaseous samples. II. Analytical performance, *Spectrochim Acta, Part B* 39: 625–636.

451 UDEN P. **(1992)** *Element-specific chromatographic detection by atomic emission spectroscopy*, American Chemical Society, Washington.

452 LOBINSKI R. **(1997)** Elemental speciation and coupled techniques, *Appl Spectrosc* 51: 260A–278A.

453 HUBERT J., MOISAN M. and RICARD A. **(1979)** A new microwave plasma at atmospheric pressure, *Spectrochim Acta, Part B* 33: 1–10.

454 RICHTS U., BROEKAERT J. A. C., TSCHÖPEL P. and TÖLG G. **(1991)** Comparative study of a Beenakker cavity and a surfatron in combination with electrothermal evaporation from a tungsten coil for microwave plasma optical emission spectrometry (MIP-AES), *Talanta* 38: 863–869.

455 YANG J., SCHICKLING C., BROEKAERT J. A. C., TSCHÖPEL P. and TÖLG G. **(1995)** Evaluation of continuous hydride generation combined with helium and argon microwave induced plasmas using a surfatron for atomic emission spectrometric determination of arsenic, antimony and selenium, *Spectrochim Acta, Part B* 50: 1351–1363.

456 JIN Q., ZHU C., BORER M. and HIEFTJE G. M. **(1991)** A microwave plasma torch assembly for atomic emission spectrometry, *Spectrochim Acta, Part B* 46: 417–430.

457 PROKISCH C., BILGIC A. M., VOGES E., BROEKAERT J. A. C., JONKERS J., VAN SANDE M. and VAN DER MULLEN J. A. M. **(1999)** Photographic plasma images and electron number density as well as electron temperature mappings of a plasma sustained with a modified argon microwave torch (MPT) measured by spatially resolved Thomson scattering, *Spectrochim Acta, Part B* 54: 1253–1266.

458 ENGEL U., PROKISCH C., VOGES E., HIEFTJE G. M. and BROEKAERT J. A. C. **(1998)** Spatially resolved measurements and plasma tomography with respect to the rotational temperatures for a microwave plasma torch, *J Anal At Spectrom* 13: 955–961.

459 PROKISCH C. and BROEKAERT J. A. C. **(1998)** Elemental determinations in aqueous and acetonitrile containing solutions by atomic spectrometry using a microwave plasma torch, *Spectrochim Acta, Part B* 53: 1109–1119.

460 PEREIRO R., WU M., BROEKAERT J. A. C. and HIEFTJE G. M. **(1994)** Direct coupling of continuous hydride generation with microwave plasma torch atomic emission spectrometry for the determination of arsenic, antimony and tin, *Spectrochim Acta, Part B* 49: 59–73.

461 BILGIC A. M., PROKISCH C., BROEKAERT J. A. C. and VOGES E. **(1998)** Design and modelling of a modified 2.45 GHz coaxial plasma torch for atomic spectrometry, *Spectrochim Acta, Part B* 53: 773–777.

462 BILGIC A. M., ENGEL U., VOGES E., KÜCKELHEIM M. and BROEKAERT J. A. C. **(2000)** A new low-power microwave plasma source using microstrip technology for atomic emission spectrometry, *Plasma Sources Sci Technol* 9: 1–4.

463 ENGEL U., BILGIC A. M., HAASE O., VOGES E. and BROEKAERT J. A. C. **(2000)** A microwave-induced plasma based on microstrip technology and its use for the atomic emission spectrometric determination of mercury with the aid of the cold-vapor technique, *Analytical Chemistry* 72: 193–197.

464 BILGIC A. M., VOGES E., ENGEL U. and BROEKAERT J. A. C. **(2000)** A low-

power 2.45 GHz microwave induced helium plasma source at atmospheric pressure based on microstrip technology, *J Anal At Spectrom* 15: 579–580.

465 Bogaerts A., Donko Z., Kutasi K., Bano G., Pinhao N. and Pinheiro M. (2000) Comparison of calculated and measured optical emission intensities in a direct current argon–copper glow discharge, *Spectrochim Acta, Part B* 55: 1465–1479.

466 Bogaerts A. and Gijbels R. (2000) Description of the argon-excited levels in a radio-frequency and direct current glow discharge, *Spectrochim Acta, Part B* 55: 263–278.

467 Broekaert J. A. C. (1979) Some observations on sample volatilisation in a hot-type hollow cathode, *Spectrochim Acta, Part B* 34: 11–17.

468 Zil'bershtein Kh. I. (1977) *Spectrochemical analysis of pure substances*, Adam Hilger, Bristol, 173–227.

469 Thelin B. (1981) The use of a high temperature hollow cathode lamp for the determination of trace elements in steels, nickel-base alloys and ferroalloys by emission spectrometry, *Appl Spectrosc* 35: 302–307.

470 Schepers C. and Broekaert J. A. C. (2000) The use of a hollow cathode glow discharge (HCGD) as an atomic emission spectrometric element specific detector for chlorine and bromine in gas chromatography, *J Anal At Spectrom* 15: 51–65.

471 Broekaert J. A. C. (1976) Application of hollow cathode excitation coupled to vidicon detection to the simultaneous multielement determination of toxic elements in airborne dust – A unique sampling-analysis procedure for lead and cadmium, *Bull Soc Chim Belg* 85: 755–761.

472 Dempster M. A. and Marcus R. K. (2000) Optimization of hollow cathode diameter for particle beam/hollow cathode glow discharge atomic emission spectrometry, *Spectrochim Acta, Part B* 55: 599–610.

473 Caroli S. (1985) *Improved hollow cathode lamps for atomic spectroscopy*, Ellis Horwood, Chichester.

474 Falk H., Hoffmann E. and Lüdke Ch. (1984) A comparison of furnace atomic nonthermal excitation spectrometry (FANES) with other atomic spectroscopic techniques, *Spectrochim Acta, Part B* 39: 283–294.

475 Grimm W. (1968) Eine neue Glimmentladungslampe für die optische Emissionsspektralanalyse, *Spectrochim Acta, Part B* 23: 443–454.

476 Ko J. B. (1984) New designs of glow discharge lamps for the analysis of metals by atomic emission spectrometry, *Spectrochim Acta, Part B* 39: 1405–1423.

477 Steers E. B. M. (1997) Charge transfer excitation in glow discharge sources: the spectra of titanium and copper with neon, argon and krypton as the plasma gas, *J Anal At Spectrom* 12: 1033–1040.

478 Steers E. B. M. and Leis F. (1991) Excitation of the spectra of neutral and singly ionized atoms in the Grimm-type discharge lamp with and without supplementary microwave excitation, *Spectrochim Acta, Part B* 46: 527–537.

479 Leis F., Broekaert J. A. C. and Laqua K. (1987) Design and properties of a microwave boosted glow discharge lamp, *Spectrochim Acta, Part B* 42: 1169–1176.

480 Ko J. B. and Laqua K. (1975) *Vergleich verschiedener Anregungsverfahren bei der emissionsspektrometrischen Analyse von Aluminium-Legierungen*, Proc XVIII Coll Spectrosc Intern, GAMS, Volume II, 549–556.

481 Bengtson A. (1985) A contribution to the solution of the problem of quantification in surface analysis work using glow discharge atomic emission spectroscopy, *Spectrochim Acta, Part B* 40: 631–639.

482 Hoffmann V., Kurt R., Kämmer K., Thielsch R., Wirth Th. and Beck U. (1999) Interference phenomena at transparent layers in glow discharge optical emission spectrometry, *Appl Spectrosc* 53: 987–990.

483 Winchester M. R., Lazik C. and Marcus R. K. (1991) Characterization of a radio frequency glow discharge

emission source, *Spectrochim Acta, Part B* 46: 483–499.

484 HOFFMANN V., UHLEMANN H. J., PRÄßLER F. and WETZIG K. **(1996)** New hardware for radio frequency powered glow discharge spectroscopies and its capabilities for analytical applications, *Fresenius' J Anal Chem* 355: 826–830.

485 BELKIN M., CARUSO J. A., CHRISTOPHER S. J. and MARCUS R. K. **(2000)** Characterization of helium/argon working gas systems in a radiofrequency glow discharge atomic emission source. Part II: Langmuir probe and emission intensity studies for Al, Cu and Macor samples, *Spectrochim Acta, Part B* 53: 1197–1208.

486 LOWE R. M. **(1978)** A modified glow-discharge source for emission spectroscopy, *Spectrochim Acta, Part B* 31: 257–261.

487 FERREIRA N. P., STRAUSS J. A. and HUMAN H. G. C. **(1983)** Developments in glow discharge emission spectrometry, *Spectrochim Acta, Part B* 38: 899–911.

488 BROEKAERT J. A. C., PEREIRO R., STARN T. K. and HIEFTJE G. M. **(1993)** A gas-sampling glow discharge coupled to hydride generation for the atomic spectrometric determination of arsenic, *Spectrochim Acta, Part B* 48: 1207–1220.

489 BROEKAERT J. A. C., STARN T. K., WRIGHT L. J. and HIEFTJE G. M. **(1998)** Studies of a helium-operated gas-sampling Grimm-type glow discharge for the atomic emission spectrometric determination of chlorine in halogenated hydrocarbon vapors, *Spectrochim Acta, Part B* 53: 1723–1735.

490 McLUCKEY S. A., GLISH K. G., ASANO K. G. and GRANT B. C. **(1988)** Atmospheric sampling glow discharge ionization source for the determination of trace organic compounds in ambient air, *Anal Chem* 60: 2220–2227.

491 PEREIRO R., STARN T. K. and HIEFTJE G. M. **(1995)** Gas-sampling glow discharge for optical emission spectrometry. Part II: Optimization

and evaluation for the determination of non-metals in gas-phase samples, *Appl Spectrosc* 49: 616–622.

492 MARGETIC V., PAKULEV A., STOCKHAUS A., BOLSHOV M., NIEMAX K. and HERGENRÖDER R. **(2000)** A comparison of nanosecond and femtosecond laser-induced plasma spectroscopy of brass samples, *Spectrochim Acta, Part B* 55: 1771–1785.

493 SUGDEN T. M. in: REED R. L. (ed) **(1965)** *Mass spectrometry*, Academic Press, New York, p. 347.

494 DOUGLAS D. J. and FRENCH J. B. **(1981)** Elemental analysis with a microwave-induced plasma quadrupole mass spectrometer system, *Anal Chem* 53: 37–41.

495 HOUK R. S., FASSEL V. A., FLESH G. D., SVEC H. J., GRAY A. L. and TAYLOR Ch. E. **(1980)** Inductively coupled argon plasma as an ion source for mass spectrometric determination of trace elements, *Anal Chem* 52: 2283–2289.

496 DATE A. R. and GRAY A. L. **(1989)** *Applications of ICP-MS*, Blackie, Glasgow.

497 JAKUBOWSKI N., RAEYMAEKERS B. J., BROEKAERT J. A. C. and STUEWER D. **(1989)** Study of plasma potential effects in a 40 MHz inductively coupled plasma mass spectrometry system, *Spectrochim Acta, Part B* 44: 219–228.

498 VAUGHAN M. A. and HORLICK G. **(1990)** Ion trajectories through the input ion optics of an ICP-MS, *Spectrochim Acta, Part B* 45: 1301–1311.

499 JAKUBOWSKI N., MOENS L. and VANHAECKE F. **(1998)** Sector field mass spectrometers in ICP-MS, *Spectrochim Acta, Part B* 53: 1739–1763.

500 SOLYOM D. A., BURGOYNE T. W. and HIEFTJE G. M. **(1999)** Plasma source sector field mass spectrometry with array detection, *J Anal At Spectrom* 14: 1101–1110.

501 BROEKAERT J. A. C. **(1990)** *ICP Massenspektrometrie*, in: Günzler H., Borsdorf R., Fresenius W., Huber W., Kelker G., Lüdenwald I., Tölg G. und

Wisser H. (eds) Analytiker Taschenbuch, Band 9. Springer-Verlag, Berlin, 127–163.

502 GRAY A. L., HOUK R. S. and WILLIAMS J. G. **(1987)** Langmuir probe potential measurements in the plasma and their correlations with mass spectral characteristics in ICP-MS, *J Anal At Spectrom* 2: 13–20.

503 VAUGHAN M. A. and HORLICK G. **(1986)** Oxide, hydroxide and doubly charged analyte species in ICP mass spectrometry, *Appl Spectrosc* 40: 434–445.

504 TITTES W., Jakubowski N., STUEWER D., TOELG G. and BROEKAERT J. A. C. **(1994)** Reduction of some selected spectral interferences in inductively coupled plasma mass spectrometry, *J Anal At Spectrom* 9: 1015–1020.

505 VAUGHAN M. A., HORLICK G. and TAN S. H. **(1987)** Effect of the operating parameters on analyte signals in inductively coupled plasma mass spectrometry, *J Anal At Spectrom* 2: 765–772.

506 DATE A. R. and GRAY A. L. **(1985)** Determination of trace elements in geological samples by inductively coupled plasma source mass spectrometry, *Spectrochim Acta, Part B* 40: 115–122.

507 BROEKAERT J. A. C., BRANDT R., LEIS F., PILGER C., POLLMANN D., TSCHOEPEL P. and TOELG G. **(1994)** Analysis of aluminium oxide and silicon carbide ceramic materials by inductively coupled plasma mass spectrometry, *J Anal At Spectrom* 9: 1063–1070.

508 TIAN X., EMTEBORG H. and ADAMS F. C. **(1999)** Analytical performance of axial inductively coupled plasma time of flight mass spectrometry (ICP-TOFMS), *J Anal At Spectrom* 14: 1807–1814.

509 VANDECASTEELE C., NAGELS M., VANHOE H. and DAMS R. **(1988)** Suppression of analyte signal in ICP-MS and the use of an internal standard, *Anal Chim Acta* 211: 91–98.

510 CRAIN J. S., HOUK R. S. and SMITH F. G. **(1988)** Matrix interferences in ICP MS: some effects of skimmer orifice

diameter and ion lense voltages, *Spectrochim Acta, Part B* 43: 1355–1364.

511 FRASER M. M. and BEAUCHEMIN D. **(2000)** Effect of concomitant elements on the distribution of ions in inductively coupled plasma-mass spectroscopy. Part 1. Elemental ions, *Spectrochim Acta, Part B* 55: 1705–1731.

512 TAN S. H. and HORLICK G. **(1986)** Background spectral features in ICP mass spectrometry, *Appl Spectrosc* 40: 445–460.

513 DOUGLAS D. J. and FRENCH J. B. **(1992)** Collisional focusing effects in radio frequency quadrupoles, *J Am Soc Mass Spectrom* 3: 398–408.

514 BARINAGA C. J., EIDEN G. C., ALEXANDER M. L. and KOPPENAAL D. W. **(1996)** Analytical atomic spectroscopy using ion trap devices, *Fresenius' J Anal Chem* 355: 487–493.

515 FELDMANN I., JAKUBOWSKI N. and STUEWER D. **(1999)** Application of a hexapole collision and reaction cell in ICP-MS. Part I: Instrumental aspects and operational optimization, *Fresenius' J Anal Chem* 365: 415–421.

516 MASON P. R. D., KASPERS K. and VAN BERGEN M. J. **(1999)** Determination of sulfur isotope ratios and concentrations in water samples using ICP-MS incorporating hexapole ion optics, *J Anal At Spectrom* 14: 1067–1074.

517 RIEPE H. G., GOMEZ M., CAMARA C. and BETTMER J. **(2000)** Modification of capillaries coupled to micro-flow nebulizers: a new strategy for on-line interference removal in inductively coupled plasma mass spectrometry, *J Mass Spectrom* 35: 891–896.

518 HEUMANN K. G. **(1982)** Isotope dilution mass spectrometry for micro- and trace element determination, *Trends Anal Chem* 1: 357–361.

519 DEAN J. R., EBDON L. and MASSEY R. J. **(1987)** Selection of mode for the measurement of lead isotope ratios by ICP mass spectrometry and its application to milk powder analysis, *J Anal At Spectrom* 2: 369–374.

520 TING B. T. G. and JANGHORBANI M. **(1987)** Application of ICP-MS to

accurate isotopic analysis for human metabolic studies, *Spectrochim Acta, Part B* 42: 21–27.

521 VANHAECKE F., MOENS L., DAMS R., ALLEN L. and GEORGITIS G. **(1999)** Evaluation of the isotope ratio performance of an axial time-of-flight ICP mass spectrometer, *Anal Chem* 71: 3297–3303.

522 VANHAECKE F., MOENS L. and DAMS R. **(1997)** Applicability of high-resolution ICP-mass spectrometry for isotope ratio measurements, *Anal Chem* 69: 268–273.

523 BECKER J. S. and DIETZE H. J. **(1998)** Ultratrace and precise isotope analysis by double-focusing sector field inductively coupled plasma mass spectrometry, *J Anal At Spectrom* 13: 1057–1063.

524 HIRATA T. and YAMAGUCHI T. **(1999)** Isotopic analysis of zirconium using enhanced sensitivity-laser ablation-multiple collector-inductively coupled plasma mass spectrometry, *J Anal At Spectrom* 14: 1455–1459.

525 HAUSLER D. **(1987)** Trace element analysis of organic solutions using ICP mass spectrometry, *Spectrochim Acta, Part B* 42: 63–73.

526 ZHU G. and BROWNER R. F. **(1988)** Study of the influence of water vapor loading and interface pressure in ICP-MS, *J Anal At Spectrom* 3: 781–790.

527 JAKUBOWSKI N., FELDMANN I., STÜWER D. and BERNDT H. **(1992)** Hydraulic high pressure nebulization – application of a new nebulization system for inductively coupled plasma mass spectrometry, *Spectrochim Acta, Part B* 47: 119–129.

528 WANG X., VICZIAN M., LASZTITY A. and BARNES R. M. **(1988)** Lead hydride generation for isotope analysis by ICP-MS, *J Anal At Spectrom* 3: 821–828.

529 PARK C. J., VAN LOON J. C., ARROWSMITH P. and FRENCH J. B. **(1987)** Sample analysis using plasma source mass spectrometry with electrothermal sample introduction, *Anal Chem* 59: 2191–2196.

530 BOOMER D. W., POWELL M., SING R. L. A. and SALIN E. D. **(1986)** Application of a wire loop direct sample-insertion device for ICP-MS, *Anal Chem* 58: 975–976.

531 HALL G. E. M., PELCHAT J. C., BOOMER D. W. and POWELL M. **(1988)** Relative merits of two methods of sample introduction in ICP-MS: electrothermal vaporisation and direct sample insertion, *J Anal At Spectrom* 3: 791–797.

532 WILLIAMS J. G., GRAY A. L., NORMAN P. and EBDON L. **(1987)** Feasibility of solid sample introduction by slurry nebulisation for ICP-MS, *J Anal At Spectrom* 2: 469–472.

533 FONSECA R. W., MILLER-IHLI N. J., SPARKS C., HOLCOMBE J. A. and SHAVER B. **(1997)** Effect of oxygen ashing on analyte transport efficiency using ETV-ICP-MS, *Appl Spectrosc* 51: 1800–1806.

534 HAUPTKORN S., KRIVAN V., GERCKEN B. and PAVEL J. **(1997)** Determination of trace impurities in high-purity quartz by electrothermal vaporization inductively coupled plasma mass spectrometry using the slurry sampling technique, *J Anal At Spectrom* 12: 421–428.

535 WENDE M. C. and BROEKAERT J. A. C. **(2001)** Investigations on the use of chemical modifiers for the direct determination of trace impurities in Al_2O_3 ceramic powders by slurry electrothermal evaporation coupled to inductively coupled plasma mass spectrometry (ETV-ICP-MS), *Fresenius' J Anal Chem*, 370: 513–520.

536 JAKUBOWSKI N., FELDMANN I., SACK B. and STÜWER D. **(1992)** Analysis of conducting solids by ICP-MS with spark ablation, *J Anal At Spectrom* 7: 121–125.

537 ARROWSMITH P. **(1987)** Laser ablation of solids elemental analysis by ICP mass spectrometry, *Anal Chem* 59: 1437–1444.

538 GUENTHER D. and HEINRICH C. A. **(1999)** Comparison of the ablation behaviour of 266 nm Nd:YAG and 193 nm ArF excimer lasers for LA-ICP-MS analyses, *J Anal At Spectrom* 14: 1369–1474.

539 Borisov O. V., Mao X. L., Fernandez A., Caetano M. and Russo R. E. **(1999)** Inductively coupled plasma mass spectrometric study of non-linear calibration behavior during laser ablation of binary Cu–Zn alloys, *Spectrochim Acta, Part B* 54: 1351–1365.

540 Date A. R. and Hutchinson D. **(1987)** Determination of rare earth elements in geological samples by ICP source mass spectrometry, *J Anal At Spectrom* 2: 269–281.

541 Garbarino J. R. and Taylor H. E. **(1987)** Stable isotope dilution analysis of hydrologic samples by ICP-MS, *Anal Chem* 59: 1568–1575.

542 Park C. J. and Hall G. E. M. **(1988)** Analysis of geological materials by ICP-MS with sample introduction by electrothermal vaporisation, *J Anal At Spectrom* 3: 355–361.

543 Gregoire D. C. **(1988)** Determination of platinum, ruthenium and iridium in geological materials by ICP-MS with sample introduction by electrothermal vaporization, *J Anal At Spectrom* 3: 309–314.

544 Audetat A., Guenther D. and Heinrich Ch. A. **(1998)** Formation of a magmatic-hydrothermal ore deposit: insight with LA-ICP-MS analysis of liquid inclusions, *Science* 279: 2091–2094.

545 McLeod C. W., Date A. R. and Cheung Y. Y. **(1986)** Metal ions in ICP-mass spectrometric analysis of nickel-base alloys, *Spectrochim Acta, Part B* 41: 169–174.

546 Pilger C., Leis F., Tschöpel P., Broekaert J. A. C. and Toelg G. **(1995)** Analysis of silicon carbide powders with ICP-MS subsequent to sample dissolution without and with matrix, *Fresenius' J Anal Chem* 351: 110–116.

547 Jiang S. J. and Houk R. S. **(1986)** Arc nebulisation for elemental analysis of conducting solids by ICP mass spectrometry, *Anal Chem* 58: 1739–1743.

548 Arrowsmith P. and Hughes S. K. **(1988)** Entrainment and transport of laser ablated plumes for subsequent elemental analysis, *Appl Spectrosc* 42: 1231–1239.

549 Arrowsmith P. **(1988)** in: Barnes R. M. (ed) *Abstracts of the 1988 Winter Conference on Plasma Spectrochemistry, San Diego*, The University of Massachusetts, p. 54.

550 Lyon T. D. B., Fell G. S., Hutton R. C. and Eaton A. N. **(1988)** Evaluation of ICAP MS for simultaneous multi-element trace analysis in clinical chemistry, *J Anal At Spectrom* 3: 265–272.

551 Delves H. T. and Campbell M. J. **(1988)** Measurements of total lead concentrations and of lead isotope ratios in whole blood by use of ICP source MS, *J Anal At Spectrom* 3: 343–348.

552 Serfass R. E., Thompson J. J. and Houk R. S. **(1986)** Isotope ratio determination by ICP MS for zinc bioavailability studies, *Anal Chim Acta* 188: 73–84.

553 Dean J. R., Ebdon L., Crews H. M. and Massey R. C. **(1988)** Characteristics of flow injection ICP MS for trace metal analysis, *J Anal At Spectrom* 3: 349–354.

554 Dean J. R., Munro S., Ebdon L., Crews H. M. and Massey R. C. **(1987)** Studies of metalloprotein species by directly coupled high-performance liquid chromatography ICP-MS, *J Anal At Spectrom* 2: 607–610.

556 Baker S. A. and Miller-Ihli N. J. **(1999)** Comparison of a cross-flow and microconcentric nebulizer for chemical speciation measurements using CZE-ICP-MS, *Appl Spectrosc* 53: 471–478.

557 Van Holderbeke M., Zhao Y., Vanhaecke F., Moens L., Dams R. and Sandra P. **(1999)** Speciation of six arsenic compounds using capillary electrophoresis–inductively coupled plasma mass spectrometry, *J Anal At Spectrom* 14: 229–234.

558 McLean J. A., Zhang H. and Montaser A. **(1998)** A direct injection high-efficiency nebulizer for inductively coupled plasma mass spectrometry, *Anal Chem* 70: 1012–1020.

559 HERZOG R. and DIETZ F. **(1987)** Anwendung der ICP MS in der Wasseranalytik, *Gewässerschutz, Wasser, Abwasser*, 92: 109–141.

560 BEAUCHEMIN D., MCLAREN J. W., MYKYTIUK A. P. and BERMAN S. S. **(1988)** Determination of trace metals in an open ocean water reference material by ICP-MS, *J Anal At Spectrom* 3: 305–308.

561 MCLAREN J. W., BEAUCHEMIN D. and BERMAN S. S. **(1987)** Application of isotope dilution ICP mass spectrometry to the analysis of marine sediments, *Anal Chem* 59: 610–613.

562 MCLAREN J. W., BEAUCHEMIN D. and BERMAN S. S. **(1987)** Determination of trace metals in marine sediments by ICP mass spectrometry, *J Anal At Spectrom* 2: 277–281.

563 RIDOUT P. S., JONES H. R. and WILLIAMS J. G. **(1988)** Determination of trace elements in a marine reference material of lobster hepatopancreas (TORT-1) using ICP-MS, *Analyst* 113: 1383–1386.

564 BEAUCHEMIN D., MCLAREN J. W., WILLIE S. N. and BERMAN S. S. **(1988)** Determination of trace metals in marine biological reference materials by ICP MS, *Anal Chem* 60: 687–691.

565 BARBANTE C., COZZI G., CAPODAGLIO G., VAN DE VELDE K., FERRARI C., VEYSSEYSE A., BOUTRON C. F., SCARPONI G. and CESCON P. **(1999)** Determination of Rh, Pd, and Pt in polar and alpine snow and ice by double-focusing ICP-MS with microconcentric nebulization, *Anal Chem* 71: 4125–4133.

566 BECKER J. S., DIETZE H. J., MCLEAN J. A. and MONTASER A. **(1999)** Ultratrace and isotope analysis of long-lived radionuclides by inductively coupled plasma quadrupole mass spectrometry using a direct injection high-efficiency nebulizer, *Anal Chem* 71: 3077–3084.

567 ANDRLE C. M., JAKUBOWSKI N. and BROEKAERT J. A. C. **(1997)** Speciation of chromium using reversed phase-high performance liquid chromatography coupled to different spectro-metric detection methods, *Spectrochim Acta, Part B* 52: 189–200.

568 BARNOWSKI C., JAKUBOWSKI N., STUEWER D. and BROEKAERT J. A. C. **(1997)** Speciation of chromium by direct coupling of ion exchange chromatography with inductively coupled plasma mass spectrometry, *J Anal At Spectrom* 12: 1155–1161.

569 PRANGE A. and JANTZEN E. **(1995)** Determination of organometallic species by gas chromatography ICP MS, *J Anal At Spectrom* 10: 105–109.

570 HEISTERKAMP M., DE SMAELE T., CANDELONE J. P., MOENS L., DAMS R. and ADAMS F. C. **(1997)** Inductively coupled plasma mass spectrometry hyphenated to capillary gas chromatography as a detection system for the speciation analysis of environmental samples, *J Anal At Spectrom* 12: 1077–1081.

571 RODRIGUEZ J., MOUNICON S., LOBINSKI R., SIDELNIKOV V., PATRUSHEV Y. and YAMANAKU M. **(1999)** Species-selective analysis by microcolumn multicapillary gas chromatography with inductively coupled plasma mass spectrometric detection, *Anal Chem* 71: 4534–4543.

572 LEACH A. M., HEISTERKAMP M., ADAMS F. C. and HIEFTJE G. M. **(2000)** Gas chromatography-inductively coupled plasma time-of-flight mass spectrometry for the speciation analysis of organometallic compounds, *J Anal At Spectrom* 15: 151–155.

573 MICHAUD-POUSSEL E. and MERMET J. M. **(1986)** Influence of the torch frequency and the plasma-gas inlet area on torch design in ICP AES, *Spectrochim Acta, Part B* 41: 125–132.

574 LAM J. W. H. and HORLICK G. **(1990)** A comparison of argon and mixed gas plasmas for ICP MS, *Spectrochim Acta, Part B* 45: 1313–1325.

575 CHAN S. K. and MONTASER A. **(1987)** Characterization of an annular helium ICP generated in a low-gas-flow torch, *Spectrochim Acta, Part B* 42: 591–597.

576 NAM S. H., MASAMBA W. R. L. and MONTASER A. **(1993)** Investigation of helium inductively coupled plasma-mass spectrometry of metals and nonmetals in aqueous solutions, *Anal Chem* 65: 2784–2790.

577 SATZGER R. D., FRICKE F. L., BROWN P. G. and CARUSO J. A. **(1987)** Detection of halogens as positive ions using a He microwave induced plasma as an ion source for mass spectrometry, *Spectrochim Acta, Part B* 42: 705–712.

578 WILSON D. A., VICKERS G. H. and HIEFTJE G. M. **(1987)** Use of microwave-induced nitrogen discharge at atmospheric pressure as an ion source for elemental mass spectrometry, *Anal Chem* 59: 1664–1670.

579 OKAMOTO Y. **(1994)** High-sensitivity microwave-induced plasma mass spectrometry for trace element analysis, *J Anal At Spectrom* 9: 745–749.

580 BLADES M. W. **(1994)** Atmospheric-pressure, radiofrequency, capacitively coupled helium plasmas, *Spectrochim Acta, Part B* 49: 47–57.

581 STEWART I. I., GUEVREMONT R. and STURGEON R. E. **(1999)** The use of a sampler–skimmer interface for ion sampling in furnace atomization plasma ionization mass spectrometry, *Anal Chem* 71: 5146–5156.

582 BINGS N. H., COSTA-FERNANDEZ J. M., GUZOWSKI Jr. J. P., LEACH A. M. and HIEFTJE G. M. **(2000)** Time-of-flight mass spectrometry as a tool for speciation analysis, *Spectrochim Acta, Part B* 55: 767–778.

583 MILGRAM K. E., WHITE F. M., WHITE M., GOODNER K. L., WATSON C. H., KOPPENAAL D. W., BARINAGA C. W., SMITH B. H., WINFORDNER J. G., MARSHALL A. G., HOUK R. S. and EYLER J. R. **(1997)** High-resolution inductively coupled plasma Fourier transform ion cyclotron resonance mass spectrometry, *Anal Chem* 69: 3714–3721.

584 HARRISON W. W. and BENTZ B. L. **(1988)** Glow discharge mass spectrometry, *Prog Anal Spectrosc* 11: 53–110.

585 RATLIFF P. H. and HARRISON W. W. **(1994)** The effects of water vapour in glow discharge mass spectrometry, *Spectrochim Acta, Part B* 49: 1747–1757.

586 RATLIFF P. H. and HARRISON W. W. **(1995)** Time-resolved studies of the effects of water vapor in glow discharge mass spectrometry, *Appl Spectrosc* 49: 863–871.

587 VIETH W. and HUNEKE J. C. **(1990)** Studies on ion formation in a glow discharge mass spectrometry ion source, *Spectrochim Acta, Part B* 45: 941–949.

588 TAYLOR W. S., DULAK J. G. and KETKAR S. N. **(1990)** Characterization of a glow discharge plasma as a function of sampling orifice potential, *J Am Soc Mass Spectrom* 1: 448.

589 GOODNER K. L., EYLER J. R., BARSHICK C. M. and SMITH D. H. **(1995)** Elemental quantification based on argides, dimers, and doubly charged glow discharge ions, *Int J Mass Spectrom Ion Processes* 146/147: 65–73.

590 DOUGLAS Y. M., DUCKWORTH C., CABLE P. R. and MARCUS R. K. **(1994)** Effects of target gas in collision-induced dissociation using a double quadrupole mass spectrometer and radiofrequency glow discharge, *J Am Soc Mass Spectrom* 5: 845–851.

591 MOLLE C., WAUTELET D., DAUCHOT J. P. and HECQ M. **(1995)** Characterization of a magnetron radiofrequency glow discharge with a glass cathode using experimental design and mass spectrometry, *J Anal At Spectrom* 10: 1039–1045.

592 JAKUBOWSKI N., STÜWER D. and TÖLG G. **(1986)** Improvement of ion source performance in glow discharge mass spectrometry, *Int J Mass Spectrom Ion Processes* 71: 183–197.

593 MILTON D. M. P., HUTTON R. C. and RONAN G. A. **(1992)** Optimization of discharge parameters for the analysis of high-purity silicon wafers by magnetic sector glow discharge mass

spectrometry, *Fresenius' J Anal Chem* 343: 773.

594 SHAO Y. and HORLICK G. **(1991)** Design and characterization of glow discharge devices as complementary sources for an "ICP" mass spectrometer, *Spectrochim Acta, Part B* 46: 165–174.

595 KLINGLER J. A. and HARRISON W. W. **(1991)** Glow discharge mass spectrometry using pulsed ual cathodes, *Anal Chem* 63: 2982–2984.

596 MARCUS R. K., HARVILLE T. R., MEI Y. and SHICK Jr. C. R. **(1994)** R.f. powered glow discharges elemental analysis across the solids spectrum, *Anal Chem* 66: 902A–911A.

597 SAPRYKIN A. I., MELCHERS F. G., BECKER J. S. and DIETZE H. J. **(1995)** Radio-frequency glow discharge ion source for high resolution mass spectrometry, *Fresenius' J Anal Chem* 353: 570–574.

598 MARCUS R. K. **(1994)** Radiofrequency powered glow discharges for emission and mass spectrometry: operating characteristics, figures of merit and future prospects, *J Anal At Spectrom* 9: 1029–1037.

599 MYERS D. P., HEINTZ M. J., MAHONEY P. P., LI G. and HIEFTJE G. M. **(1994)** Characterization of a radio-frequency glow discharge/time-of-flight mass spectrometer, *Appl Spectrosc* 48: 1337–1346.

600 HANG W., WALDEN W. O. and HARRISON W. W. **(1996)** Microsecond pulsed glow discharge as an analytical spectroscopic source, *Anal Chem* 68: 1148–1152.

601 HEINTZ M. J. and HIEFTJE G. M. **(1995)** Effect of driving frequency on the operation of a radiofrequency glow discharge emission source, *Spectrochim Acta, Part B* 50: 1125–1141.

602 DENG R. C. and WILLIAMS P. **(1994)** Suppression of cluster ion interferences in glow discharge mass spectrometry by sampling high-energy ions from a reversed hollow cathode ion source, *Anal Chem* 66: 1890–1896.

603 YOU J., FANNING J. C. and MARCUS R. K. **(1994)** Particle beam aqueous sample introduction for hollow cathode atomic emission spectroscopy, *Anal Chem* 66: 3916–3924.

604 YOU J., DEMPSTER M. A. and MARCUS R. K. **(1997)** Studies of analyte particle transport in a particle beam-hollow cathode atomic emission spectrometry system, *J Anal At Spectrom* 12: 807–815.

605 GUZOWSKI J. P. Jr., BROEKAERT J. A. C. and HIEFTJE G. M. **(2000)** Electrothermal vaporization for sample introduction into a gas sampling glow discharge time-of-flight mass spectrometer, *Spectrochim Acta, Part B* 55: 1295–1314.

606 CREED J. T., DAVIDSON T. M., SHEN W. and CARUSO J. A. **(1990)** Low-pressure helium microwave-induced plasma mass spectrometry for the detection of halogenated gas chromatographic effluents, *J Anal At Spectrom* 5: 109–113.

607 EIJKEL J. C. T., STOERI H. and MANZ A. **(1999)** A molecular emission detector on a chip employing a direct current microplasma, *Anal Chem* 71: 2600–2606.

608 MICLEA M., KUNZE K., MUSA G., FRANZKE J. and NIEMAX K. **(2001)** The dielectric barrier discharge – a powerful microchip plasma for diode laser spectrometry, *Spectrochim Acta, Part B* 56: 37–43.

609 VIETH W. and HUNEKE J. C. **(1991)** Relative sensitivity factors in glow discharge mass spectrometry, *Spectrochim Acta, Part B* 46: 137–154.

610 CHARALAMBOUS P. M. **(1987)** The application of a glow discharge mass spectrometer in the steel industry, *Steel Res* 5/87: 197–203.

611 JAKUBOWSKI N. and STÜWER D. **(1988)** Comparison of Ar and Ne as working gases in analytical glow discharge mass spectrometry, *Fresenius' Z Anal Chem* 335: 680–686.

612 CAROLI S., SENOFONTE O. and MODESTI G. **(1999)** Glow discharge atomic spectrometry, *Adv At Spectrom* 5: 173–234.

613 Jakubowski N. and Stuewer D. (1992) Application of glow discharge mass spectrometry with low mass resolution for in-depth analysis of technical layers, *J Anal At Spectrom* 7: 951–958.

614 Smithwick R. W., Lynch D. W. and Franklin J. C. (1993) Relative ion yields measured with a high-resolution glow discharge mass spectrometer operated with an argon/hydrogen mixture, *J Am Soc Mass Spectrom* 4: 178–285.

615 Takahashi T. and Shimamura T. (1994) Determination of zirconium, niobium and molybdenum in steel by glow discharge mass spectrometry, *Anal Chem* 66: 3274–3280.

616 Jakubowski N., Stuewer D. and Vieth W. (1987) Performance of a glow discharge mass spectrometer for simultaneous multielement analysis of steel, *Anal Chem* 59: 1825–1830.

617 *Technical Information GD012*, VG Instruments, Manchester, U.K.

618 *Application Note 02.681*, VG Instruments, Manchester, U.K.

619 Mykytiuk A. P., Semeniuk P. and Berman S. (1990) Analysis of high purity metal and semiconductor materials by glow discharge mass spectrometry, *Spectrochim Acta Rev* 13: 1–10.

620 Oksenoid K. G., Liebich V. and Pietsch G. (1996) Improvement of analyte to interferent line intensity ratio in direct current glow discharge mass spectrometry, *Fresenius' Z Anal Chem* 355: 863–865.

621 Wilhartitz P., Ortner H. M., Krismer R. and Krabichler D. (1990) Classical analysis involving trace-matrix separation versus solid state mass spectrometry: a comparative study for the analysis of high-purity Mo, W and Cr, *Mikrochim Acta* 2: 259–271.

622 Vassamillet L. F. (1989) Use of glow discharge mass spectrometers for quality-control of high-purity aluminium, *J Anal At Spectrom* 4: 451–455.

623 Feng X. and Horlick G. (1994) Analysis of aluminium alloys using inductively coupled plasma and glow discharge mass spectrometry, *J Anal At Spectrom* 9: 823–831.

624 Venczago C. and Weigert M. (1994) Application of glow discharge mass spectrometry (GDMS) for the multielement trace and ultratrace analysis of sputtering targets, *Fresenius' J Anal Chem* 350: 303–309.

625 Saito M., Hirose F. and Okochi H. (1995) Comparative study of the determination of micro-amounts of boron in high-purity molybdenum by a spectrophotometric method using curcumin, spark source mass spectrometry and glow discharge mass spectrometry, *Anal Sci* 11: 695–697.

626 *Technical Information GD701*, VG Instruments, Manchester, U.K.

627 Tigwell M., Clark J., Shuttleworth S. and Bottomley M. (1992) Analysis and characterization of rare earths by plasma source mass spectrometry, *Mater Chem Phys* 31: 23–27.

628 Ecker K. H. and Pritzkow W. (1994) Measurement of isotope ratios in solids by glow discharge mass spectrometry, *Fresenius' J Anal Chem* 349: 207–208.

629 Donohue D. L. and Petek M. (1991) Isotopic measurements of palladium metal containing protium and deuterium by glow discharge mass spectrometry, *Anal Chem* 63: 740–744.

630 Van Straeten M., Swenters K., Gijberls R., Verlinden J. and Adriaenssens E. (1994) Analysis of platinum powder by glow discharge mass spectrometry, *J Anal At Spectrom* 9: 1389–1397.

631 Held A., Taylor P., Ingelbrecht C., De Bièvre P., Broekaert J. A. C., Van Straeten M. and Gijbels R. (1995) Determination of the scandium content of high-purity titanium using inductively coupled plasma mass spectrometry and glow discharge mass spectrometry as part of its certification as a reference material, *J Anal At Spectrom* 10: 849–852.

632 Shi Z., Brewer S. and Sachs R. (1995) Application of a magnetron glow discharge to direct solid sampling for mass spectrometry, *Appl Spectrosc* 49: 1232–1238.

633 HUTTON R. C. and RAITH A. **(1992)** Analysis of pure metals using a quadrupole-based glow discharge mass spectrometer, *J Anal At Spectrom* 7: 623–627.

634 VG Instruments, Technical Information GD 701, Manchester, U.K.

635 DE GENDT S., SCHELLES W., VAN GRIEKEN R. and MÜLLER V. **(1995)** Quantitative analysis of iron-rich and other oxide-based samples by means of glow discharge mass spectrometry, *J Anal At Spectrom* 10: 681–687.

636 SCHELLES W., DE GENDT S., MULLER V. and VAN GRIEKEN R. **(1995)** Evaluation of secondary cathodes for glow discharge mass spectrometry analysis of different nonconducting sample types, *Appl Spectrosc* 49: 939–944.

637 SHIMAMURA T., TAKAHASHI T., HONDA M. and NAGAI H. **(1993)** Multi-element and isotopic analyses of iron meteorites using a glow discharge mass spectrometer, *J Anal At Spectrom* 8: 453–460.

638 TENG J., BARSHICK C. M., DUCKWORTH D. C., MORTON S. J., SMITH D. H. and KING F. L. **(1995)** Factors influencing the quantitative determination of trace elements in soils by glow discharge mass spectrometry, *Appl Spectrosc* 49: 1361–1366.

639 BETTI M., GIAMARELLI S., HIERNAUT T., RASMUSSEN G. and KOCH L. **(1996)** Detection of trace radioisotopes in soil, sediment and vegetation by glow discharge mass spectrometry, *Fresenius' J Anal Chem* 355: 642–646.

640 BARSHICK **(1996)** *Rapid Commun Mass Spectrom* 10: 341.

641 TAKAHASHI T. and SHIMAMURA T. **(1994)** Determination of zirconium, niobium and molybdenum in steel by glow discharge mass spectrometry, *Anal Chem* 66: 3274–3280.

642 SCHELLES W. and VAN GRIEKEN R. E. **(1996)** Direct current glow discharge mass spectrometric analysis of macor ceramic using a secondary cathode, *Anal Chem* 68: 3570–3574.

643 LUO F. C. H. and HUNEKE J. C. **(1991)** *Proc Intern Conf on elemental analysis of coal and its by-products*, Barren River Resort, KY, September 9–11, p. 17.

644 WOO J. C., JAKUBOWSKI N. and STUWER D. **(1993)** Analysis of aluminium oxide powder by glow discharge mass spectrometry with low mass resolution, *J Anal At Spectrom* 8: 881–889.

645 DE GENDT S., VAN GRIEKEN R., HANG W. and HARRISON W. W. **(1995)** Comparison between direct current and radiofrequency glow discharge mass spectrometry for the analysis of oxide-based samples, *J Anal At Spectrom* 10: 689–695.

646 SAPRYKIN A. I., BECKER J. S. and DIETZE H. J. **(1996)** Optimization of an rf-powered magnetron glow discharge for the trace analysis of glasses and ceramics, *Fresenius' J Anal Chem* 355: 831–835.

647 SCHELLES W., DE GENDT S., MÜLLER V. and VAN GRIEKEN R. **(1995)** Evaluation of secondary cathodes for glow discharge mass spectrometry analysis of different nonconducting sample types, *Appl Spectrosc* 49: 939–944.

648 SHICK C. R. Jr., DE PALMA P. A. Jr. and MARCUS R. K. **(1996)** Radiofrequency glow discharge mass spectrometry for the characterization of bulk polymers, *Anal Chem* 68: 2113–2121.

649 BARSHICK C. M. and HARRISON W. W. **(1989)** The laser as an analytical probe in glow discharge mass spectrometry, *Mikrochim Acta* 3: 169–177.

650 MEI Y. and HARRISON W. W. **(1991)** Getters and plasma reagents in glow discharge mass spectrometry, *Spectrochim Acta, Part B* 46: 175–182.

651 HECQ M., HECQ A. and FONTIGNIES M. **(1983)** Analysis of thin metallic films by glow discharge mass spectrometry, *Anal Chim Acta* 155: 191–198.

652 BEHN U., GERBIG F. A. and ALBRECHT H. **(1994)** Depth profiling of frictional brass coated steel samples by glow discharge mass spectrometry, *Fresenius' J Anal Chem* 349: 209–210.

653 Mattson W. A., Bentz B. L. and Harrison W. W. **(1976)** Coaxial cathode ion source for solids mass spectrometry, *Anal Chem* 48: 489–491.

654 Foss G. O., Svec H. J. and Conzemius R. J. **(1983)** The determination of trace elements in aqueous media without preconcentration using a cryogenic hollow-cathode ion source, *Anal Chim Acta* 147: 151–162.

655 Jakubowski N., Stuewer D. and Toelg G. **(1991)** Microchemical determination of platinum and iridium by glow discharge mass spectrometry, *Spectrochim Acta, Part B* 46: 155–163.

656 Barshick C. M., Duckworth D. C. and Smith D. H. **(1993)** Analysis of solution residues by glow discharge mass spectrometry, *J Am Soc Mass Spectrom* 4: 47–53.

657 Evetts I., Milton D. and Mason R. **(1991)** Trace element analysis in body fluids by glow discharge mass spectrometry: a study of lead mobilization by the drug cis-platin, *Biol Mass Spectrom* 20: 153–159.

658 Barshick C. M., Smith D. H., Wade J. W. and Bayne C. K. **(1994)** Isotope dilution glow discharge mass spectrometry as applied to the analysis of lead in waste oil samples, *J Anal At Spectrom* 9: 83–87.

659 Barshick C. M., Smith D. H., Hackney J. H., Cole B. A. and Wade J. W. **(1994)** Glow discharge mass spectrometric analysis of trace metals in petroleum, *Anal Chem* 66: 730–734.

660 Wetzel W., Broekaert J. A. C. and Hieftje G. M., Determination of arsenic by hydride generation coupled to time-of-flight mass spectrometry with a gas sampling glow discharge, spectrochimica Acta, Part B, submitted for publication.

661 Olson L. K., Belkin M. and Caruso J. A. **(1996)** Radiofrequency glow discharge mass spectrometry for gas chromatographic detection: a new departure for elemental speciation studies, *J Anal At Spectrom* 11: 491–496.

662 Omenetto N. and Winefordner J. D. **(1979)** Atomic fluorescence spectrometry: basic principles and applications, *Prog Anal At Spectrom* 2: 1–183.

663 Omenetto N., Human H. G. C., Cavalli P. and Rossi G. **(1984)** Laser excited atomic and ionic non-resonance fluorescence detection limits for several elements in an argon inductively coupled plasma, *Spectrochim Acta, Part B* 39: 115–117.

664 Omenetto N., Smith B. W., Hart L. P., Cavalli P. and Rossi G. **(1985)** Laser induced double-resonance fluorescence in an inductively coupled plasma, *Spectrochim Acta, Part B* 40: 1411–1422.

665 Bolshov M. A., Zybin A. V. and Smirenkina I. I. **(1981)** Atomic fluorescence spectrometry with laser sources, *Spectrochim Acta, Part B* 36: 1143–1152.

666 Smith B. W., Farnsworth P. B., Cavalli P. and Omenetto N. **(1990)** Optimization of laser excited atomic fluorescence in a graphite furnace for the determination of thallium, *Spectrochim Acta, Part B* 45: 1369–1373.

667 Grazhulene S., Khvostikov V. and Sorokin M. **(1991)** The possibilities of glow discharge cathode sputtering for laser atomic-fluorescence analysis of microelectronics materials, *Spectrochim Acta, Part B* 46: 459–465.

668 Bolshov M. A., Rudniev S. N., Rudnieva A. A., Boutron C. F. and Hong S. **(1997)** Determination of heavy metals in polar snow and ice by laser excited atomic fluorescence spectrometry with electrothermal atomization in a graphite cup, *Spectrochim Acta, Part B* 52: 1535–1544.

669 Sdorra W. and Niemax K. **(1990)** Temporal and spatial distribution of analyte atoms and ions in micro-plasmas produced by laser ablation of solid samples, *Spectrochim Acta, Part B* 45: 917–926.

670 Turk G. C., Travis J. C., DeVoe J. R. and O'Haver T. C. **(1979)** Laser enhanced ionization spectrometry in analytical flames, *Anal Chem* 51: 1890–1896.

671 HAVRILLA G. J., WEEKS S. J. and TRAVIS J. C. **(1982)** Continuous wave excitation in laser enhanced ionization spectrometry, *Anal Chem* 54: 2566–2570.

672 NIEMAX K. **(1985)** Spectroscopy using thermionic diode detectors, *Appl Phys, Part B* 38: 147–157.

673 SMITH B. W., QUENTMEIER A., BOLSHOV M. and NIEMAX K. **(1999)** Measurement of uranium isotope ratios in solid samples using laser ablation and diode laser-excited atomic fluorescence spectrometry, *Spectrochim Acta, Part B* 54: 943–958.

674 OMENETTO N., BERTHOUD Th., CAVALLI P. and ROSSI G. **(1985)** Analytical laser enhanced ionization studies of thallium in the air–acetylene flame, *Anal Chem* 57: 1256–1261.

675 BABIN F. J. and GAGNÉ J. M. **(1992)** Hollow cathode discharge (HCD) dark space diagnostics with laser photo-ionisation and galvanic detection, *Appl Phys, Part B* 54: 35–45.

676 YOUNG J. P., SHAW R. W. and SMITH D. H. **(1989)** Resonance ionization mass spectrometry, *Anal Chem* 61: 1271A–1279A.

677 KOCH K. H. **(1999)** *Process analytical chemistry*, Springer, Berlin.

678 TSCHOEPEL P. and TOELG G. **(1982)** Comments on the accuracy of analytical results in nanogram and picogram trace analysis of the elements, *J Trace Microprobe Techn* 1: 1–77.

679 KINGSTON H. M. and HASWELL S. J. **(1997)** *Microwave enhanced chemistry – fundamentals, sample preparation and applications*, American Chemical Society, Washington DC.

680 PICHLER U., HAASE A., KNAPP G. and MICHAELIS M. **(1999)** Microwave enhanced flow system for high-temperature digestion of resistant organic materials, *Anal Chem* 71: 4050.

681 ZISCHKA M., KETTISCH P., SCHALK A. and KNAPP G. **(1998)** Closed microwave-assisted wet digestion with simultaneous control of pressure and temperature in all vessels, *Fresenius' J Anal Chem* 361: 90–95.

682 KNAPP G., RAPTIS S. E., KAISER G., TOELG G., SCHRAMEL P. and SCHREIBER B. **(1981)** A partially mechanized system for the combustion of organic samples in a stream of oxygen with quantitative recovery of the trace elements, *Fresenius' Z Anal Chem* 308: 97–103.

683 JACOB E. **(1989)** Inorganic multiele-mental analysis by fluorine volatiliza-tion systems (FV/FTIR and FV/MS), *Fresenius' Z Anal Chem* 333: 761–762.

684 JACOB E. **(1989)** A novel micro ele-mental analyzer for the determination of non-metals and some metals, *Fresenius' Z Anal Chem* 334: 641–642.

685 KAISER G., MAYER A., FRIESS M., RIEDEL R., HARRIS M., JACOB E. and TOELG G. **(1995)** Critical comparison of ICP-OES, XRF and fluorine volatilization–FT-IR spectrometry for the reliable determination of the silicon main constituent in ceramic materials, *Fresenius' J Anal Chem* 352: 318–326.

686 KIPPHARDT H., GARTEN R. P. H., JACOB E., BROEKAERT J. A. C. and TOELG G. **(1997)** Mass spectrometric analysis of ceramics after decom-position with elemental fluorine, *Mikrochim Acta* 125: 101–105.

687 GERBERSMANN C., HEISTERKAMP M., ADAMS F. C. and BROEKAERT J. A. C. **(1997)** Two methods for the speciation analysis of mercury in fish involving microwave-assisted digestion and gas chromatography-atomic emission spectrometry, *Anal Chim Acta* 350: 273–285.

688 KLOCKENKÄMPER R. **(1997)** *Total-reflection x-ray fluorescence analysis*, Wiley, New York.

689 PAPPERT E., FLOCK J. and BROEKAERT J. A. C. **(1999)** Speciation of chromium in solid materials with the aid of soft-x-ray spectrometry, *Spectrochim Acta, Part B* 54: 299–310.

Index